EUROPEAN MINERALOGICAL NOTES IN MINERALOGY

Commissioning Editor: M. Plötze

Volume 21

CHEMICAL GEODYNAMICS OF THE EARTH'S MANTLE: NEW PARADIGMS

UNIVERSITY TEXTBOOK
Edited by
COSTANZA BONADIMAN AND ELISABETTA RAMPONE

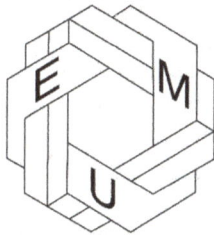

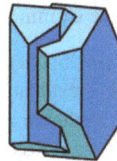

Mineralogical Society
of the UK and Ireland

Published by the European Mineralogical Union and the Mineralogical Society of the
United Kingdom & Ireland, London, 2024

The publication of this textbook is supported by the European Mineralogical Union

EMU Notes in Mineralogy

A series published under the auspices of the European Mineralogical Union (EMU).

Initiator of the EMU Schools and the EMU Notes in Mineralogy
Giovanni Ferraris, Torino, President of the EMU 1992–1996
Commissioning Editor: Michal Plötze, Zurich (previous editors: Roberta Oberti, Pavia, Giovanni Ferraris, Torino, Tamás G. Weiszburg and Gábor Papp, Budapest)

Editors of this Volume
Costanza Bonadiman, University of Ferrara, Italy
Elisabetta Rampone, University of Genoa, Italy

Managing Editor and Indexer: Kevin Murphy, London

Front cover design: Michel H. Guay
Executive Committee of the EMU (2020–2024)

Ewa Słaby, Poland (President), Michael Carpenter, UK (Past-President), Isabelle Daniel, Poland (Vice President, Acting President), Alessandro Pavese, Italy (Vice President), Helen King, The Netherlands (Treasurer) Public Information Officers: Simona Krmíčková, Czech Republic and Vadim Kovrugin, France

On the front cover: (left) Pyroxenite veins with dunitic reaction rim in peridotite (Trinity ophiolite, California) witnessing the process of infiltration melting in the Earth's mantle. (right) A passive heterogeneity (to the right of the plume axis) is deformed readily into a filament. An active (i.e. affecting the flow) heterogeneity 20 times more viscous than the surroundings maintains a 'blob-like' shape and does not deform. The figure shows the same 'blob' at various times. The plume conduit necking at 660 km depth is due to a 30× viscosity jump between the lower and upper mantle.

ISSN: 1417 2917
ISBN: 9780903056670

Published by the European Mineralogical Union and the Mineralogical Society of the United Kingdom and & Ireland (12, Baylis Mews, Amyand Park Road, Twickenham TW1 3HQ, UK)

The EMU Notes in Mineralogy Series

Published volumes

Volume	Year	Editors	Title
1	1997	S. Merlino	*Modular aspects of minerals*
2	2000	D.J. Vaughan R. Wogelius	*Environmental mineralogy*
3	2001	C.A. Geiger	*Solid Solutions in Silicate and Oxide Systems*
4	2002	C. Gramaccioli	*Energy Modelling in Minerals*
5	2003	D.A. Carswell R. Compagnoni	*Ultrahigh Pressure Metamorphism*
6	2004	A. Beran E. Libowitzky	*Spectroscopic Methods in Mineralogy*
7	2005	R. Miletich	*Mineral Behaviour at Extreme Conditions*
8	2010	F.E. Brenker G. Jordan	*Nanoscopic Approaches in Earth and Planetary Sciences*
9	2011	G. Christidis	*Advances in the Characterization of Industrial Minerals*
10	2010	M. Prieto	*Ion-partitioning in Ambient-temperature Aqueous Systems*
11	2011	M.F. Brigatti A. Mottana	*Layered Mineral Structures and their Application in Advanced Technologies*
12	2012	J. Dubessy M.-C. Caumon F. Rull	*Applications of Raman Spectroscopy to Earth Sciences and Cultural Heritage*
13	2013	D.J. Vaughan R.A. Wogelius	*Environmental Mineralogy II*
14	2013	F. Nieto K.J.T. Livi	*Minerals at the Nanoscale*
15	2015	M. Lee H. Leroux	*Planetary Mineralogy*
16	2017	W. Heinrich R. Abart	*Mineral Reaction Kinetics: Microstructures, Textures, Chemical and Isotopic Signatures*
17	2017	I.A.M. Ahmed K.A. Hudson-Edwards	*Redox-reactive minerals: Properties, reactions and applications in natural systems and clean technologies*
18	2017	A.F. Gualtieri	*Mineral fibres: Crystal chemistry, chemical-physical properties, Isotopic biological interaction and toxicity*
19	2017	J. Plášil J. Majzlan S. Krivovichev	*Mineralogical Crystallography*
20	2019	G. Artioli R. Oberti	*The contribution of mineralogy to cultural heritage*
21	2024	C. Bonadiman E. Rampone	*Chemical evolution and dynamics of the Earth's mantle*

Copies of the EMU Notes (volumes 1–7) are distributed in Europe by the larger member societies of the European Mineralogical Union:

Società Italiana di Mineralogia e Petrologia:
www.socminpet.it

Mineralogical Society of the United Kingdom & Ireland:
www.minersoc.org

Société Française de Minéralogie et de Cristallographie:
www.sfmc-fr.org

in America by the Mineralogical Society of America:
www.minsocam.org

Institutional orders as well as individual requests from outside Europe and America should be sent to Mineralogical Society of the United Kingdom & Ireland.

For volumes 8 onwards, the Mineralogical Society of the United Kingdom and Ireland acts as co-publisher and copies may be ordered from www.minersoc.org

or from:
Mineralogical Society
12 Baylis Mews, Amyand Park Road
Twickenham TW1 3HQ
UK

E-mail: admin@minersoc.org
Tel. +44 (0)20 8891 6600
Fax: +44 (0)20 8891 6599

Contents

Chapter 5. The shallow mantle as a reactive filter: a hypothesis inspired and supported by field observations...111
by **Georges Ceuleneer, Mathieu Rospabé, Michel Grégoire and Mathieu Benoit**

Chapter 6. The role of H_2O in the deformation and microstructural evolution of the upper mantle 155
by Károly Hidas and José Alberto Padrón-Navarta

Preface

Creation of this volume of the EMU Notes in Mineralogy Series was motivated by at least two important factors. Firstly, we wished to celebrate the success of the two editions of the International winter school 'MElting and fluid/melt-rock REactions in the MAntle -MEREMA'. Despite the great uncertainties due to the global pandemic (2020–2021), the school attracted many young researchers from all over the world thanks to the stimulating scientific program. The excellent intellectual milieu led a group of top-level lecturers to accept the invitation to publish their cutting-edge research in this volume. Although finalizing the chapters took longer than expected, we have ultimately compiled an outstanding selection of contributions that provides readers who have broad interests with information on how petrology and geochemistry, combined with geophysics, can shed light on mantle dynamics. Secondly, we aimed to present the current debated questions concerning mantle dynamics with key studies, which trace an ideal path from the bulk Earth composition (chapter 1) to mantle heterogeneity (chapters 2 and 3), and how these aspects are reflected in the shallow portion of the mantle and derived melts (chapters 4–7). We extend our sincere thanks to the authors of these chapters.

The chapters

The following section outlines the arrangement of chapters to illustrate the different approaches covered.

The first chapter (**McDonough 2024**) is a fascinating journey through the geochemical processes forming the primordial Earth's mantle. New geochemical models are proposed which establish the modern mantle composition in the context of the global energy budget (heat flux circulation), and account for the fundamental processes of mantle heating and melting.

The Earth's mantle is chemically and isotopically heterogeneous. Mantle plumes capture this compositional variability, and chapter 2 (**Farnetani, 2024**), with a pedagogical approach, provides a detailed analysis of the elements of this heterogeneity. The discussion of the role of remnants of subducted slabs at the base of the mantle *vs.* the presence of primordial melt blebs (magma ocean or core/mantle reaction melts) as repositories of mantle geochemical anomalies is particularly intriguing.

Closer to upper mantle domains, chapter 3 (**Gregoire *et al.*, 2024**) presents an overview of upper mantle geochemistry as depicted by a large collection of worldwide mantle xenoliths carried to the Earth's surface by Phanerozoic lavas. Partial melting events and the circulation of deep mantle-derived melts/fluids account for the significant variability in magma source mantle domains.

How can we model the melting of heterogeneous mantle sources? Chapter 4 (**Liang, 2024**) proposes new numerical models by revisiting the widely used batch, fractional, continuous, and two-porosity melting models of homogeneous mantle sources, in the context of decompression melting of a two-lithology mantle. Through worked examples, the author shows that it is possible to produce partial melts with a range of REE compositions (from LREE-enriched to LREE-depleted spectra), similar to those observed in mid-ocean ridge basalts (MORB), by decompression melting of a two-lithology mantle source.

The MORB type melts described in chapter 4, migrate from the mantle source to the surface. Chapter 5 (**Ceuleneer *et al.*, 2024**) investigates the complex system of melt migration mechanism by analysing the mantle section of the Oman ophiolite sequence. Field work observation provides evidence of melts that becomes rheologically independent from the source providing constraints for geochemical modelling and numerical simulation.

The ability of melt to migrate also depends on the structural and thermal conditions of the mantle rocks. Chapter 6 (**Hidas and Padrón-Navarta 2024**) presents various natural cases that illustrate the key mechanisms of H_2O incorporation in the main mantle minerals and how these mechanisms, rather than the chemical components, control the ductile deformation of the upper mantle.

Finally, it is important to mention that compositional heterogeneities on different scales exist in mantle rocks because of incomplete equilibration. Therefore, a kinetic evaluation is necessary before applying geothermometers to evaluate upper mantle thermal conditions. The final chapter (**Zhao and Chakraborty, 2024**) explains the importance of determining the equilibrium partitioning of temperature-dependent elements before using geothermometers. To illustrate the application of these theoretical aspects and to test the partitioning equilibrium in natural mineral equilibria, the cooling histories of mantle rocks from the ophiolite suite of Xigaze Ophiolites (Tibet), is thus modelled.

Acknowledgments

The editors are grateful to EMU President, Eva Słaby and the Commissioning Editor, Michael Plötze, for their enduring support of this initiative. Special thanks to Kevin Murphy for his assistance, patience, and advice throughout the editorial process. Thanks also to all of the reviewers whose constructive comments contributed significantly to the scientific standard of this volume.

References

Ceuleneer, G., Rospabé, M., Grégoire, M. and Benoit, M. (2024) The shallow mantle as a reactive filter. A hypothesis inspired and supported by field observations. Pp. 111–154 in: *Chemical Geodynamics of the Earth's Mantle: New Paradigms* (Costanza Bonadiman and Elisabetta Rampone, editors). EMU Notes in Mineralogy, **21**. European Mineralogical Union and the Mineralogical Society of the United Kingdom and Ireland, London.

Farnetani, C. (2024) Plumes from the heterogeneous Earth's mantle. Pp. 19–38 in: *Chemical Geodynamics of the Earth's Mantle: New Paradigms* (Costanza Bonadiman and Elisabetta Rampone, editors). EMU Notes in Mineralogy, **21**. European Mineralogical Union and the Mineralogical Society of the United Kingdom and Ireland, London.

Grégoire, M., Delpech, G., Moine, B. and Cottin, J.-Y. (2024). Nature and origin of heterogeneities in the lithospheric mantle in the context of asthenospheric upwelling and mantle wedge zones: What do mantle xenoliths tell us? Pp. 39–56 in: *Chemical Geodynamics of the Earth's Mantle: New Paradigms* (Costanza Bonadiman and Elisabetta Rampone, editors). EMU Notes in Mineralogy, **21**. European Mineralogical Union and the Mineralogical Society of the United Kingdom and Ireland, London.

Hidas, K. and Padrón-Navarta, J.A. (2024) The role of H$_2$O in the deformation and microstructural evolution of the upper mantle. Pp. 155–188 in: *Chemical Geodynamics of the Earth's Mantle: New Paradigms* (Costanza Bonadiman and Elisabetta Rampone, editors). EMU Notes in Mineralogy, **21**. European Mineralogical Union and the Mineralogical Society of the United Kingdom and Ireland, London.

Liang, Y. (2024) Simple models for trace element fractionation during decompression melting in a two-lithology mantle. Pp. 57–110 in: *Chemical Geodynamics of the Earth's Mantle: New Paradigms* (Costanza Bonadiman and Elisabetta Rampone, editors). EMU Notes in Mineralogy, **21**. European Mineralogical Union and the Mineralogical Society of the United Kingdom and Ireland, London.

McDonough, W.F. (2024) Composition of the Earth and implications for geodynamics. Pp. 1–18 in: *Chemical Geodynamics of the Earth's Mantle: New Paradigms* (Costanza Bonadiman and Elisabetta Rampone, editors). EMU Notes in Mineralogy, **21**. European Mineralogical Union and the Mineralogical Society of the United Kingdom and Ireland, London.

Zhao, L. and Chakraborty, S. (2024) Kinetic controls on the thermometry of mantle rocks: A case study from the Xigaze Ophiolites, Tibet. Pp. 189–222 in: *Chemical Geodynamics of the Earth's Mantle: New Paradigms* (Costanza Bonadiman and Elisabetta Rampone, editors). EMU Notes in Mineralogy, **21**. European Mineralogical Union and the Mineralogical Society of the United Kingdom and Ireland, London.

EMU Notes in Mineralogy, Vol. 21 (2024), Chapter 1, 1–18

Composition of the Earth and implications for geodynamics

WILLIAM F. McDONOUGH[1,2,3]

[1]*Department of Geology, University of Maryland, College Park,
MD 20742, USA (ORCID number: 0000-0001-9154-3673)*
[2]*Department of Earth Science, Tohoku University, Sendai,
Miyagi 980-8578, Japan*
[3]*Research Center for Neutrino Science, Tohoku University, Sendai,
Miyagi 980-8578, Japan*
e-mail: mcdonoug@umd.edu

The composition of the bulk silicate Earth (BSE) is the product of planetary accretion, core differentiation and Moon formation. By establishing the composition of the BSE, one can determine the composition of the bulk Earth and by subtraction, calculate the core's composition. The BSE represents the bulk Earth minus the core, which in today's terms equals the modern mantle, the continental crust, and the hydrosphere-atmosphere systems. The modern mantle can be framed in terms of two compositionally distinct components, an enriched and a depleted mantle, with the latter as the MORB (mid-ocean ridge basalt) source and the former as the OIB (ocean island basalt) source.

The Earth's surface heat flux is 46 ± 3 TW (terrawatts, 10^{12} watts). Some fraction of this flux ($\sim$40% or $\sim$20 TW) is derived from radioactive heat produced by the decay of U, Th and K. Some 40% of the Earth's budget of these elements is stored in the continental crust ($\sim$7.7 TW) and does not contribute to mantle heating. The remaining energy is primordial derived from accretion and core separation. The heat flux through the mantle also includes a core contribution ($\sim$10 $\pm$ 5 TW; i.e. bottom heating of the mantle).

The rich record of seismic tomography documents ocean slabs stagnating at the base of the transition zone, others stagnating at $\sim$1000 km depth, and others plunge directly into the deep mantle. Collectively, these images reveal mass exchange between the upper and lower mantle and are consistent with whole-mantle convection. The Mantle Transition Zone (MTZ) plays a major role in the differentiation of mantle. Mantle dynamics is controlled by its viscosity, which in turn is controlled by its water content and temperature. Evidence for core–mantle exchange remains elusive. Documented Hadean $^{182}W/^{184}W$ isotopic anomalies coupled with primordial noble gas signatures in OIBs present a preservation challenge for mantle convection models.

DOI: 10.1180/EMU-notes.21.1

1. Structure of the Earth

1.1. Definition of its domains (core – mantle – crust)

Earth's structure is composed of compositionally distinct, concentric shells: a metallic core surrounded by a silicate shell and this is surrounded by a hydrosphere/atmosphere. Our understanding of the chemical and isotopic compositions of these domains decreases with depth, resulting in significant uncertainties in the composition of the deep mantle and core. All terrestrial planets share this differentiated structure of a core, mantle and crust.

1.2. PREM

Dziewonski and Anderson (1981) presented a 1D seismic model for the Earth. This model, based on body waves and free oscillation data for the Earth, is used to define the depth to the core mantle boundary (CMB), documents the liquid outer and solid inner core, and provides a density gradient for the Earth's interior. Importantly, PREM documents jumps in the velocity of the compressional (V_p) and shear (V_s) waves at different depths in the mantle, which correspond to observed phase changes in the abundant silicate minerals.

Figure 1 and Table 1 provide a reference state for Earth's interior. The temperatures reported are for the major phase changes in the mantle and represent the post-phase change in temperature. Major phase changes in the mantle occur at 410 km, 520 km and 660 km for a pyrolite composition having an Mg# of 0.89 (where Mg# = atomic proportions of Mg/(Mg + Fe)). At 410 and 520 km depths, the isochemical (($Mg, Fe)_2SiO_4$) phase transitions are olivine → wadsleyite and wadsleyite → ringwoodite, respectively. At a depth of 660 km, ringwoodite disproportionates into bridgmanite (($Mg, Fe)SiO_3$) and ferropericlase (($Mg, Fe)O$).

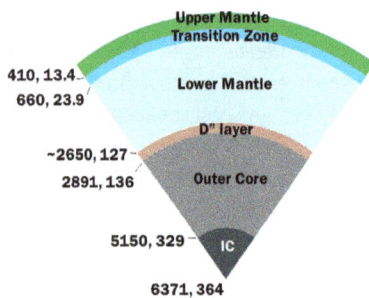

Fig. 1. The depth and pressure (km, GPa) of phase changes in the mantle for an assumed pyrolite composition (McDonough and Sun, 1995). Similarly, conditions for the core are also presented. Depth and pressure values are taken from the PREM model (Dziewonski and Anderson, 1981). See Table 1 for further information.

The seismically defined density of Earth's core is ~10% less than that of an Fe,Ni alloy at comparable P–T conditions, indicating the presence of light element(s). An estimate of the amount of light element in the outer core (i.e. 'X'%) is typically estimated to be between 8 and 10 wt. %, or twice that in atomic %, but these numbers are not fixed. The proportion of light elements depends on the temperature of the outer core and its Δ density relative to pure Fe (solid Fe, as liquid iron measurements have not been accomplished at core P and T conditions).

The temperature profile for the Earth is best constrained by the mantle potential temperature at the surface (~1640 K) and at the phase change at 410 km (~1860 K) (Sarafian *et al.*, 2017; Katsura, 2022). The temperature of the

Table 1. Properties and minerals of the Earth*.

Depth (km)	Pressure (GPa)	Temperature (K)	Phase transitions	Mineralogy
100	3.32	1672	None	Ol, Opx, Cpx,Gt
410	13.4	1860	Ol → Wa	Wa, Maj
520	17.9	1942	Wa → Ri	Ri, Maj
670	23.9	1960	Ri → Br + FeP	Br, FeP, Dm
2650	127	2540	post-perovskite	post-perovskite
2891	136	4000	liquid metal alloy	FeNi + X% light element
5150	329	6000	solid metal alloy	FeNi + 0.5*X% light element

*See Fig. 1 for further details. Ol = olivine, Opx = orthopyroxene, Cpx = clinopyroxene, Gt = garnet, Wa = wadsleyite, Maj = majorite, Ri = ringwoodite, Br = Bridgmanite, FeP = Ferropericlase, Dm = Davidmaoite. Mantle temperatures for the underside of the seismically defined boundaries are from Katsura (2022). Core temperatures are weakly defined (± 500 K for OC and ± 1000 K for IC), with the inner core boundary (ICB) providing a limit based upon the solidification of iron at these pressures.

appearance of bridgmanite at 660 km depth (1960 K) is reasonably well constrained, though some consider this depth to represent a compositional change in the mantle (e.g. Ballmer *et al.*, 2017). If so, then there should be a marked temperature discontinuity at this conductive boundary. In the model considered here, however, the upper and lower mantle have approximately the same major element composition and there is no temperature discontinuity. The lack of chemical and isotopic evidence for upper-lower mantle differentiation and seismic images of oceanic lithosphere penetrating the mantle transition zone and plunging deep into the mantle provide justification for this assumption.

Figure 2 presents a temperature profile for the Earth. The adiabatic temperature (T) gradient relative to depth (z) in the core and mantle is given as $(dT/dz) = \alpha g T/C_p$, where α is thermal expansivity (decreasing from 4×10^{-5} K^{-1} at the top of the mantle to 1×10^{-5} K^{-1} at D″ ($\sim$2700 km depth)) (Katsura, 2022), g is gravitational acceleration (approximately constant for the mantle at $\sim$9.8 m s^{-2}), and C_p is the isobaric heat capacity ($\sim$1000 J kg^{-1} s^{-1}). Values for α, T and C in the metallic core are less well-known, with gravitational acceleration going to zero at the center of the Earth.

On average, the adiabatic gradient for the mantle is $\sim$0.3 K/km (from $\sim$0.5 at the top of the mantle to $\sim$0.25 at the bottom) and for the core is between 0.5 and 1.0 K/km. The core mantle boundary (CMB) represents a major temperature discontinuity and reflects conductive cooling across this interface. The core's temperature at the inner core boundary (ICB) and its adiabatic gradient sets the temperature at the core side of the CMB. Establishing the ICB temperature depends upon knowing the composition of the inner and outer core and crystallization temperature at 330 GPa. There continues to remain considerable uncertainty on the crystallization temperature of pure iron at ICB pressures (Williams *et al.*, 1987; Tsuchiya *et al.*, 2016; Kraus *et al.*, 2022), which is the simplest analogue composition for the inner core, albeit without a light element component.

1.3. The core, CMB, D″, LLSVP and ULVZ

The core and its boundary region represent an intriguing environment that is poorly understood. The core–mantle boundary (CMB) at 2891 km depth presents a sharp seismic discontinuity, defining the major attributes of the core (its size, density,

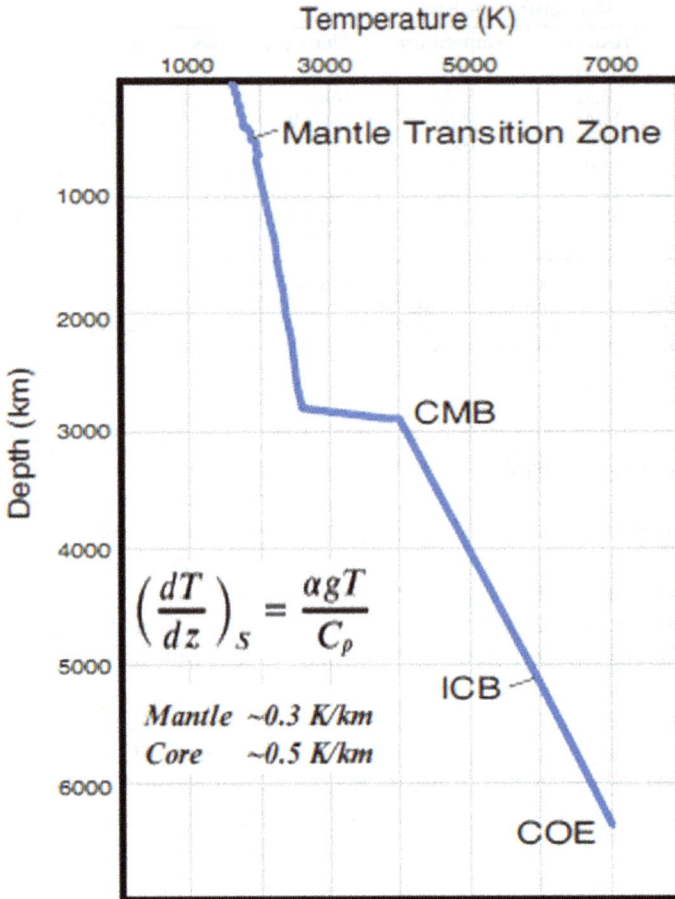

Fig. 2. The temperature profile for the Earth.

state: liquid outer solid inner core). Birch (1952) showed that the outer core's density is ~10% less than that of an Fe,Ni alloy at equivalent conditions, thus, documenting the core's density deficit. He concluded that the core must contain a proportion of one or more light elements, such as S, Si, O, C, and H. The outer core's density deficit depends on its temperature (Anderson and Isaak, 2002; Shanker *et al.*, 2004), which is probably between 3500 and 6000 K, with 4000 ± 500 K (McDonough, 2017) as the value often cited for the top of the outer core. The solid inner core represents only 4% by volume and 5% of the mass of the core, and it contains about half the amount of light element relative to the outer core (i.e. ~3–4% density deficit relative to pure Fe (Fei *et al.*, 2016; Sakamaki *et al.*, 2016)).

Earth's shape, mass, volume, bulk density, moment of inertia, and inner core libration are defined by geodetic studies, while geomagnetic studies show that the outer core has maintained a geodynamo, of approximately its current state (variation in strength and polar stability), throughout most of its history. These observations constrain the structure and nature of the core. The presence of a PKJKP seismic wave and the absence of a shear wave demonstrates the outer core is liquid and the inner core is solid. A density increase ($+820$ kg/m^3) across the inner and outer core boundary is documented (Masters and Gubbins, 2003).

The D" region is a seismically defined (a reduction in V_p waves) region of the mantle that extends from the CMB upwards for $\sim$200 km and likely represents a conductive thermal boundary layer at the bottom of the mantle. Also sitting on the CMB are two other seismically identified entities: LLSVP (large low shear-wave velocity province, also called LLVP: large low velocity province) (Cottaar and Lekic, 2016; Kim *et al.*, 2020) and ULVZ (ultra-low velocity zone) (Yu and Garnero, 2018). The two LLSVPs structures are antipodally positioned at near equatorial latitudes (beneath the central Pacific Ocean and south central Atlantic Ocean) and extend upwards some 1500 km. Their sharp vertical seismic definitions are most consistent with these structures being chemical anomalies. It is also likely that this region is thermally distinct from the surrounding mantle. Many ocean islands, associated with hotspot upwellings, are believed to have source regions emanating from these LLSVPs (Jackson *et al.*, 2021). Several ULVZ domains have been identified, which are much smaller pancake-like structures only a couple tens of kilometers across. Their distribution appears to be associated with the distal basal regions of the LLSVPs (Garnero *et al.*, 2016).

1.4. The Mantle: 410 and 660 km deep seismic discontinuities

The Earth's mantle has two prominent, seismic discontinuities at 410 and 660 km depths, which define the Mantle Transition Zone (MTZ), and separates the upper and lower mantle. These seismic discontinuities are the product of mineralogical phase changes (see Section 1.2, Figure 1 and Table 1). The MTZ is considered to play a major role in the differentiation cycle in the mantle (Birch, 1952; Ringwood, 1994).

When entering the MTZ, subducting oceanic lithosphere commonly bends and is transported laterally at or above the 660 km discontinuity. Some of these lithospheric domains later continue subducting into the lower mantle, while others may experience long-term residence in the MTZ. Importantly, the colder and markedly distinctive compositions of the basaltic and underlying harzburgitic layers of the subducting lithosphere means that it undergoes different depth dependent phase changes than that of the surrounding pyrolite-like mantle. Consequently, subducting lithosphere is distinctively more brittle and denser than the surrounding mantle, with the basaltic layer being denser than pyrolite in the MTZ and the harzburgite layer being denser than pyrolite just below the MTZ (Ringwood, 1994). These distinctive properties of subducted oceanic lithosphere play a significant role in mantle dynamics.

1.5. The lithosphere: crust and lithospheric mantle

The lithosphere is the thermal conductive layer at the top of the silicate Earth and is composed of crust and underlying lithospheric mantle. There are two different domains of lithospheres, oceanic and continental, and they have markedly different chemical, physical and temporal attributes. The boundary between the crust and the mantle is reasonably well defined by the Moho seismic discontinuity, whereas the base of the lithosphere is poorly defined physically or chemically.

In the oceanic environment the average crust is basaltic, 50 million years old (with residual domains being up to 200 million years old), and typically 7 to 10 km thick. However, in some parts of the ocean the basaltic layer is absent (e.g. central Atlantic and southwest Indian Oceans) (Bonatti *et al.*, 2001; Seyler *et al.*, 2011). In contrast, the continental crust is, on average, andesitic in composition, 2+ billion years old (with residual domains being up to 4 billion years old), and typically ~40 km thick (Rudnick and Gao, 2014; Sammon *et al.*, 2022).

The thickness of the oceanic lithospheric mantle varies from zero at the mid-ocean ridge spreading center to ~90 km thick beneath thermally mature oceanic crust. Compositionally, it varies from a highly depleted peridotite (i.e. dunite to harzburgite) at the top to a fertile lherzolite at its base. These compositional attributes reflect its generation via melt extraction at the ridges followed by off-ridge conductive cooling at its base. The thickness of the continental lithospheric mantle (CLM*) is also markedly variable depending upon its age and tectonic history. [*I pointedly avoid using the inaccurate term "subcontinental lithospheric mantle", as it implies the asthenosphere.] On average, the continental lithospheric mantle appears to be ~100 km thick beneath post-Archaean terrains and ≥200 km thick beneath Archaean terrains. Growth of the CLM is intimately linked with extraction of melt (crust formation) and the production of refractory peridotite. However, in contrast to the oceanic setting, the CLM was formed and evolved in multiple tectono-magmatic events.

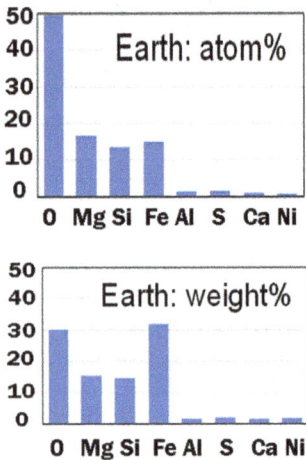

Fig. 3. The eight most abundant elements in the Earth.

2. Compositionally distinctive domains

2.1. The major elements (O, Fe, Mg, Si)

The Earth, the terrestrial planets, and chondrites have ~93% of their masses and atomic proportions being composed of O, Fe, Mg and Si (Fig. 3) (Wasson and Kallemeyn, 1988; McDonough and Yoshizaki, 2021). The relative proportions of these four elements in ordinary, carbonaceous, enstatite, and G chondrites are not fixed. McDonough and Yoshizaki (2021) noted the variation in Mg/Si (~25%), Fe/Si (factor of ≥2), and Fe/O (factor of ≥3) between chondrite groups. Variability in these four elements is consistent with them being non-refractory elements (i.e. having 50% (half-mass) condensation temperatures <1350 K).

2.2. The big eight elements

In addition to the main group elements (i.e. O, Mg, Si and Fe), there are four minor group elements (i.e. Al, Ca, S and Ni) that together constitute 99% the mass and atomic proportions of the Earth and terrestrial planets (Fig. 3). Both Al and Ca are refractory and present in chondritic proportions (Ca/Al = 1.07 ± 0.04). Setting the abundance of either Al or Ca in the planet allows us to establish Earth's absolute abundances of all 37 refractory elements (McDonough and Sun, 1995). The value of Fe/Ni in chondrites is observed to be constant (17.4 ± 0.5). Additional constraints come from the mass fraction of metallic core (32.5) and silicate Earth (67.5), the Mg# of the bulk silicate Earth (0.89) and the Ni content of the mantle. Thus, these rules allow us to put constraints on the bulk Earth composition.

To develop a model Earth composition requires additional constraints, specifically the sulfur content of, and the Fe/O and Mg/Si values for, the bulk Earth and bulk silicate Earth. The core and mantle mass fractions constrain the Earth's Fe/O value, as the core contains >90% of Earth's iron and the mantle contains >90% of Earth's oxygen. The mantle is estimated to have only a few hundred ppm sulfur, whereas the core is estimated to have as much as 6 wt.%. The remaining big unknown is the Mg/Si value for the bulk Earth and bulk silicate Earth.

Constraining the Mg/Si value for the bulk Earth and bulk silicate Earth has been a long-standing problem (Ringwood, 1989). There is no fixed Mg/Si ratio for chondrites, despite the often-quoted concept of assuming a chondritic Mg/Si. Values of this ratio for chondritic averages range from ~0.65 in EH (hi-iron enstatite chondrites) to ~0.95 in the Tagish Lake carbonaceous chondrite. This variation reflects the proportion of olivine (2:1 molar Mg/Si) to pyroxene (1:1 molar Mg/Si) and metallic Si in chondritic parent bodies. Active accretion disks in our galaxy show rapid grain growth in the inner disk region and radial variations in the relative proportions of olivine to pyroxene (van Boekel *et al.*, 2004; Bouwman *et al.*, 2010; D'Alessio *et al.*, 2001).

Predicting a planet's bulk composition requires us to establish the proportion of olivine to pyroxene in the bulk silicate shell. For the Earth this was best established on a plot of Mg/Si versus Al/Si, with mantle peridotites forming a linear melt depletion trend (Jagoutz *et al.*, 1979; Hart and Zindler, 1986). In this diagram the fertile end-member is at the intersection of the mantle (geochemical) and chondritic (cosmochemical) fractionation trends. These authors established the bulk silicate Earth's Mg/Si (1.1, metal weight ratio) and Al/Si (0.11) values.

The final step to establishing concentrations of these elements in the bulk silicate Earth and bulk Earth required us to determine the absolute concentrations of the refractory lithophile elements in the mantle. McDonough and Sun (1995), using melt-depletion trends, determined that the bulk silicate Earth had 2.7 times CI carbonaceous chondrite abundances of the refractory lithophile elements.

Table 2 provides a preferred model composition of the eight most abundant elements in the bulk Earth, core, and BSE. This model uses a pyrolite composition for the bulk silicate Earth and is comparable to the models developed by McDonough and Sun (1995) and Palme and O'Neill (2014). The core composition represents a revised estimate

Table 2. Composition of the Earth and its parts.

	Bulk Earth		Core		BSE	
	Wt.%	At.%	Wt.%	At.%	Wt.%	At.%
O	30.0	49.4	1.0	3.3	44.0	58.4
Mg	15.4	16.7	0	0	22.8	19.9
Si	14.5	13.6	1.0	1.9	21.0	15.9
Fe	31.9	15.0	85.3	79.7	6.3	2.4
Al	1.59	1.55	0	0	2.35	1.85
S	2.11	1.73	6.45	10.5	0.03	0.02
Ca	1.71	1.12	0	0	2.53	1.34
Ni	1.82	0.82	5.21	4.63	0.20	0.07
other	0.69	0.00	1.21	0.01	0.54	0.00
	99.8	100	100.1	100	99.9	100

Details given by McDonough (2014), which assume a pyrolite model composition for the BSE. Wt. = weight, At. = atomic. The At.% columns were calculated from the Wt.% values and normalized to 100%. 'other' includes Co, Cr and P for the bulk Earth and core, and Na, Cr and P for the BSE.

from McDonough (2017), now with a greater sulfur content and smaller silicon and oxygen contents.

A simple compositional model for the Earth, assuming a CI chondrite composition, is presented in Table 3. This model is often justified by forcing a Mg/Si match to the solar photosphere's composition. This alternative model is not preferred as it requires a gross distinction between the upper and lower mantle compositions. Note that all models accept that the upper mantle has a pyrolite composition, which has been documented thoroughly. To maintain a pyrolite upper mantle, there has to be limited mass exchange between the upper and lower mantles and a discontinuity in the mantle geothermal across this conductive boundary layer. These latter two observations appear to be at odds with seismic evidence (Fukao and Obayashi, 2013) and a mid-mantle viscosity

Table 3. CI composition of the Earth and its parts.

CI model →	Bulk Earth		Core		BSE		Upper mantle		Lower mantle	
	Wt.%	At.%	Wt.%	At.%	Wt.%	At.%	Wt.%	At.%	Wt.%	At.%
O	30.6	50.0	0	0	45.4	60.4	44.0	58.4	45.8	61.1
Mg	13.5	14.5	0	0	20.0	17.6	22.8	19.9	19.1	16.7
Si	15.3	14.2	0	0	22.7	17.2	21.0	15.9	23.3	17.7
Fe	26.4	12.3	75.5	62.5	5.9	2.3	6.3	2.4	5.8	2.2
Al	1.19	1.15	0	0	1.76	1.39	2.35	1.85	1.55	1.23
S	7.63	6.21	23.5	33.9	0.03	0.02	0.03	0.02	0.03	0.02
Ca	1.26	0.82	0	0	1.86	0.99	2.53	1.34	1.62	0.87
Ni	1.56	0.69	4.61	3.63	0.20	0.07	0.20	0.07	0.20	0.07
other	1.17	0.00	1.15	0.01	0.54	0.00	0.54	0.00	0.54	0.00
	98.7	100	104.7	100	98.5	100	99.9	100	97.9	100

Details for CI composition were given by Lodders (2020); this model represents a degassed composition with a final H and C content being ~2000 ppm in the BSE (bulk silicate Earth). The BSE was divided into a lower mantle and a combined upper mantle and transition zone (labelled 'upper mantle'). A pyrolite model composition was assumed for this upper mantle (see Table 3). The At.% columns were calculated from the Wt.% values and normalized to 100%. 'other' includes Co, Cr and P for the bulk Earth and core, and Na, Cr and P for the bulk silicate Earth. Wt. = weight, At. = atomic.

discontinuity (Rudolph *et al.*, 2015).The core composition in this model was established based on the sulfur content of CI, which adds a significant and unrealistic amount of light element (S) content to the core. It is also noteworthy that this model cannot accommodate any silicon or oxygen in the core.

This simple exercise shows the weaknesses associated with assuming a CI chondrite compositional model. Traditionally, such models are justified based solely on the CI Mg/Si value matching that of the solar photosphere, which is taken to be representative of the mass of the solar system. However, an exact match of the bulk Earth to CI chondrite is impossible given the latter's volatile content (i.e. H_2O and CO_2). The usual convention is to create a degassed composition leaving behind some *ad hoc* amount of H and C in the BSE and core. Moreover, in this example one would also have to lose some sulfur to accommodate the totals and fit the core's ~10% density deficit (Birch, 1952). Picking one element ratio to establish a planetary composition is fraught with many complications and unwelcome consequences.

It is also useful to consider the mineralogical composition of the lower mantle (Table 4). The lower mantle is primarily composed of three minerals: bridgmanite, ferropericlase, and davidmaoite. Davidmaoite is the calcium-bearing perovskite ($CaSiO_3$), with a simple chemistry. Likewise, ferropericlase (Mg, Fe)O also has a simple chemistry and is probably a host for nickel in the mantle. The most abundant mineral in the Earth is bridgmanite ((Mg, Fe^{2+}, Fe^{3+}, Al)$SiO_{3.x}$), with the coefficient of $3.x$ to account for the charge balance needed for the substitution of Fe^{3+} and Al^{3+} for Mg^{2+} and Fe^{2+}.

Presented in Table 4 are the mode proportions of bridgmanite, ferropericlase, and davidmaoite in a pyrolite model (Table 2) and a CI model (Table 3) composition for the lower mantle. The CI model assumes that the upper mantle and transition zone have a pyrolite composition. This compositional model for the mantle above 660 km is well documented in a wide range of mantle peridotite samples. The limited davidmaoite content of the lower mantle in this CI model means that this mantle composition has a lower mass fraction of refractory elements, including the heat-producing elements U & Th.

2.3. Heat-producing elements (K, Th and U)

The Earth is a hybrid engine powered by two sources of energy: residual primordial energy, evolved from accretion and core separation, and heat from radioactive decay. Radiogenic heat production in the Earth comes overwhelmingly from ^{40}K, ^{232}Th, ^{238}U and ^{235}U (i.e. 99.5% in total). The Earth's radiogenic power has been estimated at 20 ± 3 TW, which assumes a specific model composition (McDonough and Sun,

Table 4. Lower mantle mineral proportions for two compositional models.

	Bridgmanite Mg-perovskite (Mg, Fe²⁺, Fe³⁺, Al)SiO₃.ₓ	Ferropericlase (Mg, Fe)O	Davidmaoite Ca-pvk CaSiO₃
Pyrolite	74	19	7
CI-model	89	7	4

1995; Palme and O'Neill, 2014). Predictions from competing compositional models span from 10 to 33 TW; thus, there are considerable uncertainties in our understanding of the absolute amount of radiogenic power in the Earth.

Both Th and U, like Ca and Al, are refractory lithophile elements, are assumed to be concentrated in the silicate Earth, and occur in chondritic proportions to one another. This means that if we know the absolute concentration of one of these elements, then we know the absolute concentration of some 29 refractory lithophile elements.

McDonough and Sun (1995) found that cosmochemical (e.g. condensation and loss of volatiles) and geochemical (i.e. core formation) fractionation processes resulted in the bulk silicate Earth having an enrichment in refractory lithophile elements at 2.7 times that of an undegassed CI chondrite. Similarly, assuming instead a CI model for the composition of the bulk silicate Earth, this results in the refractory lithophile elements being enriched by 2.2 times relative to an undegassed CI chondrite. Consequently, these two different compositional models generate different amounts of radiogenic heat.

The amount of radiogenic heat in the Earth is now being determined at the global scale through the measurements the Earth's geoneutrino flux (Araki *et al.*, 2005). Geoneutrino measurements are being conducted by particle physicists and provide a transformative level of discovery of the current composition and heat production in the Earth. These data provide insights into the thermal evolution of Earth and independently define the silicate Earth's enrichment factor from a CI chondrite composition.

Existing data from two particle physics experiments predict that the Earth has either $14.6^{+5.2}_{-4.2}$ TW (KamLAND, Japan) (Abe *et al.*, 2022) or $38.2^{+13.6}_{-12.7}$ TW (Borexino, Italy) (Agostini *et al.*, 2020) total radiogenic power. More recently, however, the neutrino flux measurement for Borexino has been re-interpreted with a more accurate figure using a local lithospheric model. Sammon and McDonough (2022) showed that the initial geological model used by the Borexino physicists (Agostini *et al.*, 2020) failed to recognize the abundant local igneous rocks that have high K, Th and U contents. By under-predicting the regional contribution of the geoneutrino flux, the Borexino team over-predicted the global radiogenic budget of the Earth. Using a more accurate compositional model for the lithosphere surrounding Borexino results in a 20 TW Earth model (Sammon and McDonough, 2022). The SNO+, Canada (Andringa *et al.*, 2016) geoneutrino experiment is now counting and the JUNO, China (An *et al.*, 2016) experiment is being built. Results from these experiments will provide additional data in the coming years and reduce the uncertainties for the mantle's radiogenic power. These data also define the composition of the silicate Earth and bulk Earth for the refractory elements.

3. Present-day continental crust and modern mantle

The present-day architecture of the silicate Earth has continental and oceanic crust surrounding a modern mantle composed of chemically and isotopically depleted and enriched domains. The oceanic crust is relatively simple, basaltic in composition, and a product of single-stage melting of the mantle. In contrast, the continental crust is considerably more complicated. It is older, a product of multistage melting, and possesses an extraordinary range of rock compositions (e.g. sandstones to limestones and everything in between). On average, the continental crust is andesitic (~60 wt.%) in composition (Rudnick and Gao, 2014; Sammon *et al.*, 2022). The depleted domain of the modern mantle is the MORB source region, which occupies a considerable mass fraction of the mantle ($\geq$70%), while the enriched domain, the remaining fraction, is the source region for Ocean Island Basalts (OIB).

The mass of the continental crust is 2.32×10^{22} kg and represents ~0.58% of the silicate Earth (Sammon *et al.*, 2022). It covers around 44% of the surface area and is, on average, 41 km thick. The top of the crust is granitic in composition and from there, there is an SiO_2 gradient from ~61 wt.% (quartz monzonite) to ~53 wt.% (gabbronorite) from the middle to the base of the crust. Radiogenic heat production by the continental crust is estimated to be 330 picowatts/kg or 7.7 TW in total (Sammon *et al.*, 2022), with the continental crust containing ~40% of the BSE budget of K, Th and U. The remaining 60% of these elements are in the modern mantle providing 12.3 TW to drive its geodynamic engine.

Our understanding of the modern mantle remains weak. We can identify two main magma types (MORB and OIB), which we interpreted as being derived from depleted and enriched domains. The mass fraction, its distribution, and the composition of the depleted mantle is poorly constrained. It is possible that the depleted mantle is vertically stratified with more incompatible trace element depletions at the top of the mantle and more chemical fertility lower down. Such a chemical stratigraphy might explain differences between compositional models for the depleted mantle (Salters and Stracke, 2004; Workman and Hart, 2005) and the helium-heat flux paradox (O'Nions and Oxburgh, 1983).

The mass fraction, its distribution, and the composition of the enriched mantle are even less well understood than what we know about the depleted mantle. Current models for the enriched source regions of the OIB lavas envisage a significant role being played by the LLSVPs (Jackson *et al.*, 2021), which are large degree-2, antipodal structures that are near-equatorial positions and extend up to 1500 km from the core–mantle boundary. OIB lavas (e.g. some HIMU) and some continental intraplate basalts (e.g. recent basalts in eastern China) have been proposed as coming from recycled lithospheric components stored in the base of the transitions zone (Ringwood, 1982; Zhao *et al.*, 2011; Huang *et al.*, 2020).

4. Mantle geodynamics

Seismic tomography provides a rich record of 2D and 3D images of the fate of subducting oceanic lithosphere as they penetrate deeper into the mantle. Some oceanic slabs stagnate at the base of the transition zone and are transported laterally for considerable distances. Fukao and Obayashi (2013) showed that others stagnate below 600 km and reside horizontally at ~1000 km depth, while others plunge directly into the deep mantle.

The reasons for these different dynamical responses are not clear, but may be related to density barriers (Ringwood, 1982) and viscosity filters (Rudolph *et al.*, 2015) in this portion of the mantle.

These seismic images document considerable mass exchange between the upper and lower mantle and are consistent with whole-mantle convection. The MTZ and several hundred km beneath it appears to play a major role in the differentiation of mantle. Understanding the dynamics of this region is essential (Ringwood, 1994).

Water is a significant unknown in the mantle, with compositional estimates ranging by an order of magnitude (Marty, 2012; Ohtani, 2021). Moreover, we are not certain if the net water flux is greater or lesser from mantle to surface over time. Mantle dynamics is controlled by its viscosity, which in turn is controlled by its water content and temperature. The efforts of geoneutrino studies are focused on defining precisely the radiogenic heating curve for the history of the planet. This information will constrain the thermal evolution of the Earth and its viscosity structure.

There is considerable uncertainty in terms of the initial thermal condition of the proto-Earth, especially regarding the timing of Moon formation, via collision with a Mars-sized bolide. Much depends on the presence or absence of an atmosphere, which has a blanketing effect on surface heat loss. In addition, the Earth's τ accretion age is unknown. If the Earth's τ accretion age is 10 million years, then as McDonough *et al.* (2020) showed, ^{26}Al and ^{60}Fe played a significant role in heating of the planet and probably accelerated core formation. Subsequent to this early heating event was the Moon-forming, giant impact event, which is constrained to have occurred between 50 and 150 million years after t_{zero}, the age of Ca-Al inclusions which are the oldest-formed solids in the solar system.

Developing a thermal model for the Earth can provide insights into early mantle differentiation products, including the existence of primordial reservoirs that may (or may not) currently exist at the base of the mantle. The last 20 years of research into decay products of extinct short-lived nuclides (e.g. ^{142}Nd and ^{182}W) have provided many exciting new insights (Boyet and Carlson, 2005; Mundl *et al.*, 2017), but these studies have also raised more questions than they have resolved.

Over the last several decades we have made great strides in measuring trace element abundances, radiogenic isotope compositions (e.g. Sr, Nd, Pb, Hf) and noble gas isotopic compositions (e.g. He, Ne, Ar, Xe) of modern basalts, including Mid Ocean Ridge Basalt (sampling the depleted mantle) and Ocean Island basalts (sampling the enriched mantle). Insights gained from these measurements coupled with detailed tomographic images from seismology (French and Romanowicz, 2015) have advanced our understanding of the architecture of the mantle (Jackson *et al.*, 2017). More recently, we have added stable isotopes systems and extinct short-lived isotope systems to the repertoire of tools to investigate the mantle.

The ^{142}Nd isotope system (^{146}Sm $\rightarrow$ ^{142}Nd $+ \alpha + $ Q, $t_{1/2} = 103$ Ma) and ^{182}W isotope system (^{182}Hf $\rightarrow$ ^{182}W $+ 2e^- + 2v + $ Q, $t_{1/2} = 8.9$ Ma) provide insights into early crust formation and core-mantle segregation, respectively. After $\sim$6 half-lives, these isotope systems become extinct relative to our abilities of measurement. The ^{182}W isotope system records the fractionation of Hf (wholly stored in the mantle) from W (mostly ($\sim$90%) partitioned into the metallic core). It is predicted that the $(^{182}$W/^{184}W$)_{core}$ is

200 ppm lighter than that of the surrounding mantle and that this isotopic difference was frozen into the core within the first 100 million years of Earth's history (Kleine and Walker, 2017).

Significantly, none of the Ocean Island Basalts recording Hadean evidence of $^{182}W/^{184}W$ isotopic anomalies contains any evidence of anomalous $^{142}Nd/^{144}Nd$ isotopic compositions (Mundl-Petermeier *et al.*, 2020). Consequently, the source regions of basalts carrying Hadean isotopic signatures are probably much younger and are related to the recycling of oceanic crust subsequent to continent formation (Hofmann and White, 1982). In contrast, Eoarchaean continental rocks are known to have $^{142}Nd/^{144}Nd$ isotopic anomalies. This has led to the suggestion that these anomalous isotopic signatures have recently inherited ^{182}W along with primitive helium from the core (Mundl *et al.*, 2017; Mundl-Petermeier *et al.*, 2020).

Evidence for the mechanism(s) of core-mantle exchange remains elusive. Documented Hadean $^{182}W/^{184}W$ isotopic anomalies coupled with primordial noble gas signatures in OIBs present a preservation challenge for mantle convection models (Mundl *et al.*, 2017). Understanding the formation, nature, long-term stability, and distribution of these components in the mantle remains a significant area of study.

References

Abe, S., Asami, S., Eizuka, M., Futagi, S., Gando, A., Gando, Y., Gima, T., Goto, A., Hachiya, T., Hata, K., Hosokawa, K., Ichimura, K., Ieki, S., Ikeda, H., Inoue, K., Ishidoshiro, K., Kamei, Y., Kawada, N., Kishimoto, Y., Koga, M., Kurasawa, M., Maemura, N., Mitsui, T., Miyake, H., Nakahata, T., Nakamura, K., Nakamura, K., Nakamura, R., Ozaki, H., Sakai, T., Sambonsugi, H., Shimizu, I., Shirahata, Y., Shirai, J., Shiraishi, K., Suzuki, A., Suzuki, Y., Takeuchi, A., Tamae, K., Watanabe, H., Yoshida, Y., Obara, S., Ichikawa, A.K., Yoshida, S., Umehara, S., Fushimi, K., Kotera, K., Urano, Y., Berger, B.E., Fujikawa, B.K., Learned, J.G., Maricic, J., Axani, S.N., Fu, Z., Smolsky, J., Winslow, L.A., Efremenko, Y., Karwowski, H.J., Markoff, D.M., Tornow, W., Li, A., Detwiler, J.A., Enomoto, S., Decowski, M.P., Grant, C., Song, H., O'Donnell, T. and Dell'Oro, S. (2022) Abundances of uranium and thorium elements in earth estimated by geoneutrino spectroscopy. *Geophysical Research Letters*, **49**, e2022GL099566.

Agostini, M., Altenmüller, K., Appel, S., Atroshchenko, V., Bagdasarian, Z., Basilico, D., Bellini, G., Benziger, J., Biondi, R., Bravo, D., Caccianiga, B., Calaprice, F., Caminata, A., Cavalcante, P., Chepurnov, A., D'Angelo, D., Davini, S., Derbin, A., Di Giacinto, A., Di Marcello, V., Ding, X.F., Di Ludovico, A., Di Noto, L., Drachnev, I., Formozov, A., Franco, D., Galbiati, C., Ghiano, C., Giammarchi, M., Goretti, A., Göttel, A.S., Gromov, M., Guffanti, D., Ianni, A., Ianni, A., Jany, A., Jeschke, D., Kobychev, V., Korga, G., Kumaran, S., Laubenstein, M., Litvinovich, E., Lombardi, P., Lomskaya, I., Ludhova, L., Lukyanchenko, G., Lukyanchenko, L., Machulin, I., Martyn, J., Meroni, E., Meyer, M., Miramonti, L., Misiaszek, M., Muratova, V., Neumair, B., Nieslony, M., Nugmanov, R., Oberauer, L., Orekhov, V., Ortica, F., Pallavicini, M., Papp, L., Pelicci, L., Penek, Pietrofaccia, L., Pilipenko, N., Pocar, A., Raikov, G., Ranalli, M.T., Ranucci, G., Razeto, A., Re, A., Redchuk, M., Romani, A., Rossi, N., Schönert, S., Semenov, D., Settanta, G., Skorokhvatov, M., Singhal, A., Smirnov, O., Sotnikov, A., Suvorov, Y., Tartaglia, R., Testera, G., Thurn, J., Unzhakov, E., Villante, F.L., Vishneva, A., Vogelaar, R.B., von Feilitzsch, F., Wojcik, M., Wurm, M., Zavatarelli, S., Zuber, K. and Zuzel, G. (2020) Experimental evidence of neutrinos produced in the CNO fusion cycle in the Sun. *Nature*, **587**, 577–582. doi:10.1038/s41586-020-2934-0.

An, F., An, G., An, Q., Antonelli, V., Baussan, E., Beacom, J., Bezrukov, L., Blyth, S., Brugnera, R., Avanzini, M.B., Busto, J., Cabrera, A., Cai, H., Cai, X., Cammi, A., Cao, G., Cao, J., Chang, Y., Chen, S., Chen, S.,

Chen, Y., Chiesa, D., Clemenza, M., Clerbaux, B., Conrad, J., D'Angelo, D., Kerret, H.D., Deng, Z., Deng, Z., Ding, Y., Djurcic, Z., Dornic, D., Dracos, M., Drapier, O., Dusini, S., Dye, S., Enqvist, T., Fan, D., Fang, J., Favart, L., Ford, R., Göger-Neff, M., Gan, H., Garfagnini, A., Giammarchi, M., Gonchar, M., Gong, G., Gong, H., Gonin, M., Grassi, M., Grewing, C., Guan, M., Guarino, V., Guo, G., Guo, W., Guo, X.H., Hagner, C., Han, R., He, M., Heng, Y., Hsiung, Y., Hu, J., Hu, S., Hu, T., Huang, H., Huang, X., Huo, L., Ioannisian, A., Jeitler, M., Ji, X., Jiang, X., Jollet, C., Kang, L., Karagounis, M., Kazarian, N., Krumshteyn, Z., Kruth, A., Kuusiniemi, P., Lachenmaier, T., Leitner, R., Li, C., Li, J., Li, W., Li, W., Li, X., Li, X., Li, Y., Li, Y., Li, Z.B., Liang, H., Lin, G.L., Lin, T., Lin, Y.H., Ling, J., Lippi, I., Liu, D., Liu, H., Liu, H., Liu, J., Liu, J., Liu, J., Liu, Q., Liu, S., Liu, S., Lombardi, P., Long, Y., Lu, H., Lu, J., Lu, J., Lu, J., Lubsandorzhiev, B., Ludhova, L., Luo, S., Vladimir Lyashuk, Möllenberg, R., Ma, X., Mantovani, F., Mao, Y., Mari, S.M., McDonough, W.F., Meng, G., Meregaglia, A., Meroni, E., Mezzetto, M., Miramonti, L., Thomas Mueller, Naumov, D., Oberauer, L., Ochoa-Ricoux, J.P., Olshevskiy, A., Ortica, F., Paoloni, A., Peng, H., Jen-Chieh Peng, Previtali, E., Qi, M., Qian, S., Qian, X., Qian, Y., Qin, Z., Raffelt, G., Ranucci, G., Ricci, B., Robens, M., Romani, A., Ruan, X., Ruan, X., Salamanna, G., Shaevitz, M., Valery Sinev, Sirignano, C., Sisti, M., Smirnov, O., Soiron, M., Stahl, A., Stanco, L., Steinmann, J., Sun, X., Sun, Y., Taichenachev, D., Tang, J., Tkachev, I., Trzaska, W., van Waasen, S., Volpe, C., Vorobel, V., Votano, L., Wang, C.H., Wang, G., Wang, H., Wang, M., Wang, R., Wang, S., Wang, W., Wang, Y., Wang, Y., Wang, Y., Wang, Z., Wang, Z., Wang, Z., Wang, Z., Wei, W., Wen, L., Wiebusch, C., Wonsak, B., Wu, Q., Wulz, C.E., Wurm, M., Xi, Y., Xia, D., Xie, Y., Zhi-zhong Xing, Xu, J., Yan, B., Yang, C., Yang, C., Yang, G., Yang, L., Yang, Y., Yao, Y., Yegin, U., Yermia, F., You, Z., Yu, B., Yu, C., Yu, Z., Zavatarelli, S., Zhan, L., Zhang, C., Zhang, H.H., Zhang, J., Zhang, J., Zhang, Q., Zhang, Y.M., Zhang, Z., Zhao, Z., Zheng, Y., Zhong, W., Zhou, G., Zhou, J., Zhou, L., Zhou, R., Zhou, S., Zhou, W., Zhou, X., Zhou, Y., Zhou, Y. and Zou, J. (2016) Neutrino physics with JUNO. *Journal of Physics G: Nuclear and Particle Physics*, **43**, 030401. doi:10.1088/0954-3899/43/3/030401.

Anderson, O. and Isaak, D. (2002) Another look at the core density deficit of earth's outer core. *Physics of the Earth and Planetary Interiors*, **131**, 19–27.

Andringa, S., Arushanova, E., Asahi, S., Askins, M., Auty, D.J., Back, A.R., Barnard, Z., Barros, N., Beier, E. W., Bialek, A., Biller, S.D., Blucher, E., Bonventre, R., Braid, D., Caden, E., Callaghan, E., Caravaca, J., Carvalho, J., Cavalli, L., Chauhan, D., Chen, M., Chkvorets, O., Clark, K., Cleveland, B., Coulter, I.T., Cressy, D., Dai, X., Darrach, C., Davis-Purcell, B., Deen, R., Depatie, M.M., Descamps, F., Di Lodovico, F., Duhaime, N., Duncan, F., Dunger, J., Falk, E., Fatemighomi, N., Ford, R., Gorel, P., Grant, C., Grullon, S., Guillian, E., Hallin, A.L., Hallman, D., Hans, S., Hartnell, J., Harvey, P., Hedayatipour, M., Heintzelman, W.J., Helmer, R.L., Hreljac, B., Hu, J., Iida, T., Jackson, C.M., Jelley, N.A., Jillings, C., Jones, C., Jones, P.G., Kamdin, K., Kaptanoglu, T., Kaspar, J., Keener, P., Khaghani, P., Kippenbrock, L., Klein, J. R., Knapik, R., Kofron, J.N., Kormos, L.L., Korte, S., Kraus, C., Krauss, C.B., Labe, K., Lam, I., Lan, C., Land, B.J., Langrock, S., LaTorre, A., Lawson, I., Lefeuvre, G.M., Leming, E.J., Lidgard, J., Liu, X., Liu, Y., Lozza, V., Maguire, S., Maio, A., Majumdar, K., Manecki, S., Maneira, J., Marzec, E., Mastbaum, A., McCauley, N., McDonald, A.B., McMillan, J.E., Mekarski, P., Miller, C., Mohan, Y., Mony, E., Mottram, M.J., Novikov, V., O'Keeffe, H.M., O'Sullivan, E., Orebi Gann, G.D., Parnell, M.J., Peeters, S.J.M., Pershing, T., Petriw, Z., Prior, G., Prouty, J.C., Quirk, S., Reichold, A., Robertson, A., Rose, J., Rosero, R., Rost, P.M., Rumleskie, J., Schumaker, M.A., Schwendener, M.H., Scislowski, D., Secrest, J., Seddighin, M., Segui, L., Seibert, S., Shantz, T., Shokair, T.M., Sibley, L., Sinclair, J.R., Singh, K., Skensved, P., Sörensen, A., Sonley, T., Stainforth, R., Strait, M., Stringer, M.I., Svoboda, R., Tatar, J., Tian, L., Tolich, N., Tseng, J., Tseung, H.W.C., Van Berg, R., Vázquez-Jáuregui, E., Virtue, C., von Krosigk, B., Walker, J.M.G., Walker, M., Wasalski, O., Waterfield, J., White, R.F., Wilson, J.R., Winchester, T.J., Wright, A., Yeh, M., Zhao, T. and Zuber, K. (2016) Current status and future prospects of the SNO+ experiment. *Advances in High Energy Physics*, 2016, 6194250. doi:10.1155/2016/6194250.

Araki, T., Enomoto, S., Furuno, K., Gando, Y., Ichimura, K., Ikeda, H., Inoue, K., Kishimoto, Y., Koga, M., Koseki, Y., Maeda, T., Mitsui, T., Motoki, M., Nakajima, K., Ogawa, H., Ogawa, M., Owada, K., Ricol, J.S., Shimizu, I., Shirai, J., Suekane, F., Suzuki, A., Tada, K., Takeuchi, S., Tamae, K., Tsuda, Y., Watanabe, H., Busenitz, J., Classen, T., Djurcic, Z., Keefer, G., Leonard, D., Piepke, A., Yakushev, E., Berger, B.E., Chan, Y.D., Decowski, M.P., Dwyer, D.A., Freedman, S.J., Fujikawa, B.K., Goldman,

J., Gray, F., Heeger, K.M., Hsu, L., Lesko, K.T., Luk, K.B., Murayama, H., O'Donnell, T., Poon, A.W.P., Steiner, H.M., Winslow, L.A., Mauger, C., McKeown, R.D., Vogel, P., Lane, C.E., Miletic, T., Guillian, G., Learned, J.G., Maricic, J., Matsuno, S., Pakvasa, S., Horton-Smith, G.A., Dazeley, S., Hatakeyama, S., Rojas, A., Svoboda, R., Dieterle, B.D., Detwiler, J., Gratta, G., Ishii, K., Tolich, N., Uchida, Y., Batygov, M., Bugg, W., Efremenko, Y., Kamyshkov, Y., Kozlov, A., Nakamura, Y., Karwowski, H.J., Markoff, D. M., Nakamura, K., Rohm, R.M., Tornow, W., Wendell, R., Chen, M.J., Wang, Y.F. and Piquemal, F. (2005) Experimental investigation of geologically produced antineutrinos with KamLAND. *Nature*, **436**, 499–503. doi:10.1038/nature03980.

Ballmer, M.D., Houser, C., Hernlund, J.W., Wentzcovitch, R.M. and Hirose, K. (2017) Persistence of strong silica-enriched domains in the Earth's lower mantle. *Nature Geoscience*, **10**, 236–240. doi:10.1038/ngeo2898.

Birch, F. (1952) Elasticity and constitution of the Earth's interior. *Journal of Geophysical Research*, **57**, 227–286. doi:10.1029/JZ057i002p00227.

van Boekel, R.J.H.M., Min, M., Leinert, C., Waters, L.B.F.M., Richichi, A., Chesneau, O., Dominik, C., Jaffe, W., Dutrey, A., Graser, U., Henning, T., de Jong, J., Köhler, R., de Koter, A., Lopez, B., Malbet, F., Morel, S., Paresce, F., Perrin, G., Preibisch, T., Przygodda, F., Schöller, M. and Wittkowski, M. (2004) The building blocks of planets within the 'terrestrial' region of protoplanetary disks. *Nature*, **432**, 479–482. doi:10. 1038/nature03088.

Bonatti, E., Brunelli, D., Fabretti, P., Ligi, M., Portaro, R.A. and Seyler, M. (2001) Steady-state creation of crust-free lithosphere at cold spots in mid-ocean ridges. *Geology*, **29**, 979–982.

Bouwman, J., Lawson, W.A., Juhász, A., Dominik, C., Feigelson, E.D., Henning, T., Tielens, A.G.G.M. and Waters, L.B.F.M. (2010) The protoplanetary disk around the M4 star RECX 5: witnessing the influence of planet formation? *The Astrophysical Journal Letters*, **723**, L243–L247. doi:10.1088/2041-8205/723/2/L243.

Boyet, M. and Carlson, R.W. (2005) [142]Nd evidence for early (>4.53 Ga) global differentiation of the silicate Earth. *Science*, **309**, 576–581. doi:10.1126/science.1113634.

Cottaar, S. and Lekic, V. (2016) Morphology of seismically slow lower-mantle structures. *Geophysical Supplements to the Monthly Notices of the Royal Astronomical Society*, **207**, 1122–1136.

D'Alessio, P., Calvet, N. and Hartmann, L. (2001) Accretion disks around young objects. III. Grain growth. *The Astrophysical Journal*, **553**, 321. doi:10.1086/320655.

Dziewonski, A.M. and Anderson, D.L. (1981) Preliminary reference Earth model. *Physics of the Earth and Planetary Interiors*, **25**, 297–356. doi:10.1016/0031-9201(81)90046-7.

Fei, Y., Murphy, C., Shibazaki, Y., Shahar, A. and Huang, H. (2016) Thermal equation of state of hcp-iron: Constraint on the density deficit of Earth's solid inner core. *Geophysical Research Letters*, **43**, 6837–6843.

French, S.W. and Romanowicz, B. (2015) Broad plumes rooted at the base of the Earth's mantle beneath major hotspots. *Nature*, **525**, 95–99.

Fukao, Y. and Obayashi, M. (2013) Subducted slabs stagnant above, penetrating through, and trapped below the 660 km discontinuity. *Journal of Geophysical Research: Solid Earth*, **118**, 5920–5938.

Garnero, E.J., McNamara, A.K. and Shim, S.H. (2016) Continent-sized anomalous zones with low seismic velocity at the base of Earth's mantle. *Nature Geoscience*, **9**, 481–489.

Hart, S.R. and Zindler, A. (1986) In search of a bulk-Earth composition. *Chemical Geology*, **57**, 247–267. doi:10.1016/0009-2541(86)90053-7.

Hofmann, A.W. and White, W.M. (1982) Mantle plumes from ancient oceanic crust. *Earth and Planetary Science Letters*, **57**, 421–436.

Huang, S., Tschauner, O., Yang, S., Humayun, M., Liu, W., Corder, S.N.G., Bechtel, H.A. and Tischler, J. (2020) HIMU geochemical signature originating from the transition zone. *Earth and Planetary Science Letters*, **542**, 116323.

Jackson, M., Becker, T. and Steinberger, B. (2021) Spatial characteristics of recycled and primordial reservoirs in the deep mantle. *Geochemistry, Geophysics, Geosystems*, **22**, e2020GC009525.

Jackson, M., Konter, J. and Becker, T. (2017) Primordial helium entrained by the hottest mantle plumes. *Nature*, **542**, 340–343.

Jagoutz, E., Palme, H., Baddenhausen, H., Blum, K., Cendales, M., Dreibus, G., Spettel, B., Lorenz, V. and Wänke, H. (1979) The abundances of major, minor and trace elements in the Earth's mantle as derived from primitive ultramafic nodules. *Lunar and Planetary Science Conference Proceedings*, pp. 2031–2050.

Katsura, T. (2022) A revised adiabatic temperature profile for the mantle. *Journal of Geophysical Research: Solid Earth*, **127**, e2021JB023562.

Kim, D., Lekić, V., Ménard, B., Baron, D. and Taghizadeh-Popp, M. (2020) Sequencing seismograms: A panoptic view of scattering in the core-mantle boundary region. *Science*, **368**, 1223–1228.

Kleine, T. and Walker, R.J. (2017) Tungsten isotopes in planets. *Annual Review of Earth and Planetary Sciences*, **45**, 389–417.

Kraus, R.G., Hemley, R.J., Ali, S.J., Belof, J.L., Benedict, L.X., Bernier, J., Braun, D., Cohen, R., Collins, G.W., Coppari, F., *et al.* (2022) Measuring the melting curve of iron at super-earth core conditions. *Science*, **375**, 202–205.

Lodders, K. (2020) Solar Elemental Abundances. Pp. 1–68 in: *Oxford Research Encyclopedia of Planetary Science* (P. Read, editor). Oxford University Press, Oxford, UK. doi:10.1093/acrefore/9780190647926.013.145.

Marty, B. (2012) The origins and concentrations of water, carbon, nitrogen and noble gases on Earth. *Earth and Planetary Science Letters*, **313**, 56–66. doi:10.1016/j.epsl.2011.10.040.

Masters, G. and Gubbins, D. (2003) On the resolution of density within the Earth. *Physics of the Earth and Planetary Interiors*, **140**, 159–167. doi:10.1016/j.pepi.2003.07.008.

McDonough, W.F. (2014) Compositional model for the Earth's core. Pp. 559–577 in: *The Mantle and Core* (R.W. Carlson, editor). Treatise on Geochemistry, editors-in-chief: H.D. Holland and K.K. Turekian (second edition), Elsevier, Oxford, UK.. doi:10.1016/B978-0-08-095975-7.00215-1.

McDonough, W.F. (2017) Earth's core. Pp. 1–13 in: *Encyclopedia of Geochemistry: A Comprehensive Reference Source on the Chemistry of the Earth* (W.M. White, editor). Springer International Publishing, doi:10.1007/978-3-319-39193-9_258-1.

McDonough, W.F. and Sun, S.S. (1995) The composition of the Earth. *Chemical Geology*, **120**, 223–253. doi:10.1016/0009-2541(94)00140-4.

McDonough, W.F., Šrámek, O. and Wipperfurth, S.A. (2020) Radiogenic power and geoneutrino luminosity of the Earth and other terrestrial bodies through time. *Geochemistry, Geophysics, Geosystems*, **21**, e2019GC008865. doi:10.1029/2019GC008865.

McDonough, W.F. and Yoshizaki, T. (2021) Terrestrial planet compositions controlled by accretion disk magnetic field. *Progress in Earth and Planetary Science*, **8**, 39. doi:10.1186/s40645-021-00429-4.

Mundl, A., Touboul, M., Jackson, M.G., Day, J.M., Kurz, M.D., Lekic, V., Helz, R.T. and Walker, R.J. (2017) Tungsten-182 heterogeneity in modern ocean island basalts. *Science*, **356**, 66–69.

Mundl-Petermeier, A., Walker, R., Fischer, R., Lekic, V., Jackson, M. and Kurz, M. (2020) Anomalous ^{182}W in high ^{3}He/^{4}He ocean island basalts: Fingerprints of Earth's core? *Geochimica et Cosmochimica Acta*, **271**, 194–211.

Ohtani, E. (2021) Hydration and dehydration in Earth's interior. *Annual Review of Earth and Planetary Sciences*, **49**, 253–278.

O'Nions, R. and Oxburgh, E. (1983) Heat and helium in the Earth. *Nature*, **306**, 429–431.

Palme, H. and O'Neill, H.S.C. (2014) Cosmochemical estimates of mantle composition. Pp. 1–39 in: *The Mantle and Core* (R.W. Carlson, editor). Volume 3 of Treatise on Geochemistry – Editors-in-chief: H.D. Holland and K.K. Turekian (second edition), Elsevier, Oxford, UK. doi:10.1016/B978-0-08-095975-7.00201-1.

Ringwood, A. (1989) Significance of the terrestrial Mg/Si ratio. *Earth and Planetary Science Letters*, **95**, 1–7.

Ringwood, A. (1994) Role of the transition zone and 660 km discontinuity in mantle dynamics. *Physics of the Earth and Planetary Interiors*, **86**, 5–24.

Ringwood, A.E. (1982) Phase transformations and differentiation in subducted lithosphere: Implications for mantle dynamics, basalt petrogenesis, and crustal evolution. *The Journal of Geology*, **90**, 611–643.

Rudnick, R.L. and Gao, S. (2014) Composition of the continental crust. Pp. 1–51 in: *Treatise on Geochemistry* (Second Edition). Editors-in-Chief: H.D. Holland and K.K. Turekian (editors), Elsevier, Oxford, UK. doi:10.1016/B978-0-08-095975-7.00301-6.

Rudolph, M.L., Lekić, V. and Lithgow-Bertelloni, C. (2015) Viscosity jump in Earth's mid-mantle. *Science*, **350**, 1349–1352.

Sakamaki, T., Ohtani, E., Fukui, H., Kamada, S., Takahashi, S., Sakairi, T., Takahata, A., Sakai, T., Tsutsui, S., Ishikawa, D., *et al.* (2016) Constraints on Earth's inner core composition inferred from measurements of the sound velocity of hcp-iron in extreme conditions. *Science Advances*, **2**, e1500802.

Salters, V.J. and Stracke, A. (2004) Composition of the depleted mantle. *Geochemistry, Geophysics, Geosystems,* **5**. doi:10.1029/2003GC000597

Sammon, L.G. and McDonough, W.F., 2022. Quantifying Earth's radiogenic heat budget. *Earth and Planetary Science Letters*, **593**, 117684.

Sammon, L.G., McDonough, W.F. and Mooney, W.D. (2022) Compositional attributes of the deep continental crust inferred from geochemical and geophysical data. *Journal of Geophysical Research: Solid Earth*, **127**, e2022JB024041.

Sarafian, E., Gaetani, G.A., Hauri, E.H. and Sarafian, A.R. (2017) Experimental constraints on the damp peridotite solidus and oceanic mantle potential temperature. *Science*, **355**, 942–945. doi:10.1126/science.aaj2165.

Seyler, M., Brunelli, D., Toplis, M.J. and Mével, C. (2011) Multiscale chemical heterogeneities beneath the eastern southwest Indian ridge (52°E–68°E): Trace element compositions of along-axis dredged peridotites. *Geochemistry, Geophysics, Geosystems* **12**. https://doi.org/10.1029/2011GC003585

Shanker, J., Singh, B. and Srivastava, S. (2004) Volume–temperature relationship for iron at 330 GPa and the Earth's core density deficit. *Physics of the Earth and Planetary Interiors*, **147**, 333–341.

Tsuchiya, T., Kawai, K., Wang, X., Ichikawa, H. and Dekura, H. (2016) Temperature of the lower mantle and core based on ab initio mineral physics data. Deep Earth. *Physics and Chemistry of the Lower Mantle and Core*, 13–30.

Wasson, J.T. and Kallemeyn, G.W. (1988) Compositions of chondrites. *Philosophical Transactions of the Royal Society of London A: Mathematical, Physical and Engineering Sciences*, **325**, 535–544. doi:10.1098/rsta.1988.0066.

Williams, Q., Jeanloz, R., Bass, J., Svendsen, B. and Ahrens, T.J. (1987) The melting curve of iron to 250 gigapascals: A constraint on the temperature at earth's center. *Science*, **236**, 181–182.

Workman, R.K. and Hart, S.R. (2005) Major and trace element composition of the depleted MORB mantle (DMM). *Earth and Planetary Science Letters*, **231**, 53–72.

Yu, S. and Garnero, E.J. (2018) Ultralow velocity zone locations: A global assessment. *Geochemistry, Geophysics, Geosystems*, **19**, 396–414.

Zhao, D., Yu, S. and Ohtani, E. (2011) East Asia: Seismotectonics, magmatism and mantle dynamics. *Journal of Asian Earth Sciences*, **40**, 689–709.

EMU Notes in Mineralogy, Vol. 21 (2024), Chapter 2, 19–38

Plumes from the heterogeneous Earth's mantle

Cinzia G. FARNETANI

*Institut de Physique du Globe de Paris and Université de Paris-Cité,
1, rue Jussieu, 75238 Paris, Cedex 05 France
e-mail: cinzia@ipgp.fr*

The spectrum of geochemical compositions of Oceanic Island Basalts (OIBs) and their systematic differences from Mid-Ocean Ridge Basalts (MORBs) reveal that the Earth's mantle is chemically and isotopically heterogeneous. Two main processes, both related to plate tectonics, contribute to the creation of mantle heterogeneities: (1) partial melting generates melts enriched in incompatible elements and leaves a depleted residual rock; and (2) subduction of the oceanic lithosphere injects heterogeneous material at depth, in particular, altered oceanic crust and continental/oceanic sediments. Moreover, delamination and foundering of metasomatized subcontinental lithospheric mantle might have been important in the early Earth history, when plate tectonics did not operate as today. The fate of the subducted plate is still a matter of debate; presumably some of it is stirred by convection and some may segregate at the base of the mantle, in particular the oceanic crust, which is compositionally denser than the pyrolitic mantle. The view of the lower mantle as a "graveyard" of subducted crust prevailed for decades and was supported by the Hofmann and White (1982) observation that the geochemical fingerprint of most OIB reveals the presence of ancient recycled crust. However, recent geochemical data on short-lived systems (*e.g.* $^{182}Hf \rightarrow ^{182}W$ has a half-life of 8.9 My) showed that some hotspots, namely Hawaii, Samoa, Iceland and Galápagos, have a negative $\mu^{182}W$ anomaly. This discovery prompted a change in our view of the deep mantle because anomalies in short-lived systems require additional processes, which include, but are not limited to, the preservation of 'pockets' of melt from a primordial magma ocean, and/or chemical reactions between the metallic core and the silicate mantle. Exchanges at the core-mantle boundary would cause a negative $\mu^{182}W$ anomaly, and might also add ^3He to mantle material later entrained by plumes. It is now clear that some plumes probe the deepest mantle and are highly heterogeneous, as revealed by isotope ratios from long-lived radiogenic systems, noble gases and short-lived isotope systems. Here I will focus on the dynamics of plumes carrying compositional and rheological heterogeneities. This contribution attempts to be pedagogic and multi-disciplinary, spanning from seismology to geochemistry and geodynamics.

1. Hotspots and mantle plumes

Hotspots are regions of intraplate volcanism, or of excessive volcanism along a portion of a spreading ridge (Koppers *et al.*, 2021). Their long-lasting magmatic activity can form an age-progressive volcanic chain, the best example being the Hawaiian hotspot: volcanoes younger than 1 Ma are still active (*i.e.* Kilauea and Loihi), whereas older, extinct, volcanoes form the 6000 km-long Hawaiian–Emperor chain (Fig. 1). The age of the oldest volcano (80 Ma) attests to the long-lived activity of the hotspot, whereas any information about previous volcanism is missing since the Pacific plate subducts at the

DOI: 10.1180/EMU-notes.21.2

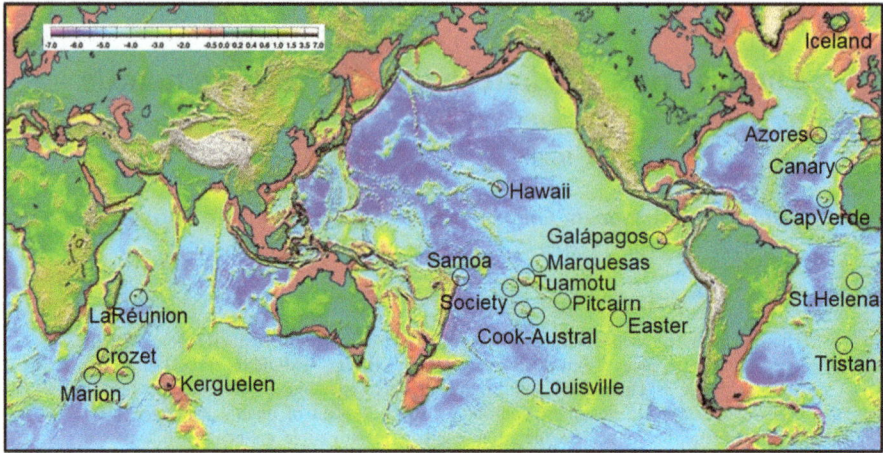

Fig. 1. Global map of sea-floor depth (Smith and Sandwell, 1997) and continental elevation. Only some hotspots are indicated; for a map of all hotspots see Ito and van Keken (2007).

Aleutians subduction zone. Ito and van Keken (2007) indicated that hotspots younger than 100 Ma are generally active, although their vigour may vary considerably, whereas older hotspots are either waning or inactive.

Morgan (1971) first suggested that hotspots are the surface expression of plumes rising from the deep mantle. Ever since, the existence of plumes and their depth of origin have been hotly debated. Courtillot *et al.* (2003) proposed five criteria to assess whether a plume has a deep origin: (1) a hotspot fed by deep, relatively fixed, plume should form an age progressive volcanic chain the linear trend of which is consistent with the direction of plate motion. According to Ito and van Keken (2007) only 13 long-lived (> 50 Ma) and eight short-lived (<20 Ma) volcanic chains are age-progressive, whereas all other volcanic chains lack a clear age progression. (2) The onset of hotspot magmatism should be marked by the emplacement of a Large Igneous Province (LIP), a term including continental flood basalts and oceanic plateaus, formed in continental and oceanic settings, respectively. Note that both criteria are associated with the plume shape constrained by pioneering laboratory experiments (Whitehead and Luther, 1975), namely a large "head", the partial melting of which generates a LIP, and a narrow, long-lasting, plume conduit or "tail", the melting of which generates a volcanic chain (Richards *et al.*, 1989). (3) A deep plume is expected to have a large buoyancy flux. This parameter is evaluated from the topographic swell of a hotspot. The swell is a region with an anomalously high topography (swell heights range from 500 to 1200 m) over lateral widths of 1000–1500 km, in the direction perpendicular to the volcanic chain (Crough, 1983; Wessel, 1993). Sleep (1990) estimated the buoyancy flux of Hawaii, $B_{Hawaii} = 8.7 \times 10^3$ kg s^{-1} and the total buoyancy flux of all hotspots $B_{Total} = 55 \times 10^3$ kg s^{-1}. Recent estimates propose $B_{Total} = 22 \times 10^3$ kg s^{-1} (King and Adam, 2014) and $B_{Total} = 46 \times 10^3$ kg s^{-1} (Hoggard *et al.*, 2020). The last two criteria, proposed by Courtillot *et al.* (2003), rely on geochemical observations, namely large ^{3}He/^{4}He for hotspot lavas fed by a deep mantle plume, and on seismological observations, namely a low shear wave velocity zone in the underlying mantle. Today there is a general agreement that 18 mantle plumes originate from the deepest regions of the Earth's mantle (Koppers *et al.*, 2021).

One important parameter not mentioned above is the plume excess temperature with respect to the surrounding mantle. Following Putirka (2005) and Bao *et al.* (2022), the plume excess temperature for the most vigorous hotspots ranges between 100 and 250°C. This excess temperature enables plumes to melt even below a thick lithosphere (*e.g.* Hawaii) or to generate excess magmatism along some parts of a spreading ridge (*e.g.* Iceland). Moreover, the excess temperature provides the buoyancy force necessary to ascend from the deep mantle. Last but not least, the excess temperature poses a first-order constraint on the location of the root zone of plumes (Stacey and Loper, 1983), likely to be the core–mantle boundary region, at 2700–2900 km depth, where the estimated temperature increment is 800–1000°C (Jeanloz and Morris, 1986).

2. Plumes and the Earth's mantle structure: seismic observations

Ever since the plume concept was proposed, the questions addressed to seismologists have been: "Can seismic tomography detect mantle plumes?" "Is there a deep seismic velocity anomaly beneath hotspots?"

Because of the uneven distribution of earthquakes and of seismic stations, and because of wavefront healing effects (Nolet and Dahlen, 2000) it is difficult to resolve narrow, low-velocity structures (Goes *et al.*, 2004). Using finite frequency tomography, Montelli *et al.* (2006) showed that zones with a negative shear-wave velocity anomaly were broader (*i.e.* 600–800 km diameter) than the plume conduits hypothesized by Morgan (1971). Moreover, conduits had a variable depth extent and sometimes lacked a clear depth continuity. This is in agreement with resolution analysis, indicating that plumes would not be visible if their diameter was <~600 km (Montelli *et al.*, 2006; French and Romanowicz, 2015). By using travel times of seismic core waves recorded by the dense USArray seismic network in North America, Nelson and Grand (2018) were able to detect a narrow, cylindrically shaped, slow-velocity anomaly, ~350 km in diameter, interpreted as the deep mantle plume beneath the Yellowstone hotspot.

Global tomography models have shown the existence of two low-velocity structures in the lower mantle below Africa and the central Pacific. The Large Low Shear Velocity Provinces (LLSVP) observed in both shear- and compressional-wave tomography, cover 25–30% of the CMB surface and rise 500–1000 km above it (Ritsema *et al.*, 1999; Masters *et al.*, 2000; Romanowicz and Gung, 2002). LLSVPs are frequently associated with mantle plumes beneath active hotspots (French and Romanowicz, 2015). Moreover, the reconstructed LIP position, obtained by calculating the plate motion back in time to the location where the LIP erupted, often corresponds to the edges of the LLSVPs (Burke and Torsvik, 2004). These reconstructions, together with other geophysical observations (*e.g.* Dziewonski *et al.*, 2010), suggest that the LLSVPs might have been stable during the last 250 My. LLSVPs are probably hotter than the surrounding mantle, but there is still a debate about their nature, either purely thermal or thermo-chemical. To better constrain their nature, Su and Dziewonski (1997) looked at relative perturbations in the shear velocity ($V_s = (\mu/\rho)^{1/2}$, where μ is the modulus of rigidity, ρ is density) and in bulk sound velocity ($V_\Phi = (K/\rho)^{1/2}$, where K is the bulk modulus) and found that in the lowermost mantle the two are negatively correlated. Interestingly, thermal variations alone do not generate anti-correlation; therefore, both Su and Dziewonski (1997) and Masters *et al.* (2000) suggested that compositional anomalies are required to explain seismic observations. Although the chemical composition of LLSVPs is unknown, the hypothesis of ancient subducted oceanic crust was proposed decades ago. This is not surprising because the subducted crust probably represents 10–15% the mass of the mantle (Stracke *et al.*, 2003) and because the oceanic crust remains compositionally denser even in the lowermost mantle (Ricolleau *et al.*, 2010; Stixrude and Lithgow-Bertelloni, 2012). Geodynamic models (Christensen and Hofmann, 1996; Brandenburg *et al.*, 2008) confirmed that dense recycled crust piles up at the base of the mantle, forming LLSVP-like structures. However, Deschamps *et al.* (2012) proposed that LLSVPs have a distinct, more primitive, composition enriched in iron by 3.0% and in (Mg,Fe)-perovskite by 20%, compared to regular mantle. The last ten years have seen the bloom of thermo-chemical geodynamic models, combining recycled crust and relatively primitive material (Li *et al.*, 2014; Nakagawa and Tackley, 2014; Tucker *et al.*, 2020; Jones at al., 2021). The results show that the LLSVPs may indeed be compositionally heterogeneous and that plumes may be fed by LLSVP material (see Fig. 2 for a schematic representation). However, even the latest geodynamic models do not capture the complex shape observed by the most recent tomographic studies, *e.g.* Tsekhmistrenko *et al.* (2021) found that mantle upwellings of the African LLSVP are arranged in a tree-like structure: from a central, compact

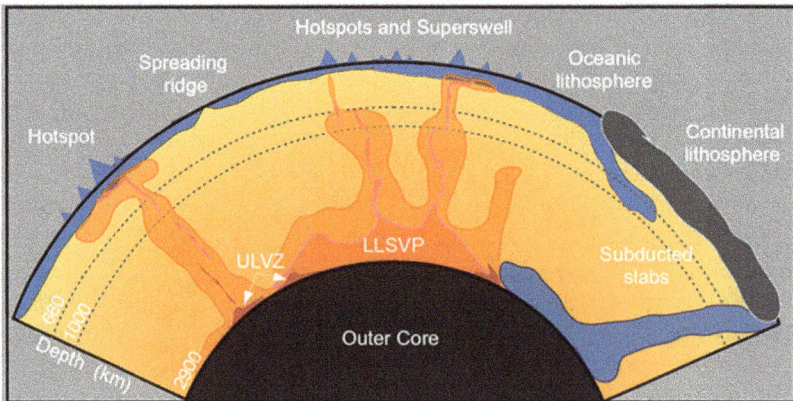

Fig. 2. Cartoon showing, in a simplistic way, various plume morphologies. For the LLSVP and the ULVZ the thickness is exaggerated for graphical reasons. Complexities such as plume–ridge interactions are not shown. The fate of subducted lithosphere is illustrated schematically. Refer to Fukao and Obayashi (2013) for tomographic images of slabs and to French and Romanowicz (2015) for tomographic images of LLSVP.

trunk below ~1500 km depth, three branches tilt outwards and up towards various Indo-Austral hotspots. Tomographic images by Wamba *et al.* (2023) of the African LLSVP in the Indian ocean, form the Comores, La Réunion and Crozet hotspots and indicate the presence of slow-velocity anomalies in the lower mantle, forming broad conduit-like structures with a diameter of ~900 km at 1500 km depth. At 1000 km depth the vertical continuity is disrupted and horizontal spreading seems to prevail between 1000 and 660 km depth. For the Pacific LLSVP, French and Romanowicz (2015) and Davaille and Romanowicz (2020) showed the presence of well-separated, low-velocity conduits that extend vertically throughout most of the lower mantle.

Seismologists have also discovered the Ultra Low Velocity Zones (ULVZ), patches on the core-mantle boundary, that exhibit reductions of S-wave velocity by as much as 30% and of P-wave velocity by up to 10% (*e.g.* Garnero and Helmberger, 1996; Rost and Revenaugh, 2003; Thorne and Garnero, 2004). ULVZs are 10–40 km thick, have a lateral extent of hundreds of km, and up to 1000 km for the Mega-ULVZs (Cottaar and Romanowicz, 2012; Kim *et al.*, 2020). Their density increase may be up to ~10% (Rost *et al.*, 2005). Several mechanisms have been proposed to explain their high density; just two are mentioned here: core–mantle reaction; and the presence of dense silicate melts or of iron-enriched oxides (see the review by McNamara (2019) and references therein).

3. Plumes and the Earth's mantle structure: geochemical observations

Oceanic island basalts have a different chemical composition in both trace and major elements with respect to Mid-Ocean-Basalts (Schilling, 1973). These differences are due, in part, to a lower degree of melting in plumes than at spreading ridges; however, the isotopic differences between OIBs and MORBs suggest an isotopically distinct source (Hart *et al.*, 1973; Hofmann and White, 1982; Hofmann, 2003; White, 2010, 2015b). The present study will focus on key isotopic ratios, starting with long-lived isotopes and moving on to short-lived isotopes.

3.1. The message from the long-lived radioactive decay systems

Isotopic systems with long-lived radioactive parent elements (*i.e.* half lives of billions of years) shed light on mantle processes that occurred over long time-scales, *e.g.*: (1) partial melting and melt extraction, leading to a residual mantle depleted in incompatible elements; (2) the generation of the crust, enriched in incompatible elements; (c) recycling of the oceanic crust and of portions of continental crust, in the form of marine sediments and/or delaminated lower crust; and (4) dynamic instabilities of the continental lithosphere (Cottrell *et al.*, 2004; Fourel *et al.*, 2013) leading to foundering of the sub-continental lithospheric mantle. All of these compositionally and isotopically distinct materials, possibly with distinct density and viscosity, are variably stirred by mantle convection and/or partially segregate in the deep mantle over long time-scales (1–2 Gy), before being entrained in mantle plumes.

Since the studies of Zindler *et al.* (1982) and Hart *et al.* (1992), radiogenic isotope ratios such as $^{143}Nd/^{144}Nd$, $^{87}Sr/^{86}Sr$, $^{206}Pb/^{204}Pb$ have been the 'backbone' of the definition of mantle end-member isotopic compositions (*i.e.* the Depleted MORB Mantle (DMM), the Enriched Mantle (EM-1 and EM-2), the High-μ (HIMU), where μ is the $^{238}U/^{204}Pb$ ratio). Clearly, these end-members should not be taken as physical entities that actually exist in the mantle. For a thorough discussion, which goes beyond the classical isotopic variability in 2D isotope ratio, see Stracke (2021) and Stracke *et al.* (2022). Here I will follow a more traditional, albeit simplistic, approach. In Fig. 3 is a pedagogic way to understand how, over time, the residual mantle and the crust develop distinct $^{143}Nd/^{144}Nd$ and $^{87}Sr/^{86}Sr$ ratios. Two aspects are noteworthy: first, strontium isotope ratios are negatively correlated with neodymium. Second, there is a complementarity between the residual mantle and the crust. Figure 3c shows that partial melting does not modify an isotopic ratio; this is important because, by measuring the isotopic ratios in basalts we obtain information about the source rock, assumed to be isotopically homogeneous.

MORB and OIB are isotopically variable, and the isotopic variation is non-random. For example, OIBs are systematically more radiogenic in strontium and less radiogenic in neodymium than MORBs, and they extend to more extreme values of $^{206}Pb/^{204}Pb$. OIBs do not have the isotopic composition of a pure end-member but form arrays between end-members. Detailed discussions of geochemical end-members can be found in many articles (Zindler *et al.*, 1982; Hart *et al.*, 1992; Hofmann, 1997; Stracke *et al.*, 2005; White 2015a, 2015b, and references therein); here, just a brief description is provided.

The HIMU end-member has a very radiogenic Pb isotopic composition (*i.e.* high $^{206}Pb/^{204}Pb$) which requires a mantle source with high U/Pb for time periods of the

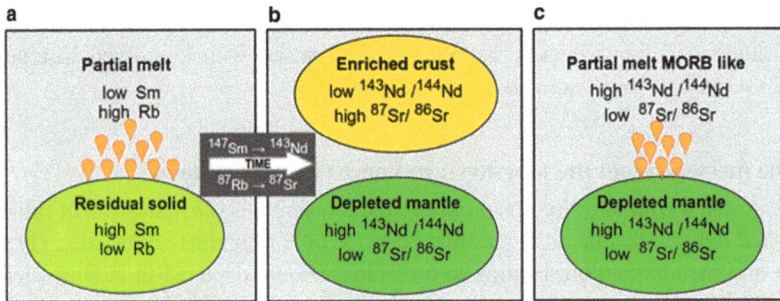

Fig. 3. (*a*) Partial melting in geologic times. For the system $^{147}Sm \rightarrow ^{143}Nd$, the parent element, samarium, is more compatible than neodymium. Thus, the residual solid has a high Sm (*i.e.* high Sm/Nd ratio) while the liquid magma has a low Sm. For the system $^{87}Rb \rightarrow ^{87}Sr$, the parent element, rubidium, is less compatible than strontium. Thus, the residual solid has a low Rb (*i.e.* low Rb/Sr ratio), while the liquid magma has a high Rb. (*b*) Over time, the parent element decays into the daughter element (^{147}Sm has a half-life of 106 Gy; ^{87}Sr has a half-life of 48.8 Gy). The residual solid, assimilated to present day depleted mantle, has a high $^{143}Nd/^{144}Nd$ and low $^{87}Sr/^{86}Sr$. The crust, formed from the crystallization of magmas, has low $^{143}Nd/^{144}Nd$ and high $^{87}Sr/^{86}Sr$. (*c*) Partial melting does not change an isotopic ratio.

order of ~1.5–2 Gy since the parent isotope ^{238}U has a long half-life (4.469 Gy). Only a few hotspots have a strong HIMU component, namely St. Helena in the Atlantic, the Cook-Austral islands (Polynesia) in the South Pacific, and some of the Galápagos islands (Harpp and Weis, 2020). However, Homrighausen *et al.* (2018) proposed that the HIMU end-member has a more global distribution than previously thought. What is the origin of the HIMU component? The classical view is that HIMU reflects the presence of recycled oceanic crust enriched in continental ^{238}U carried by sediments; alternatively, it might reflect the presence of recycled basaltic crust that lost fluid-mobile trace elements, such as Pb, because of hydrothermal alteration at ridges or dehydration during subduction (Chauvel *et al.*, 1992; Hofmann, 1997). Anomalous sulphur isotope signatures, indicating mass-independent fractionation (MIF), were found by Cabral *et al.* (2013) in sulfides from Mangaia's OIBs (Cook Islands). This is important, because MIF processes require photo-chemical reactions that occurred during the Archaean, when the atmosphere was oxygen-poor and relatively transparent to solar ultraviolet radiation. Cabral *et al.* (2013) proposed that the source of Mangaia's lavas carries ancient (>2.45 Gy old), hydrothermally altered oceanic crust. There is also a long-standing debate about the role of metasomatized lithosphere in HIMU basalts. Weiss *et al.* (2016) pointed out that olivine phenocrysts in HIMU lavas do not support melting of recycled crust but indicate melting of peridotite metasomatized by carbonatite fluids. The metasomatism probably occurred during the Archaean and affected subcontinental lithospheric mantle (Weiss *et al.*, 2016; Homrighausen *et al.*, 2018). According to this model, ancient, metasomatized subcontinental lithospheric mantle delaminated and cycled in the deep mantle before being entrained in plumes.

The enriched EM-1 end-member is present in hotspots such as Kerguelen, Tristan, Hawaii and Pitcairn (Eisele *et al.*, 2003; Weis *et al.*, 2011 and references therein). This component indicates the contribution of recycled ancient pelagic sediments and/ or of recycled delaminated subcontinental lithosphere. The enriched EM-2 end-member is observed in hotspots such as Society, Marquesas, Samoa (White and Hofmann, 1982; Jackson *et al.*, 2007; Chauvel *et al.*, 2012 and references therein) and is associated with a small fraction (~2%) of subducted terrigenous-continental sediments. Globally, the enriched components (*i.e.* high ^{87}Sr/^{86}Sr, low ^{143}Nd/^{144}Nd, unradiogenic Pb isotopic ratios) indicate that plumes carry subducted material, either continental sediments or oceanic crust, as first suggested by Hofmann and White (1982).

3.2. The message from the short-lived radioactive decay systems

During the last ten years, geochemists were able to define heterogeneities by measuring isotopic ratios (*e.g.* ^{182}W/^{184}W, ^{129}Xe/^{136}Xe, ^{142}Nd/^{144}Nd) from short-lived radioactive parent elements (Touboul *et al.*, 2012; Horan *et al.*, 2018; Rizo *et al.*, 2019). The parent element has such a short half-life that it became extinct soon after the formation of the solar system. In other words, only fractionation processes that occurred very early in the Earth's history could form these isotopic anomalies. The hafnium-tungsten systematics offer a clear example: ^{182}Hf has a half-life of 8.9 My, so that after 50 My no more daughter isotope ^{182}W could be produced. Today the observed μ^{182}W (*i.e.* ppm deviation relative terrestrial standard) in MORBs and most OIBs is close to zero, but hotspots such as Hawaii, Samoa, Iceland and Galápagos have negative $\mu^{182}W$, often associated with high $^3He/^4He$ (Mundl *et al.*, 2017; Jackson *et al.*, 2020; Mundl-Petermeier *et al.*, 2020; Peters *et al.*, 2021).

Short-lived isotope systems prompt us to consider processes that occurred very early in the Earth's history. The physical conditions that prevailed during the first 50 My of the Earth's life are still debated, but there is a general consensus that this period was highly energetic. Impacts of planetesimals and embryos generated enormous amounts of heat and the giant Moon-forming impact probably melted the entire Earth, forming a global silicate magma ocean (Stevenson, 1987; Canup, 2008; Elkins-Tanton, 2012; Bolrão *et al.*, 2021 and references therein). The possibility that heterogeneous domains could form during crystallization of the magma ocean is supported by first-principles molecular dynamics (Deng and Stixrude, 2021) showing that Hf is more compatible than W (see Fig. 4a). Even though it remains unclear if the solidification of a magma ocean proceeded "bottom up" (Andrault *et al.*, 2011) or from mid-depth (Stixrude *et al.*, 2009; Boukaré *et al.*, 2015), the residual liquids of a magma ocean are progressively enriched in incompatible elements; therefore, they are expected to have a high iron content and a sub-chondritic Hf/W (Brown *et al.*, 2014). This 'Enriched Magma-Ocean Residue', if preserved from 4 Gy of convective stirring, would still have a slightly negative $\mu^{182}W$ (see Fig. 4c, magenta colour).

Alternatively, the metallic core could be a reservoir with a strongly negative $\mu^{182}W$ (Touboul *et al.*, 2012; Rizo *et al.*, 2019), as W is a moderately siderophile element, whereas Hf is lithophile (see Fig. 4b,c). A key issue with this model is how to transfer a 'core signature' to the lowermost mantle. High-pressure experiments by Yoshino *et al.* (2020) suggest that is possible to transfer W from the core to the mantle because grain boundary diffusion of W is a fast and strongly temperature-dependent process. This supports the idea that a fraction of a lowermost mantle might be thermally and chemically equilibrated with the core. Mundl-Petermeier *et al.* (2020) proposed the existence of a "Core-Mantle equilibrated reservoir", with a core-like negative $\mu^{182}W$, and show that

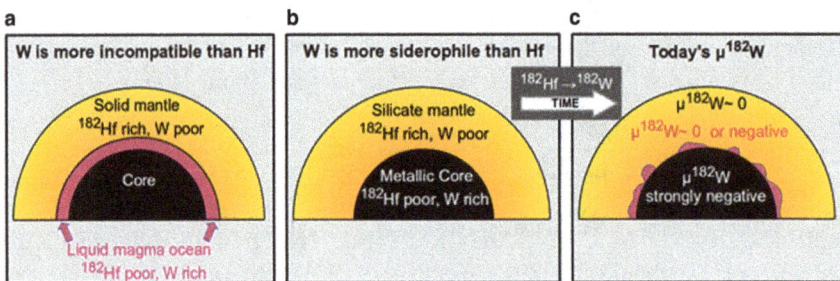

a W is more incompatible than Hf
Solid mantle
^{182}Hf rich, W poor
Core
Liquid magma ocean
^{182}Hf poor, W rich

b W is more siderophile than Hf
$^{182}Hf \rightarrow ^{182}W$
TIME
Silicate mantle
^{182}Hf rich, W poor
Metallic Core
^{182}Hf poor, W rich

c Today's $\mu^{182}W$
$\mu^{182}W \sim 0$
$\mu^{182}W \sim 0$ or negative
$\mu^{182}W$ strongly negative

Fig. 4. (*a–b*) At the time when the parent element ^{182}Hf was extant. (*a*) W is more incompatible than Hf, thus a residual liquid from a magma ocean is expected to be W rich and Hf poor. (*b*) W is more siderophile than Hf, thus the metallic core is W rich, whereas the silicate mantle is Hf rich. (*c*) By definition, $\mu^{182}W$ indicates the deviation of $^{182}W/^{182}W$ of the sample from the $^{182}W/^{182}W$ of a standard. The mantle has $\mu^{182}W \approx 0$, relics of the magma ocean are expected to have a slightly negative $\mu^{182}W$ and the core has a strongly negative $\mu^{182}W$.

only 0.3% of its entrainment in plumes would be sufficient to explain the most negative $\mu^{182}W$ observed in OIBs. Although highly speculative, it is tempting to associate a "Core-Mantle equilibrated reservoir" with the ULVZs, now detected beneath several hotspots.

4. Fluid dynamics of mantle plumes

Our view of mantle plumes has evolved considerably over time, moving from the 'classical' thermal plume in Newtonian rheology to the more complex thermo-chemical plumes.

4.1. Purely thermal plumes

Purely thermal plumes have one source of buoyancy, namely, their excess temperature. The morphology of a thermal plume in a Newtonian fluid depends on the viscosity contrast between the hot plume and the colder ambient mantle. For strongly temperature-dependent viscosity, the plume has a 'mushroom-shape' with a large head and a narrow tail, whereas for constant viscosity the plume has a 'spout' morphology, with a roughly constant diameter (Whitehead and Luther, 1975). The upwelling flow within the conduit is controlled by the viscosity contrast between the hot axial part and the colder periphery: the vertical velocity (Vz) is maximum at the plume axis and decreases exponentially with the square of the radial distance (Olson *et al.*, 1993). This velocity profile (Fig. 5) has two profound implications. First, Vz decreases with radial distance more rapidly than the excess temperature ΔT. Material with a low ΔT, either surrounding mantle conductively heated by the plume, or the most peripheral parts of the plume, have such a negligible upwelling velocity that they contribute very little to the plume buoyancy flux (Fig. 5) and may never reach the melting zone in the plume head. Thus, the key parameter to define entrained material should not be the excess temperature (*e.g.* Hauri *et al.*, 1994, used a ΔT value which is only 1% of the axial ΔT) but the upwelling velocity. Second, this velocity profile generates zones with high strain rates within the conduit so that passive (*i.e.* not affecting the flow) geochemical heterogeneities are stretched readily into filaments (Fig. 6) as they ascend in the plume tail (Farnetani and Hofmann, 2009). The existence of geochemically distinct 'streaks' in the Hawaiian plume was first proposed by Abouchami *et al.* (2005) in order to explain the spatio-temporal geochemical variability of volcanoes belonging to the Kea-trend. Numerical simulations later showed that a filament crosses the Hawaiian melting zone over timescales of the order of 1 My (Farnetani and Hofmann, 2010) so that two volcanoes can indeed sample the same heterogeneity. Obviously this model is very simplified, as it ignores mixing of partial melts *en route* to the surface and/or ponding of melts in a magma chamber. However, it shows that the laminar flow within the plume conduit does not induce toroidal stirring.

4.2. Thermo-chemical plumes

Thermo-chemical plumes have two sources of buoyancy: thermal and compositional. The nature of compositionally denser material is still elusive. Ancient recycled eclogitic

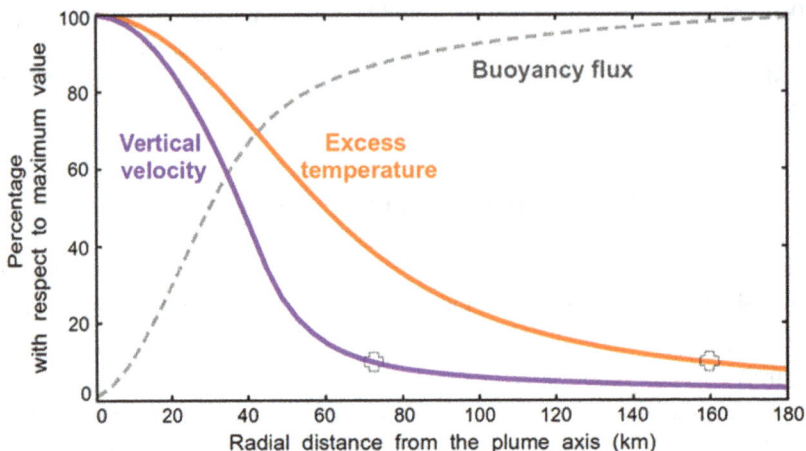

Fig. 5. Profiles of the vertical velocity, Vz, of the excess temperature ΔT, and of the buoyancy flux across the plume conduit. Both Vz_{max} and ΔT_{max} occur at the plume axis. Because of temperature-dependent viscosity, Vz decreases more rapidly than ΔT with radial distance. The open crosses indicate that Vz is 10% of Vz_{max} at 73 km distance, whereas ΔT is 10% of ΔT_{max} at 160 km distance. This implies that low excess temperature material (either from the plume periphery or entrained mantle conductively heated) cannot upwell efficiently and will probably never reach the melting zone. Note also that at the radial distance where Vz is 10% of Vz_{max}, the buoyancy flux $B \approx 87\% B_{max}$ meaning that the low-velocity plume periphery contributes very little to the total buoyancy flux. The buoyancy flux $B = \int \rho \ \alpha \ \Delta T \ Vz \ dS$, where the integral is over the horizontal surface S of the conduit, ρ is density and α the thermal expansion coefficient.

crust is denser than the pyrolitic mantle (Hirose *et al.*, 1999; Ricolleau *et al.*, 2010) while deep-seated portions of relatively primitive mantle might be iron-enriched relative to the depleted surrounding mantle. The effect of compositional density is important, *e.g.* Dannenberg and Sobolev (2015) estimated that a plume with an excess temperature of 250°C can entrain ~20% of dense eclogite before becoming neutrally buoyant. For thermo-chemical plumes the subtle balance between positive thermal buoyancy and negative compositional buoyancy can induce an oscillatory behavior (Davaille, 1999) and a complex internal dynamics, where parts of the conduit sink while others well up (Kumagai *et al.*, 2008). Moreover, thermo-chemical plumes have irregular and non-symmetric morphologies (Tackley, 1998; Farnetani and Samuel, 2005; Nakagawa and Tackley, 2014; Li *et al.*, 2014; Limare *et al.*, 2019). Another important difference is that in a purely thermal mantle a large-scale isotopic zonation across the source region translates into an isotopic zonation of the plume conduit (Farnetani *et al.*, 2012), whereas this might not be true in a thermo-chemical mantle. As pointed out by Jones *et al.* (2016) compositionally denser material rises preferentially at the plume axis, thereby disrupting any 'geochemical mapping' between the conduit and the source. As mentioned above we also do not know to which extent partial melts mix within a complex melt drainage system. In any case, the geochemical zonation of many hotspots is an observational fact, first recognized at Hawaii, where volcanoes younger than 5 Ma form two parallel and geochemically distinct chains. The difference between lavas from Kea- and Loa-trend volcanoes, clearly expressed in radiogenic lead isotope ratios, has been explained by a bilateral zonation (Abouchami *et al.*, 2005) of the Hawaiian plume. According to Weis *et al.* (2011) the southern Loa-side of the plume entrains geochemically enriched LLSVP material, whereas the northern Kea-side samples more depleted lower mantle. These geochemical considerations, taken alongside the results of Jones *et al.* (2016), indicate that the entrained LLSVP component is only slightly denser than the depleted mantle; otherwise the bilateral zonation could not be maintained. Interestingly, a bilateral zonation is observed in other hotspots, *e.g.* Samoa, Societies, Galápagos, Easter, and Marquesas Islands in the Pacific (Hoernle *et al.*, 2000; Huang *et al.*, 2011; Chauvel *et al.*, 2012; Payne *et al.*, 2012; Harpp and Weis, 2020) and the Tristan-Gough hotspot track in the Atlantic (Rohde *et al.*, 2013; Hoernle *et al.*, 2015), but see Stracke (2021) for a different interpretation.

4.3. The role of a distinct rheology

Most models consider that the mantle has a Newtonian rheology, while Davaille *et al.* (2018) explored the effect of a visco-plastic rheology. In such a case, the flow occurs only when the local deviatoric stress is greater than a critical yield stress. This rheology affects both the plume morphology, which becomes broader, and the vertical velocity profile across the conduit, which becomes smoother than shown in Fig. 5.

The Earth's mantle could also be rheologically heterogeneous, with volumes of rocks characterized by a distinct viscosity, caused by an increase in the mineral-grain size (Ammann *et al.*, 2010), by an increase in the silica content (Yamazaki *et al.*, 2000) or by variations in the water content (Hirth and Kohlstedt, 1996; Karato, 2010). Zones with an increased viscosity have been modelled at different length-scales, *e.g.* Ballmer *et al.* (2017) considered the hypothesis that the whole lower mantle might be more viscous because of a silica enrichment. Gülcher *et al.* (2020) explored a wide range of parameters (*i.e.* excess compositional density and/or viscosity) and mapped distinct regimes of thermo-chemical convection, ranging from efficient mixing to variable degrees of preservation of the heterogeneities forming piles, domes or isolated blobs. Far-netani *et al.* (2018) focused instead on mantle plumes carrying finite size (30–40 km radius) rheological heterogeneities 20–30 times more viscous than the surrounding rocks. Their numerical simulations showed that the heterogeneities do not stretch, they simply rotate as they well up in the plume conduit (Fig. 6). Intrinsically more viscous material, that resists stretching and mixing, is an ideal candidate to preserve a distinct isotopic fingerprint. In such a case, the presence of rheological heterogeneities crossing the plume melting zone, would induce 'pulses' of geochemically distinct material with a time-scale of ∼1 My.

5. Concluding remarks and future perspectives

In this brief review I tried to combine insights from seismology, geochemistry and fluid dynamics in order to constrain the existence, the depth of origin and the heterogeneous nature of plumes. I think it is now clear that some plumes have a deep origin and thus

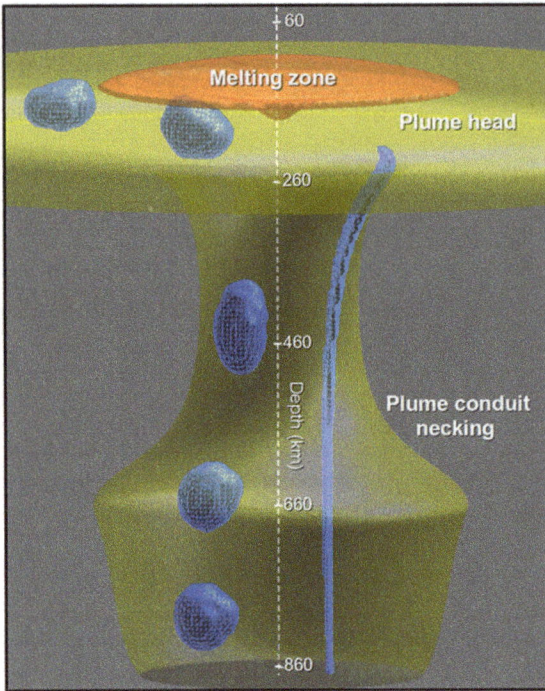

Fig. 6. A passive heterogeneity (right of the plume axis) is deformed readily into a filament. Instead, an active (*i.e.* affecting the flow) heterogeneity 20 times more viscous than the surroundings maintains a 'blob-like' shape, and does not deform. The figure shows the same 'blob' at various times. The plume conduit necking at 660 km depth is due to a 30 times viscosity jump between the lower and upper mantle.

represent a unique window into the lower mantle; however, deciphering the message carried by deep plumes has proved to be challenging. During the past decade there have been considerable advances in analytical geochemistry capabilities (*i.e.* increases in precision, sensitivity and resolving power of mass spectrometers) enabling us to detect small isotopic anomalies and to explore short-lived isotopic systems. In parallel, advances in whole-mantle seismic tomography techniques enabled us to have increasingly sharp images of the Earth's internal structure. In the previous paragraphs I focused on aspects that are generally accepted, albeit still debated. Here I will focus on aspects that still need to be elucidated; I did not not order them according to their importance, they are all important, but following a depth, top-down, criteria.

- Lithospheric and sub-lithospheric processes include, but are not limited to, partial melting of distinct lithologies carried by plumes, the subsequent melt transport within the lithosphere, and the hypothetical mixing of distinct batches of melt in a magma chamber. The key issue will be to quantitatively understand to which extent all these processes can modify the geochemical fingerprint of plume-derived magmas. We also need to understand how to explain geochemical variations occurring on short-time scales (*e.g.* 10^3 years or less), as models based on filaments/blobs carried by plumes only provide a framework to explain isotopic variations occurring on time-scales of order 10^6 years.

- Dynamic processes within an 'enlarged' mantle transition zone. For decades the mantle transition zone was bounded by two mineralogical solid-state phase transitions, at depths of 410 and 660 km, which led to density and viscosity variations. However, seismic tomography has clearly shown that both subducted slabs (Fukao and Obayashi, 2013) and mantle plumes (French and Romanowicz, 2015) are 'perturbed' also at a depth of 1000 km. In particular, plumes become narrower and start to be more tilted by the global mantle circulation. A viscosity change (Rudolph *et al.*, 2015) and/or a compositional change (Ballmer *et al.*, 2017) have been invoked, but the exact nature of the 'transition' at a depth of 1000 km is still elusive.

- Linking seismic observations to the geochemical fingerprint of plumes. Plumes with anomalous μ^{182}W have large ULVZs at their base (*e.g.* Cottaar and Romanowicz, 2012; Kim *et al.*, 2020; Li *et al.*, 2022) something that is not yet explained but is an intriguing observation. Do ULVZs represent the 'Core-Mantle Equilibrated Reservoir' proposed by Mundl-Petermeier *et al.* (2020)? Is this consistent with the observation that plumes with a negative μ^{182}W, coupled to high ^{3}He/^{4}He ratio, also show a weak geochemical fingerprint of recycled material (Jackson *et al.*, 2020)? How would this dense material be transported by plumes? LLSVPs have also been considered as a heterogeneous reservoir feeding mantle plumes (*e.g.* Weis *et al.*, 2011; McNamara, 2019; Williams *et al.*, 2019). Resolving uncertainties about the heterogeneous nature (recycled *vs.* more primitive materials) of LLSVPs and their internal structure might help to explain spatial differences in OIB compositions.
- Interactions between the core and the mantle. The metallic core is a major reservoir for siderophile elements (Carlson *et al.*, 2014 and references therein). The core can also be a major reservoir for ^{3}He (Olson and Sharp, 2022) as it was preserved from degassing caused by impacts and by plate cycling. But, is it possible for elements to 'leak out' of the core? We clearly need more metal-silicate partitioning experiments to understand over which time-scales and length-scales such exchanges might occur. This will enable us to assess if core–mantle processes do explain the negative μ^{182}W and the high ^{3}He/^{4}He observed in some plumes.

I conclude by saying that understanding mantle plumes requires understanding large-scale planetary processes responsible for the chemical evolution of the Earth over billions of years.

Acknowledgments

I gratefully acknowledge Barbara Romanowicz and Andreas Stracke for their thoughtful reviews. I wish to thank Albrecht Hofmann for comments on an early version of the manuscript. Finally, I thank Costanza Bonadiman and Elisabetta Rampone for the editorial assistance and for the organization of the MEREMA meeting in Sestri Levante.

References

Abouchami, W., Hofmann, A.W., Galer, S.J.G., Frey, F., Eisele, J. and Feigenson, M. (2005) Pb isotopes reveal bilateral asymmetry and vertical continuity in the Hawaiian plume. *Nature,* **434**, 851–856.

Ammann, M.W., Brodholt, J.P., Wookey, J. and Dobson, D.P. (2010) First-principles constraints on diffusion in lower-mantle minerals and a weak D″ layer. *Nature,* **465**, 462–465, doi.org/10.1038/nature09052.

Andrault, D., Bolfan-Casanova, N., Nigro, G.L., Bouhifd, M.A., Garbarino, G. and Mezouar, M. (2011) Solidus and liquidus profiles of chondritic mantle: implication for melting of the Earth across its history. *Earth and Planetary Science Letters*, **304**, 251–259. doi.org/10.1016/j.epsl.2011.02.006.

Ballmer, M.D., Houser, C., Hernlund, J.W., Wentzcovitch, R.M. and Hirose, K. (2017) Persistence of strong silica-enriched domains in the Earth's lower mantle. *Nature Geoscience*, 236–241. https://doi.org/10.1038/NGEO2898.

Bao, X., Lithgow-Bertelloni, C.R., Jackson, M.G. and Romanowicz, B. (2022) On the relative temperatures of Earth's volcanic hotspots and mid-ocean ridges. *Science*, **375**, 57–61.

Bolrão, D.P., Ballmer, M.D., Morison, A., Rozel, A.B., Sanan, P., Labrosse, S. and Tackley, P.J. (2021) Time-scales of chemical equilibrium between the convecting solid mantle and over- and underlying magma oceans. *Solid Earth*, **12**, 421–437, doi.org/10.5194/se-12-421-2021.

Boukaré, C.-E., Ricard, Y. and Fiquet, G. (2015) Thermodynamics of the MgO-FeO-SiO$_2$ system up to 140 GPa: application to the crystallization of Earth's magma ocean, *Journal of Geophysical Research Solid Earth*, **120**, 6085–6101, doi.org/10.1002/2015JB011929.

Brown, S.M., Elkins-Tanton, L.T. and Walker, R.J. (2014) Effects of magma ocean crystallization and overturn on the development of 142Nd and 182W isotopic heterogeneities in the primordial mantle. *Earth and Planetary Science Letters*, **408**, 319–330.

Brandenburg, J.P., Hauri, E.H., van Keken, P.E. and Ballentine, C.J. (2008) A multiple-system study of the geochemical evolution of the mantle with force-balanced plates and thermochemical effects. *Earth and Planetary Science Letters*, **276**, 1–13.

Burke, K. and Torsvik, H. (2004) Derivation of Large Igneous Provinces of the past 200 million years from long-term heterogeneities in the deep mantle. *Earth and Planetary Science Letters*, **227**, 531–538.

Cabral, R.A., Jackson, M.G., Rose-Koga, E.F., Koga, K.T., Whitehouse, M.J. *et al.* (2013) Anomalous sulphur isotopes in plume lavas reveal deep mantle storage of Archaean crust. *Nature*, **496**, 490–493.

Canup, R.M. (2008) Accretion of the Earth. *Philosophical Transactions of the Royal Society A*, **336**, 4061–4075.

Carlson, R.W., Garnero, E., Harrison, T.M., Li, J., Manga, M., McDonough, W.F., Mukhopadhyay, S., Romanowicz, B. et al. (2014) How did Early Earth become our modern world? *Annual Reviews of Earth and Planetary Sciences*, **42**, 151–178.

Chauvel, C., Hofmann, A.W. and Vidal, P. (1992) HIMU-EM: the French Polynesian connection. *Earth and Planetary Science Letters*, **110**, 99–119.

Chauvel, C., Maury, R.C., Blais, S., Lewin, E., Guillou, H., Guille, G., Rossi, P. and Gutscher, M.-A. (2012) The size of plume heterogeneities constrained by Marquesas isotopic stripes. *Geochemistry, Geophysics, Geosystems*, **13(1)**, Q07005 doi:10.1029/2012GC004123.

Christensen, U.R. and Hofmann, A.W. (1994) Segregation of subducted oceanic crust in the convecting mantle. *Journal of Geophysical Research*, **99**, 19867–19884.

Cottaar, S. and Romanowicz, B. (2012) An unusually large ULVZ at the base of the mantle near Hawaii. *Earth and Planetary Science Letters*, **355**, 213–222.

Cottrell, E., Jaupart, C. and Molnar, P. (2004) Marginal stability of thick continental lithosphere. *Geophysical Research Letters*, **31**, L18612, doi:10.1029/2004GL020332.

Courtillot, V., Davaille, A., Besse, J. and Stock, J. (2003) Three distinct types of hotspots in the Earth's mantle. *Earth and Planetary Science Letters*, **205**, 295–308.

Crough, S.T. (1983) Hotspot swells. *Annual Reviews in Earth and Planetary Science*, **11**, 165–193.

Dannberg, J. and Sobolev, S.V. (2015) Low-buoyancy thermochemical plumes resolve controversy of classical mantle plume concept. *Nature Communications*, **6**, 6960.

Davaille, A. (1999) Simultaneous generation of hotspots and superswells by convection in a heterogeneous planetary mantle. *Nature*, **402**, 756–760.

Davaille, A., Carrez, Ph. and Cordier, P. (2018) Fat plumes may reflect the complex rheology of the lower mantle. *Geophysical Research Letters*, **45**, 1349–1354. doi.org/10.1002/2017GL076575.

Davaille, A. and Romanowicz, B. (2020) Deflating the LLSVPs: bundles of mantle thermochemical plumes rather than thick stagnant "piles". *Tectonics*, **39**, e2020TC006265.

Deng, J. and Stixrude, L. (2021) Deep fractionation of Hf in a solidifying magma ocean and its implications for tungsten isotopic heterogeneities in the mantle. *Earth and Planetary Science Letters*, **562**, 116873.

Deschamps, F., Cobden, L. and Tackley, P.J. (2012) The primitive nature of large low shear-wave velocity provinces. *Earth and Planetary Science Letters*, **349–350**, 198–208.

Dziewonski, A.M., Lekic, V. and Romanowicz, B.A. (2010) Mantle anchor structure: An argument for bottom up tectonics. *Earth and Planetary Science Letters*, **299**, 69–79.

Eisele, J., Abouchami, W., Galer, S.J.G. and Hofmann, A.W. (2003) The 320 kyr Pb isotope evolution of Mauna Kea lavas recorded in the HSDP-2 drill core. *Geochemistry, Geophysics, Geosystems*, **4(5)**, 8710, doi:10.1029/2002GC000339.

Elkins-Tanton, L.T. (2012) Magma oceans in the inner solar system. *Annual Reviews in Earth and Planetary Science*, **40**,113–139, doi:10.1146/annurev-earth-042711-105503.

Farnetani, C.G. and Hofmann, A.W. (2009) Dynamics and internal structure of a lower mantle plume conduit. *Earth and Planetary Science Letters*, **282**, 314–322.

Farnetani, C.G. and Hofmann, A.W. (2010) Dynamics and internal structure of the Hawaiian plume. *Earth and Planetary Science Letters*, **295**, 231–240.

Farnetani, C.G. and Samuel, H. (2005) Beyond the thermal plume paradigm. *Geophysical Research Letters*, **32**, L07311.

Farnetani, C.G., Hofmann, A.W. and Class, C. (2012) How double volcanic chains sample geochemical anomalies from the lowermost mantle. *Earth and Planetary Science Letters*, **359-360**, 240–247.

Farnetani, C.G., Hofmann, A.W., Duvernay, T. and Limare, A. (2018) Dynamics of rheological heterogeneities in mantle plumes. *Earth and Planetary Science Letters*, **499**, 74–82.

Fourel, L., Milelli, L., Jaupart, C. and Limare, A. (2013) Generation of continental rifts, basins, and swells by lithosphere instabilities. *Journal of Geophysical Research Solid Earth*, **118**, 3080–3100, doi:10.1002/jgrb.50218.

French, S.W. and Romanowicz, B. (2015) Broad plumes rooted at the base of the Earth's mantle beneath major hotspots. *Nature*, **525**, 95–99.

Fukao, Y. and Obayashi, M. (2013) Subducted slabs stagnant above, penetrating through, and trapped below the 660 km discontinuity. *Journal of Geophysical Research*, **118**, 5920–5938, doi:10.1002/2013JB010466.

Garnero, E.J. and Helmberger, D.V. (1996) Seismic detection of a thin laterally varying boundary layer at the base of the mantle beneath the central-Pacific. *Geophysical Research Letters*, **23**, 977–980.

Goes, S., Cammarano, F. and Hansen, U. (2004) Synthetic seismic signature of thermal mantle plumes. *Earth and Planetary Science Letters*, **218**, 403–419.

Gülcher, A.J.P., Gebhardt, D.J., Ballmer, M.D. and Tackley, P.J. (2020) Variable dynamic styles of primordial heterogeneity preservation in the Earth's lower mantle. *Earth and Planetary Science Letters*, **536**, 116160.

Harpp, K.S. and Weis, D. (2020) Insights into the origins and compositions of mantle plumes: A comparison of Galápagos and Hawai'i. *Geochemistry, Geophysics, Geosystems*, **21**, e2019GC008887 doi:10.1029/2019GC008887.

Hart, S.R., Schilling, J.G. and Powell, J.L. (1973) Basalts from Iceland and along the Reykjanes Ridge: Sr isotope geochemistry. *Nature*, **246**, 104–107.

Hart, S.R., Hauri, E.H., Oschmann, L.A. and Whitehead, J.A. (1992) Mantle plumes and entrainment: isotopic evidence. *Science*, **256**, 517–520.

Hauri, E., Whitehead, J. and Hart, S.R. (1994) Fluid dynamic and geochemical aspects of entrainment in mantle plumes. *Journal of Geophysical Research*, **99** 24275–24300.

Hirose, K., Fei, Y., Ma, Y. and Mao, H.-K. (1999) The fate of subducted basaltic crust in the Earth's lower mantle. *Nature*, **397**, 53–56.

Hirth, G. and Kohlstedt, D.L. (1996) Water in the oceanic upper mantle: implications for rheology, melt extraction and the evolution of the lithosphere. *Earth and Planetary Science Letters*, **144**, 93–108.

Hoernle, K., Rohde, J., Hauff, F., Garbe-Schönberg, D., Hornrighausen, S., Werner, R. and Morgan, G.P. (2015) How and when plume zonation appeared during the 132 My evolution of the Tristian Hotspot. *Nature Communications*, **6,** 1–10.

Hoernle, K., Werner, R., Morgan, J.P., Garbe-Schonberg, D., Bryce, J. and Mrazek, J. (2000) Existence of complex spatial zonation in the Galapagos plume for at least 14 m.y. *Geology*, **28**, 435–438.

Hofmann, A.W. (1997) Mantle geochemistry: the message from oceanic volcanism. *Nature*, **385**, 219–229.

Hofmann, A.W. (2003) Sampling Mantle Heterogeneity through Oceanic Basalts: Isotopes and Trace Elements. Pp. 61–101 in: *Treatise On Geochemistry* (R. Carlson, editor). Vol. **2**. ISBN: 0-08-044337-0.

Hofmann, A.W. and White, W.M. (1982) Mantle plumes from ancient oceanic crust. *Earth and Planetary Science Letters*, **57**, 421–436.

Hoggard, M.J., Parnell-Turner, R. and White, N. (2020) Hotspots and mantle plumes revisited: Towards reconciling the mantle heat transfer discrepancy. *Earth and Planetary Science Letters*, **542**, 116317.

Homrighausen, S., Hoernle, K., Hauff, F., Geldmacher, J., Wartho, J.A., van den Bogaard, P. and Garbe-Schönberg, D. (2018) Global distribution of the HIMU end member: Formation through Archean plume-lid tectonics. *Earth-science reviews*, **182**, 85–101.

Horan, M., Carlson, R.W., Walker, R.J., Jackson, M., Garcon, M. and Norman, M. (2018) Tracking Hadean processes in modern basalts with 142-Neodymium. *Earth and Planetary Science Letters*, **484**, 184–191, doi:10.1016/j.epsl.2017.12.017.

Huang, S., Hall, P.S. and Jackson, M.G. (2011) Geochemical zoning of volcanic chains associated with Pacific hotspots. *Nature Geoscience*, **4(12)**, 874–878.

Ito, G. and van Keken, P.E. (2007) Hotspots and melting anomalies. Pp. 371–435 in: *Treatise on Geophysics: Mantle Dynamics* (D. Bercovici, editor). Vol. **7(09)**, Elsevier, Amsterdam.

Jackson, M.G., Hart, S.R., Koppers, A.A.P., Staudigel, H., Konter, J., Blusztajn, J., Kurz, M. and Russell, J.A. (2007) The return of subducted continental crust in Samoan lavas. *Nature*, **448**, 684–687 doi:10.1038/nature06048.

Jackson, M.G., Blichert-Toft, J., Halldorsson, S.A., Mundl-Petermeier, A. *et al.* (2020) Ancient helium and tungsten isotopic signatures preserved in mantle domains least modified by crustal recycling. *Proceedings of the National Academy of Science, USA*, **117**, 30993–31001.

Jeanloz, R. and Morris, S. (1986) Temperature distribution in the crust and in the mantle. *Annual Reviews in Earth and Planetary Science*, **14**, 377–415.

Jones, T.D., Davies, D.R., Campbell, I.H., Wilson, C.R. and Kramer, S.C. (2016) Do mantle plumes preserve the heterogeneous structure of their deep-mantle source? *Earth and Planetary Science Letters*, **434**, 10–17.

Jones, T.D., Sime, N. and van Keken, P.E. (2021) Burying Earth's Primitive Mantle in the Slab Graveyard. *Geochemistry, Geophysics, Geosystems*, doi.org/10.1029/2020GC009396.

Karato, S. (2010) Rheology of the deep upper mantle and its implications for the preservation of the continental roots: a review. *Tectonophysics*, **481**, 82–98.

Kim, D., Lekic, V., Ménard, B., Baron, D. and Taghizadeh-Popp, M. (2020) Sequencing seismograms: A panoptic view of scattering in the core-mantle boundary region. *Science*, **368**, 1223–1228.

King, S.D. and Adam, C. (2014) Hotspot swells revisited. *Physics of the Earth and Planetary Interiors*, **235**, 66–83.

Koppers, A.A.P., Becker, T.W., Jackson, M.G., Konrad, K., Muller, R.D., Romanowicz, B., Steinberger, B. and Whittaker, J.M. (2021) Mantle plumes and their role in Earth processes. *Nature Reviews Earth & Environment* **2**, 382–401.

Kumagai, I., Davaille, A., Kurita, K. and Stutzmann, E. (2008) Mantle plumes: Thin, fat, successful, or failing? Constraints to explain hot spot volcanism through time and space. *Geophysical Research Letters*, **35**, L16301.

Li, M., McNamara, A.K. and Garnero, E.J. (2014) Chemical complexity of hotspots caused by cycling oceanic crust through mantle reservoirs. *Nature Geoscience*, **7**, 366–370.

Li, Z., Leng, K.D., Jenkins, J. and Cottaar, S. (2022) Kilometer-scale structure on the core-mantle boundary near Hawaii. *Nature Communications*, **13**, 2787, doi:10.1038/s41467-022-30502-5.

Limare, A., Jaupart, C., Kaminski, E., Fourel, L. and Farnetani, C.G. (2019) Convection in an internally heated stratified heterogeneous reservoir. *Journal of Fluid Mechanics*, **870**, 67–105.

Masters, G., Laske, G., Bolton, H. and Dziewonski, A. (2000) The relative behavior of shear velocity, bulk sound speed, and compressional velocity in the mantle: Implications for chemical and thermal structure. Pp. 63–87 in: *Mineral Physics and Tomography from the Atomic to the Global Scale* (S.I. Karato *et al.* editors). Vol. **117**, American Geophysical Union, Washington, DC.

McNamara, A.K. (2019) A review of large low shear velocity provinces and ultralow velocity zones. *Tectonophysics*, **760**, 199–220.

Montelli, R., Nolet, G., Dahlen, F.A. and Masters, G. (2006) A catalogue of deep mantle plumes: New results from finite-frequency tomography. *Geochemistry, Geophysics, Geosystems*, **7**, Q11007.

Morgan, W.J. (1971) Convection plumes in the lower mantle. *Nature*, **230**, 42–43.

Mundl, A., Touboul, M., Jackson, M.G., Day, J.M.D., Kurz, M.D., Lekic, V., Helz, R.T. and Walker, R.J. (2017) Tungsten-182 heterogeneity in modern ocean island basalts. *Science*, **356**, 66–69.

Mundl-Petermeier, A., Walker, R.J., Fischer, R.A., Lekic, V., Jackson, M.G. and Kurz, M.D. (2020) Anomalous ^{182}W in high ^{3}He/^{4}He ocean island basalts: Fingerprints of Earth's core? *Geochimica et Cosmochimica Acta*, **271**, 194–211.

Nakagawa, T. and Tackley, P.J. (2014) Influence of combined primordial layering and recycled MORB on the coupled thermal evolution of Earth's mantle and core. *Geochemistry, Geophysics, Geosystems*, **15**, 619–633, doi:10.1002/2013GC005128.

Nelson, P.L. and Grand, S.P. (2018) Lower-mantle plume beneath the Yellowstone hotspot revealed by core waves. *Nature Geoscience*, **11**, 280–284.

Nolet, G. and Dahlen, F.A. (2000) Wavefront healing and the evolution of seismic delay times. *Journal of Geophysical Research*, **105**, 19043–19054.

Olson, P., Schubert, G. and Anderson C. (1993) Structure of axisymmetric mantle plumes. *Journal of Geophysical Research*, **98**, 6829–6844.

Olson, P.L. and Sharp, Z.D. (2022) Primordial helium-3 exchange between Earth's core and mantle. *Geochemistry, Geophysics, Geosystems*, **23**, e2021GC009985.

Payne, J.A., Jackson, M.G. and Hall, P.S. (2012) Parallel volcano trends and geochemical asymmetry of the Society hotspot track. *Geology*, **41(1)**, 19–22.

Peters, B.J., Mundl-Petermeier, A., Carlson, R.W., Walker, R.J. and Day, J.M. D. (2021) Combined lithophile-siderophile isotopic constraints on Hadean processes preserved in ocean island basalt sources. *Geochemistry, Geophysics, Geosystems*, **22**, doi:10.1029/2020GC009479.

Putirka, K.D. (2005) Mantle potential temperatures at Hawaii, Iceland, and the mid-ocean ridge system, as inferred from olivine phenocrysts: Evidence for thermally driven mantle plumes. *Geochemistry, Geophysics, Geosystems*, **6**, doi:10.1029/2005GC000915.

Richards, M.A., Duncan, R.A. and Courtillot, V.E. (1989) Flood basalts and hotspot tracks: Plume heads and tails. *Science*, **246**, 103–107.

Ricolleau, A., Perrillat, J.-P., Fiquet, G., Daniel, I., Matas, J., Addad, A., Menguy, N., Cardon, H., Mezouar, M. and Guignot, N. (2010) Phase relations and equation of state of a natural MORB: Implications for the density profile of subducted oceanic crust in the Earth's lower mantle. *Journal of Geophysical Research*, **115**, B08202, doi:10.1029/2009JB006709.

Ritsema, H.J., van Heijst, J.H. and Woodhouse, J.H. (1999) Complex shear velocity structure beneath Africa and Iceland. *Science*, **286**, 1925–1928.

Rizo, H., Andrault, D., Bennett, N.R., Humayun, M., Brandon, A., Vlastélic, I., Moine, B.N., Poirier, A., Bouhifd, M.A. and Murphy, D.T. (2019) 182W evidence for core-mantle interaction in the source of mantle plumes. *Geochemical Perspectives Letters*, **11**, 6–11.

Rohde, J., Hoernle, K., Hauff, F., Werner, R., O'Connor, J., Class, C., Garbe-Schönberg, D. and Jokat, W. (2013) 70 Ma chemical zonation of the Tristan-Gough hotspot track. *Geology*, **41(3)**, 335–338.

Romanowicz, B. and Gung, Y.C. (2002) Superplumes from the core–mantle boundary to the lithosphere: Implications for heat flux. *Science*, **296**, 513–516.

Rost, S. and Revenaugh, J. (2003) Small-scale ultralow-velocity zone structure imaged by ScP. *Journal of Geophysical Research*, **108**, 2056, doi:10.1029/2001JB001627.

Rost, S., Garnero, E.J., Williams, Q. and Manga, M. (2005) Seismological constraints on a possible plume root at the core–mantle boundary. *Nature*, **435**, 666–669.

Rudolph, M.L., Lekic, V. and Lithgow-Bertelloni, C. (2015) Viscosity jump in Earth's mid-mantle. *Science*, **350**, 1349–1352.

Schilling, J.G. (1973) Iceland mantle plume: geochemical evidence along Reykjanes Ridge. *Nature*, **242**, 565–571.

Sleep, N.H. (1990) Hotspots and mantle plumes: Some phenomenology. *Journal of Geophysical Research*, **95**, 6715–6736.

Smith, W.H.F. and Sandwell, D.T. (1997) Global seafloor topography from satellite altimetry and ship depth soundings. *Science*, **277**, 1956–1961.

Stacey, F.D. and Loper, D.E. (1983) The thermal boundary layer interpretation of D″ and its role as a plume source. *Physics of the Earth and Planetary Interiors*, **33**, 45–55.

Stevenson, D.J. (1987) Origin of the Moon – the collision hypothesis. *Annual Reviews in Earth and Planetary Science,* **15**, 271–315.

Stixrude, L. and Lithgow-Bertelloni, C. (2012) Geophysics of Chemical Heterogeneity in the Mantle. *Annual Reviews in Earth and Planetary Science,* **40**, 569–595.

Stixrude, L., de Koker, N., Sun, N., Mookherjee, M. and Karki, B.B. (2009) Thermodynamics of silicate liquids in the deep Earth. *Earth and Planetary Science Letters,* **278**, 226–232.

Stracke, A. (2021) A process-oriented approach to mantle geochemistry. *Chemical Geology,* **579**, 120350. doi.org/10.1016/j.chemgeo.2021.120350.

Stracke, A., Bizimis, M. and Salters, V.J.M. (2003) Recycling oceanic crust: Quantitative constraints. *Geochemistry, Geophysics, Geosystems,* **4(3)**, 8003, doi:10.1029/2001GC000223.

Stracke, A., Hofmann, A.W. and Hart, S.R. (2005) FOZO, HIMU, and the rest of the mantle zoo. *Geochemistry, Geophysics, Geosystems,* **6(5)**, Q05007.

Stracke, A., Willig, M., Genske, F., Béguelin, P. and Todd, E. (2022) Chemical geodynamics insights from a machine learning approach. *Geochemistry, Geophysics, Geosystems,* **23**, e2022GC010606. doi.org/10.1029/2022GC010606.

Su, W.-J. and Dziewonski, A.M. (1997) Simultaneous inversion for 3-D variations in shear and bulk velocity in the mantle. *Physics of the Earth and Planetary Interiors*, **100**, 135–156.

Tackley, P.J. (1998) Three-dimensional simulations of mantle convection with a thermo-chemical basal boundary layer: D″? Pp. 231–253 in: *The Core–Mantle Boundary Region*. Geophysical Monograph Series, **28**, (M. Gurnis *et al.*, editors). American Geophysical Union, Washington, DC.

Thorne, M.S. and Garnero, E.J. (2004) Inferences on ultralow-velocity zone structure from a global analysis of SPdKS waves. *Journal of Geophysical Research*, **109**, B08301, doi:10.1029/2004JB003010.

Touboul, M., Puchtel, I.S. and Walker, R.J. (2012) 182W evidence for long-term preservation of early mantle differentiation products. *Science*, **335**, 1065-1069, doi:10.1126/science.1216351.

Tsekhmistrenko, M., Sigloch, K., Hosseini, K. and Barruol, G. (2021) A tree of Indo-African mantle plumes imaged by seismic tomography. *Nature Geoscience,* **14**, 612–619.

Tucker, J.M., van Keken, P.E., Jones, R.E. and Ballentine, C.J. (2020) A role for subducted oceanic crust ingenerating the depleted mid-ocean ridge basalt mantle. *Geochemistry, Geophysics, Geosystems,* **21**, doi.org/10.1029/2020GC009148.

Wamba, M.D., Montagner, J.-P. and Romanowicz, B. (2023) Imaging deep-mantle plumbing beneath La Réunion and Comores hot spots: Vertical plume conduits and horizontal ponding zones. *Science Advances,* **9**, doi:10.1126/sciadv.ade3723.

Weis, D., Garcia, M.O., Rhodes, J.M., Jellinek, M. and Scoates, J.S. (2011) Role of the deep mantle in generating the compositional asymmetry of the Hawaiian mantle plume. *Nature Geoscience,* **4**, 831–838. doi: 10.1038/NGEO1328.

Weiss, Y., Class, C., Goldstein, S.L. and Hanyu, T. (2016) Key new pieces of the HIMU puzzle from olivines and diamond inclusions. *Nature*, **537**, 666–670.

Wessel, P. (1993) Observational constraints on models of the Hawaiian hot spot swell. *Journal of Geophysical Research*, **98**, 16095–16104.

White, W.M. (2010) Oceanic island basalts and mantle plumes: the geochemical perspective. *Annual Reviews in Earth and Planetary Science*, **38**, 133–160.

White, W.M. (2015a) Probing the Earth's deep interior through geochemistry. *Geochemical Perspectives,* **4(2)**, 95–251.

White, W.M. (2015b) Isotopes, DUPAL, LLSVPs, and Anekantavada. *Chemical Geology,* **419**, 10–28.

White, W.M. and Hofmann, A.W. (1982) Sr and Nd isotope geochemistry of oceanic basalts and mantle evolution. *Nature,* **296**, 821–825.

Whitehead, J.A. and Luther, D.S. (1975) Dynamics of laboratory diapir and plume models. *Journal of Geophysical Research,* **80**, 705–717.

Williams, C.D., Mukhopadhyay, S., Rudolph, M.L. and Romanowicz, B. (2019) Primitive helium is sourced from seismically slow regions in the lowermost mantle. *Geochemistry, Geophysics, Geosystems,* **20**, 4130–4145. doi:10.1029/2019gc008437.

Yamazaki, D., Kato, T., Yurimoto, H., Ohtani, E. and Toriumi, M. (2000) Silicon self-diffusion in $MgSiO_3$ perovskite at 25 GPa. *Physics of the Earth and Planetary Interiors,* **119**, 299–309.

Yoshino, T., Makino, Y., Suzuki, T. and Hirata, T. (2020) Grain boundary diffusion of W in lower mantle phase with implications for isotopic heterogeneity in oceanic island basalts by core-mantle interactions. *Earth and Planetary Science Letters,* **530**, 115887.

Zindler, A., Jagoutz, E. and Goldstein, S. (1982) Nd, Sr and Pb isotopic systematics in a three-component mantle: a new perspective, *Nature,* **298**, 519–523.

Nature and origin of heterogeneities in the lithospheric mantle in the context of asthenospheric upwelling and mantle wedge zones: What do mantle xenoliths tell us?

M. Grégoire[1], G. Delpech[2], B. Moine[3] and J.-Y. Cottin[4]

[1]*Géosciences Environnement Toulouse, OMP, CNRS, CNES, IRD, Toulouse III University, 14 Avenue Édouard Belin, 31400 Toulouse, France,*
e-mail: michel.gregoire@get.omp.eu
[2]*Géosciences Paris Sud (GEOPS), CNRS, Université Paris-Saclay, Rue du Belvédère, Bâtiment 504, 91405 Orsay, France*
[3]*Laboratoire Magmas et Volcans, CNRS, IRD, Clermont-Ferrand University, Clermont-Ferrand, France*
[4]*Laboratoire de Géologie de Lyon: Terre, Planète, Environnement, CNRS, ENS, Université Lyon 1, Université Jean Monnet, Saint Etienne, France*

The present contribution synthesizes the main petrographic, mineralogical and chemical features of mantle xenoliths uplifted by Phanerozoic lavas. The collections of mantle xenoliths consist predominantly of peridotites but minor pyroxenites are commonly associated. Two main petrogenetic processes are responsible for the features of mantle xenoliths: partial melting and circulation of melts/fluids and associated metasomatic and magmatic processes. Partial melting processes lead to the formation of residual pieces of upper mantle while two main types of mantle metasomatism could be recognized such as LILE enrichment, the first referring to asthenosphere upwelling settings (essentially mantle plumes, rifting zones and asthenosphere window zones) and the second to mantle wedge settings. The AUZ (asthenospheric upwelling zones) metasomatism is essentially related to the migration of more or less CO_2-rich alkaline silicate melts and associated fluids while the MWZ (mantle wedge zones) metasomatism is associated with the activity of hydrated liquids (fluids) commonly SiO_2-rich.

1. Introduction

Earth's mantle rock samples come almost exclusively from a depth of <200–250 km with the exception of very rare inclusions in diamonds indicating the occurrence of majorite, calcium- and magnesium-perovskite, ringwoodite and ferropericlase (all from depths of >400 km; *e.g.* Harte and Harris, 1994; Collerson *et al.*, 2000; Pearson *et al.*, 2014; Seitz *et al.*, 2018). Therefore, mantle rock samples do not allow us to assess the nature and evolution of the Earth's mantle heterogeneities at a global scale. However, they provide very important constraints on mantle heterogeneities at the lithospheric scale, particularly in settings such as mid-oceanic ridges, intraplate rifting zones, supra-subduction zones, passive continental margins and cratonic roots (*e.g.* McInnes

DOI: 10.1180/EMU-notes.21.3

et al., 2001; Grégoire *et al.*, 2003; Python and Ceuleneer, 2003; Grégoire *et al.*, 2009; O'Reilly and Griffin, 2013; Tilhac *et al.*, 2017).

The study of the lithospheric upper mantle based on petrological, geochemical and/or petrophysical studies of mantle rocks is complementary to those of geophysics (experimental and modelling). There are four typical occurrences for mantle rock samples: mantle sections of ophiolites, peridotite orogenic massifs, abyssal mantle domains and pyroxenite and peridotite mantle xenoliths in basic lavas. Mantle rocks provide insights into the origin and evolution of the lithospheric mantle and, in particular, to the petrogenetic processes that have affected the upper mantle through geological time in various geodynamic settings.

The present contribution focuses only on mantle xenoliths uplifted by Phanerozoic lavas (Fig. 1). Similar studies of the other types of occurrences including xenoliths uplifted by kimberlites and related rocks are crucial for our understanding of the upper mantle composition and evolution. The present study aims to synthesize the main petrographic, mineralogical and chemical features of the Phanerozoic lithospheric mantle-derived peridotite xenoliths. Two main petrogenetic processes are responsible for those features: partial melting and melt/fluid circulation and associated metasomatic and magmatic processes. Also summarized briefly here is the state of knowledge on the nature and origin of mantle pyroxenites which are, although minor in terms of abundance, very important in our understanding of the history of the Earth's mantle.

2. Mantle processes recorded in peridotite xenoliths

2.1. Partial melting

Mantle peridotites defined as lherzolites, harzburgites and dunites form by far the largest volume of the lithosphere, while wehrlites are rarer. They mostly consist of three main mineral phases; olivine (Ol), orthopyroxene (Opx), clinopyroxene (Cpx) associated with a minor Al-rich phase being plagioclase (Pl), spinel (Spl) or garnet (Gt) depending on equilibration depth (*e.g.* Wyllie, 1981; Gasparik, 1984; Klemme, 2004). Because

Fig. 1. Mantle xenoliths in two basanitic dykes (emplaced a few million years ago) from Kerguelen Archipelago. The scale is given by the pencil (*a*) and the hammer (*b*).

olivine and the two pyroxenes are the most abundant minerals, the classification of peridotites is based on the modal abundances of them (Streckeisen, 1976).

Under normal geothermal conditions, the upper mantle does not undergo partial melting. However, there are a number of geodynamic cases where anomalous thermal regimes exceed the solidus of peridotite, leading to local or regional partial melting of the upper mantle. Such abnormal thermal regimes are, for instance, encountered in mantle plume settings where the temperature of the plume material is hotter than the surrounding asthenosphere (Farnetani, 2024, this volume). At mid-oceanic ridge settings, this situation occurs in response to the adiabatic decompression of the asthenosphere and in subduction settings the abnormal thermal regime is due to the influx of fluids/melts which percolate into the mantle wedge above.

In all the geodynamic settings described above, peridotites undergo partial melting. All the different constituent mineral phases taken alone have different melting temperatures, the lowest being for the Al-rich mineral phases (Pl, Spl, Gt) and then the clinopyroxene, the orthopyroxene and finally the olivine. The petrological experiments conducted to reproduce peridotite partial melting (*cf.* Hirschmann *et al.*, 1998; Bernstein *et al.*, 2007) indicate that melting reactions involve an assemblage of dominant clinopyroxene, Al-rich mineral phase and orthopyroxene or olivine depending on pressure. At low and medium pressure in the spinel stability field (from ~1 to 1.7–2 GPa), olivine crystallizes as a product of the melting reaction [Opx+Cpx+Spl → Ol+Melt; Niu, 1997]; hence the olivine modal content of the peridotite residue increases. At higher pressures in the garnet stability field (>1.7–2 GPa), orthopyroxene crystallizes in the residue of melting [Ol+Cpx+Gt → Opx+Melt; Walter, 1998]; hence the orthopyroxene content increases. Such studies also demonstrate that the eutectic composition of a melt formed by partial melting of peridotite depends on the composition of the latter before melting, on pressure, on temperature, on fluid contents (H_2O, CO_2) and on degree of melting. Therefore, an extreme diversity of melt compositions can be generated from peridotite melting. For example, the partial melting of a mantle lherzolite will first lead to the exhaustion of the Al-rich phase depending on pressure (plagioclase, spinel or garnet) and then of the clinopyroxenes and if the degree of melting is high enough, a harzburgite will form if the cpx content drops to <5% and finally a dunite will form if the sum of opx+cpx is <10% (Fig. 2). For clinopyroxene to disappear from the melting peridotite, the degree of melting needs to be >~20%, (*e.g.* Baker and Stolper, 1994; Hirschmann *et al.*, 1998). On the other hand, to form a dunite by partial melting of a lherzolite, the degree of melting has to reach as much as 40–50% (Bernstein *et al.*, 2007) and therefore this situation is probably limited to Archaean mantle melting processes only (Bernstein *et al.*, 2007).

As the mineralogy changes during increasing partial melting, the geochemical compositions of peridotites (major, minor and trace elements) also vary. While the residue becomes more and more magnesian, leading to high Mg#, its CaO, Al_2O_3, FeO_t, TiO_2, Na_2O and K_2O contents decrease significantly throughout the whole peridotite and constituent mineral phases. Minor and trace elements are also strongly affected by melting processes. Incompatible trace elements such as LILE, REE and HFSE tend to concentrate in the newly formed melt during peridotite melting while minor and trace elements of the

Fig. 2. Types of mantle peridotite (lherzolite, harzburgite, dunite) and their relationships with partial melting and mantle metasomatism. Hand specimens.

transition group especially Ni, Co and Cr behave as compatible trace elements and tend to concentrate in the residue of melting (*e.g.* Norman, 1998, Simon *et al.*, 2008).

2.2. Percolation of melts/fluids in the lithosphere and associated metasomatism

It is (very) rare to find mantle peridotite xenoliths that have been affected only by partial melting processes; the majority have been affected by two main processes – an early partial melting and subsequent metasomatic processes. Each of those processes may affect the rock multiple times. The commonly used definition of mantle metasomatism defines as physical and chemical processes that are implemented during the flow (both pervasive at grain boundaries and focused in dykes or veins) of magma and/or fluids within the upper mantle. The main changes that affect the mantle wall rocks (peridotites) are in microstructure, recrystallization, the possible formation of new minerals (and the disappearance of pristine ones) and the chemical exchanges leading to the enrichment in incompatible trace elements of the mantle peridotites. In some cases, metasomatism at large melt/rock ratio can induce the formation of a new rock such as, for example, the transformation of harzburgite into dunite (*e.g.* Kelemen, 1990; Grégoire *et al.*, 2000a) or also the formation of pyroxenites from peridotites (*e.g.* Liu *et al.*, 2020).

Depending on the geochemical and mineralogical changes they cause, three main types of metasomatism have been distinguished over the years: (1) cryptic metasomatism; (2) modal metasomatism; and (3) stealth metasomatism.

Cryptic metasomatism (Dawson, 1984) triggers only chemical modification of the mantle peridotite and of its constituent minerals without crystallization of new mineral phases. The main geochemical modifications concern the abundance of trace elements which are often associated with a decrease (or increase) in the Mg# and an increase in other basaltic components (*e.g.* Na_2O, Al_2O_3, TiO_2) of the whole rock and minerals (*e.g.* Xu *et al.*, 2000).

Modal metasomatism (Harte, 1983) describes the formation of new mineral phases which are different from the original four phases constituting a mantle peridotite, *i.e.* olivine, orthopyroxene, clinopyroxene and an Al-rich phase (spinel or garnet or plagioclase) associated with geochemical modifications such as described for cryptic metasomatism. As for cryptic metasomatism, the Mg# and other basaltic components (*e.g.* Na_2O, TiO_2) of the whole rock and mineral will also vary, possibly in greater ways than for cryptic metasomatism. The classical metasomatic phases are amphibole and phlogopite. Detailed information is given in Table 1.

A third type of metasomatism was introduced more recently by O'Reilly and Griffin (2013) and is referred to as 'stealth metasomatism'. This involves the addition of new mineral phases (*e.g.* garnet and/or clinopyroxene) to the mantle rock; this might be a 'misleading' metasomatic process as it adds similar mineral phases to those forming the original mantle peridotite mineral association.

3. Main metasomatic imprints within the Phanerozoic upper mantle

Mantle metasomatism, depending on the original melt/fluid composition and volume as well as on the duration of the metasomatic event, will result in an increase/decrease in major and trace elements of the whole rock and minerals. The nature of metasomatic mantle melts/fluids may be determined from the changes in the geochemical compositions of mantle minerals but also from the composition of fluid inclusions in minerals. The study of fluid inclusions in mantle peridotites (*e.g.* Schiano *et al.*, 1994; Hidas *et al.*, 2010; Créon *et al.*, 2017) helps to constrain the composition of the fluid phase associated

Table 1. Main differences between the two types of mantle metasomatism presented in the present study (mineral abbreviations after Whitney and Evans, 2010 and Gl for glass and Su for sulfide). Cpx: Clinopyroxene; Opx: Orthopyroxene; Ol: Olivine; Pl: Plagioclase; Spl: Spinel; Phl: Phlogopite; Fsp: Feldspar; Rt: Rutile; Ap: Apatite; Ilm: Ilmenite; Arm: Armalcolite; Cb: Carbonate; Gl: Glass; Su: Sulfide; Mag: Magnetite. Cpx1 and 2 and Opx1 and 2 refer to primary Cpx and Opx (1) and metasomatic Cpx and Opx (2), respectively.

Asthenospheric upwelling zones (AUZ)	Mantle wedge zones (MWZ)
Secondary mineralogy	**Secondary mineralogy**
Cpx-Phl-Amp-Fsp-Pl-Ol-Spl-Rt-Ap-Ilm-Arm-Cb-Gl-Su	Opx-Amp-Cpx-Ol-Phl-Spl-Su-Mag
Geochemistry	**Geochemistry**
Major elements	Major elements
– Cpx: Common enrichment in Cr-Na, sometimes only Na or Na-Ti	– Opx2 and cpx2 poorer in Al than cpx1 et opx1
– Amphibole mostly more sodic, lower in Si for a given mg# than those from MWZ metasomatism	– Phlogopite low in Ti
	– Amphibole mostly less sodic but richer in Si for a given mg# than those from AUZ metasomatism
Trace elements	**Trace elements**
U/Th wall rock variable (commonly U/Th < 1)	U/ThWR high >>1
Amphibole and phlogopite are Nb- and Ta-rich	Amphibole and phlogopite are Nb- and Ta-poor

with metasomatism (H_2O, CO_2, ...) and can also provide temperature constraints at which the fluid inclusions were trapped. A traditional way of estimating a theoretical composition of the metasomatic agent is to use the trace element compositions of mineral phases (commonly cpx) and a set of appropriate partition coefficients between melt or fluid and mineral to calculate a theoretical composition of the metasomatic agent (*e.g.* Grégoire *et al.*, 2000a). In some cases, such as mantle xenoliths from the Kerguelen Islands, it is possible to compare the composition of the metasomatic clinopyroxenes with those of clinopyroxenes occurring in deep-seated alkaline pyroxenite segregates and clinopyroxene megacrysts occurring in young alkaline lavas (Fig. 3). Metasomatic melts (fluids) in off-craton (mostly Phanerozoic) regions cover a vast

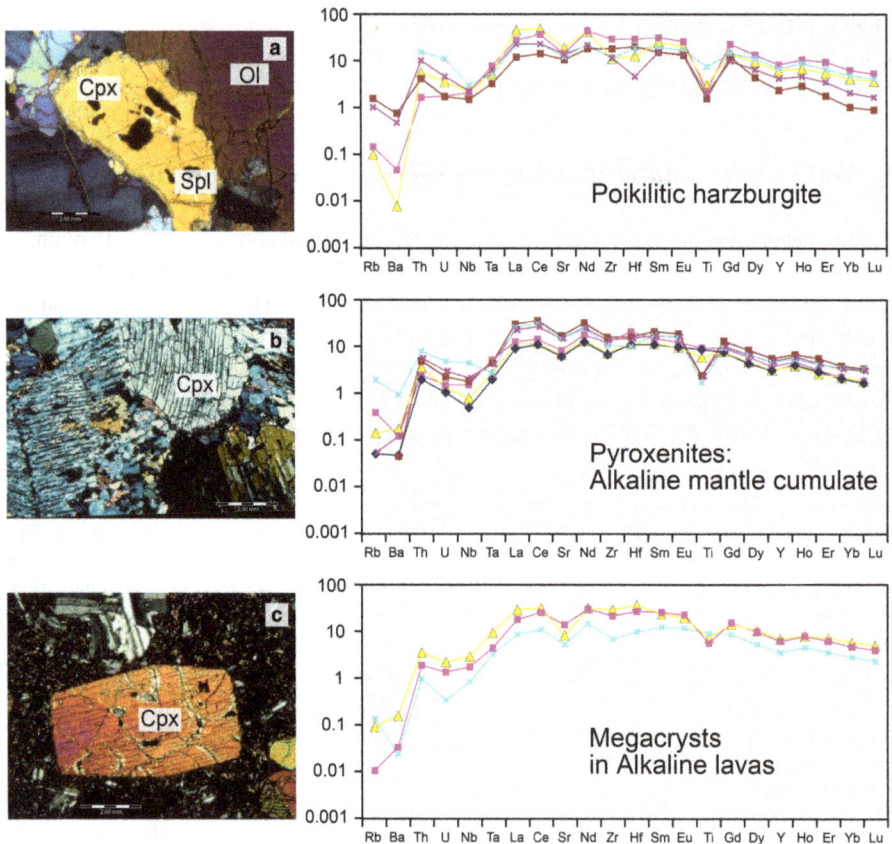

Fig. 3. Comparison between primitive mantle-normalized (McDonough and Sun, 1995) trace element patterns of clinopyroxene from a poikilitic mantle harzburgite (*a*), an alkaline deep pyroxenite cumulate (*b*) and occurring as a megacryst in an alkaline basaltic lava (*c*) from the Kerguelen Islands. The three types of patterns are similar, indicating that the metasomatic agent affecting the poikilitic harzburgite was probably an alkaline mafic silicate melt (Grégoire *et al.*, 2000a and unpublished data). All three photos (scale bar: 2 mm) were taken using an optical microscope under crossed polars.

spectrum from silicate to carbonate magmas containing variable types and abundances of dissolved fluids and solutes including brines, C-O-H species and sulfur-bearing components (O'Reilly and Griffin, 2013). It is, nevertheless, possible to discriminate between two main types of metasomatic agents, the first referring to asthenospheric upwelling settings (essentially mantle plumes, rifting zones and asthenospheric window zones) and the second to mantle wedge settings.

3.1. Metasomatism in asthenospheric upwelling settings

In this type of context, mantle xenoliths uplifted by alkaline lavas mostly prove metasomatism linked to the migration of more or less CO_2-rich alkaline silicate melts and associated fluids related to the magmatic activity in the upwelling zone. The most common metasomatic minerals associated with this type of metasomatism are clinopyroxene and interstitial hydrous minerals such as amphibole and/or phlogopite (Fig. 4). They could be associated with a great diversity of accessory secondary minerals including

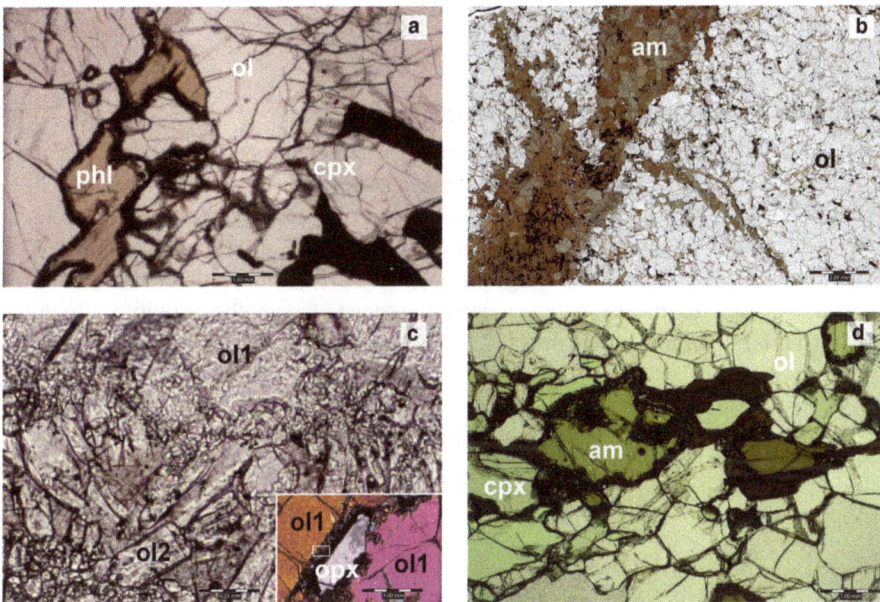

Fig. 4. (*a*) Photomicrograph of a phlogopite-bearing mantle harzburgite from the Kerguelen Archipelago (scale bar: 1mm, from the collection B.N. Moine). (*b*) Photomicrograph of an amphibole-rich dykelet cross-cutting a mantle dunite from the Kerguelen Archipelago (scale bar: 2 mm, modified after Moine *et al.*, 2001). (*c*) Photomicrograph of a contact area of a metasomatic vein (bottom) of euhedral olivine 2, interstitial feldspar and opaque oxide grains and needles replacing orthopyroxene in a mantle harzburgite xenolith from Kerguelen Archipelago (scale bar: 0.25 mm, modified after Ionov *et al.*, 1999). The inset image represents a larger view (scale bar: 3 mm) in which the detailed zone is indicated by the black rectangle. (*d*) Photomicrograph of a metasomatic amphibole and clinopyroxene mineral assemblage in a mantle peridotite xenolith from central Patagonia (scale bar: 1 mm, modified after Dantas, 2008). Phl: phlogopite; cpx: clinopyroxene; ol: olivine; am: amphibole

feldspars, olivine, spinel, rutile, apatite, ilmenite, carbonates, sulfides, armalcolite (e.g. O'Reilly and Griffin, 1988; Ionov et al., 1999; Grégoire et al., 2000a,b; Moine et al., 2001; Delpech et al., 2004). Associated geochemical modifications mostly comprise a decrease in the MgO/FeO ratio (or Mg#) and an increase in CaO, Al_2O_3, FeO_t, TiO_2, Na_2O and K_2O contents of the metasomatized peridotites. A positive correlation between the Na_2O and the Cr_2O_3 contents of primary clinopyroxenes is also observed frequently, indicating the addition of Cr to the metasomatized peridotites during metasomatism (Fig. 5). The abundance of incompatible elements, such as LILE, REE and HFSE commonly increases in metasomatized peridotites, and the more incompatible the element is, the greater the enrichment will be. Note that the same types of trace element patterns of clinopyroxenes can be observed in many localities around the world, simply indicating that the metasomatic agents were very similar (Fig. 6). The U/Th ratio is typically <1 in such metasomatized peridotites, and amphibole and phlogopite, when they occur, are Nb- and Ta-rich (see figure 2 of Coltorti et al., 2007). Much more rarely, the metasomatic agents in asthenospheric upwelling zones are pure carbonatite melts, Fe-Ti basaltic melts and tholeiitic melts (e.g. Hervig et al., 1986; Drury and van Roermund, 1988; Green and Wallace, 1988; Delpech et al., 2004; Moine et al., 2004; Dantas et al., 2009; Grégoire et al., 2010).

3.2. Metasomatism in mantle wedge settings

Mantle xenolith localities in mantle wedge settings are much less common and, therefore, the diversity of the mineral assemblages described is much less (Grégoire et al., 2001; McInnes et al., 2001; Ishimaru et al., 2007; Grégoire et al., 2008). In this context, some of the uplifted mantle xenoliths contain distinctive vein structures and mineral compositions generated by hydrofracturing and water–rock reaction within the mantle wedge. There is evidence of metasomatic processes associated with the activity of hydrated liquids (fluids) commonly rich in SiO_2 and formed during the dehydration processes

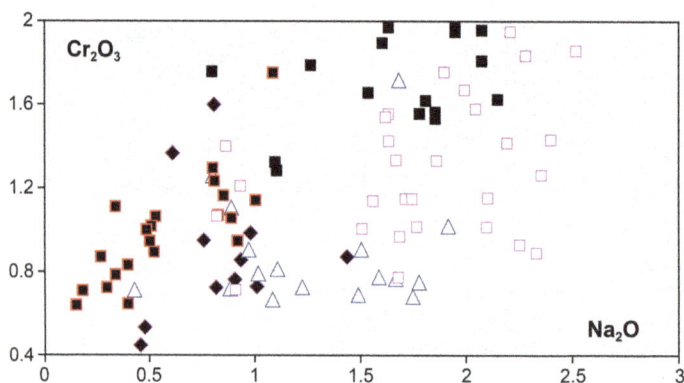

Fig. 5. Cr_2O_3 vs. Na_2O for clinopyroxenes from mantle xenoliths from Oman (diamonds), Kerguelen archipelago (black squares; Grégoire, 1994; Grégoire et al., 2001), Cameroon (triangles; Teitchou et al., 2011) and Patagonia (open squares; Dantas, 2007).

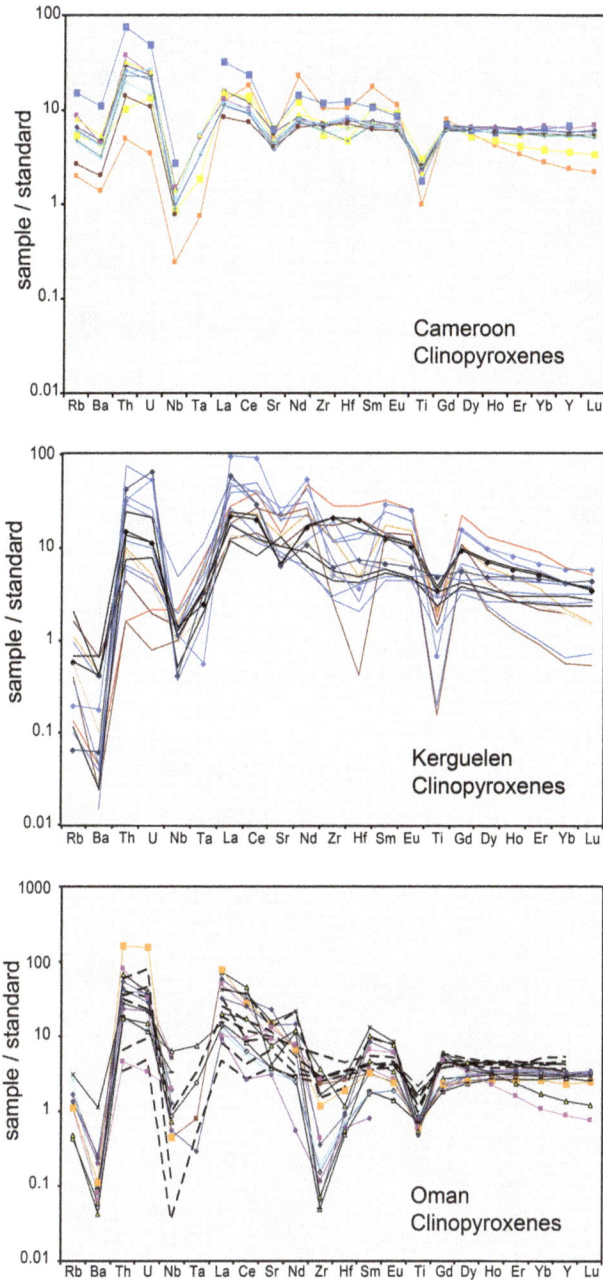

Fig. 6. Primitive mantle-normalized (McDonough and Sun, 1995) incompatible trace element patterns for clinopyroxene from mantle xenoliths analyzed by LA-ICP-MS from the Kerguelen Archipelago, Cameroon and Oman (modified from Grégoire *et al.*, 2000 and 2009 and from Teitchou *et al.*, 2011, respectively).

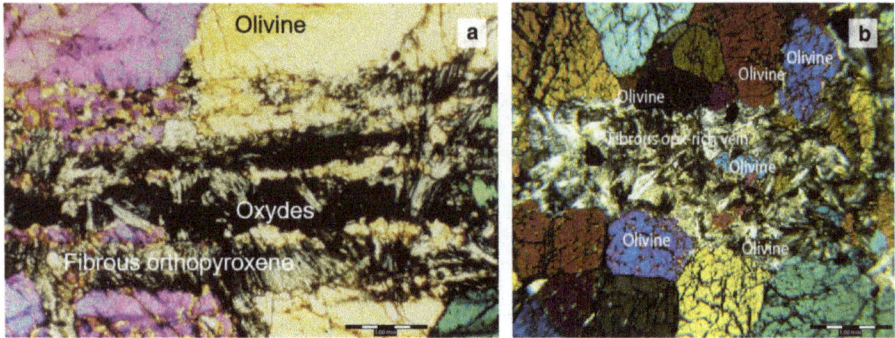

Fig. 7. (*a*) Photomicrograph of a 1 mm wide fibrous orthopyroxene-rich vein from a mantle harzburgite xenolith from Lihir (scale bar: 1mm, modified after McInnes *et al.*, 2001). (*b*) Photomicrographs of a fibrous orthopyroxene-rich vein cross-cutting a spinel dunite xenolith from Monglo (scale bar: 1 mm, modified after Grégoire *et al.*, 2008).

of subducting slabs. The typical metasomatic minerals in this case are fibrous orthopyroxene and clinopyroxene commonly associated with hydrous minerals such as amphibole and/or phlogopite as well as accessory secondary olivine, spinel, magnetite, ilmenite and sulfides (Fig. 7). The metasomatic pyroxenes (Opx, Cpx) display lower Al and Ca contents compared to their primary counterparts from the host mantle peridotite (Fig. 8). Those metasomatic orthopyroxenes and clinopyroxenes display primitive-

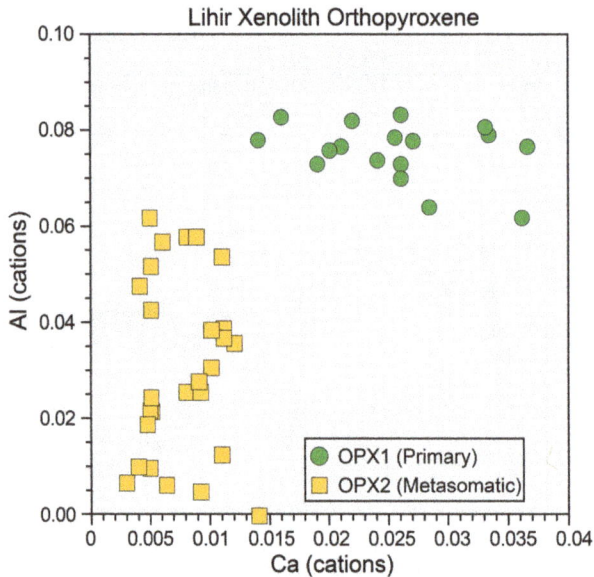

Fig. 8. Al *vs.* Ca (cations) for the two orthopyroxene generations from Lihir mantle xenoliths (Grégoire, pers. comm.). The metasomatic fibrous opx (OPX2) has smaller Al and Ca values than the primary opx (OPX1).

mantle-normalized trace element patterns characterized by strong positive U and strong negative Nb, Ta, Zr, Hf and Ti anomalies, respectively. These patterns are in good agreement with an origin of their metasomatic parental liquid linked to dehydration processes of subducting slab (*e.g.* McInnes *et al.*, 2001; Grégoire *et al.*, 2001; Fig. 9). Hydrous metasomatic minerals such as phlogopite and amphibole will, overall, display low HFSE contents (Ti, Nb and Ta). Moreover, amphibole in mantle wedge settings is less sodic and has larger Mg# values than metasomatic amphiboles from 'asthenospheric upwelling zones' (see Fig. 1 from Coltorti *et al.*, 2007). Finally, the whole-rock U/Th

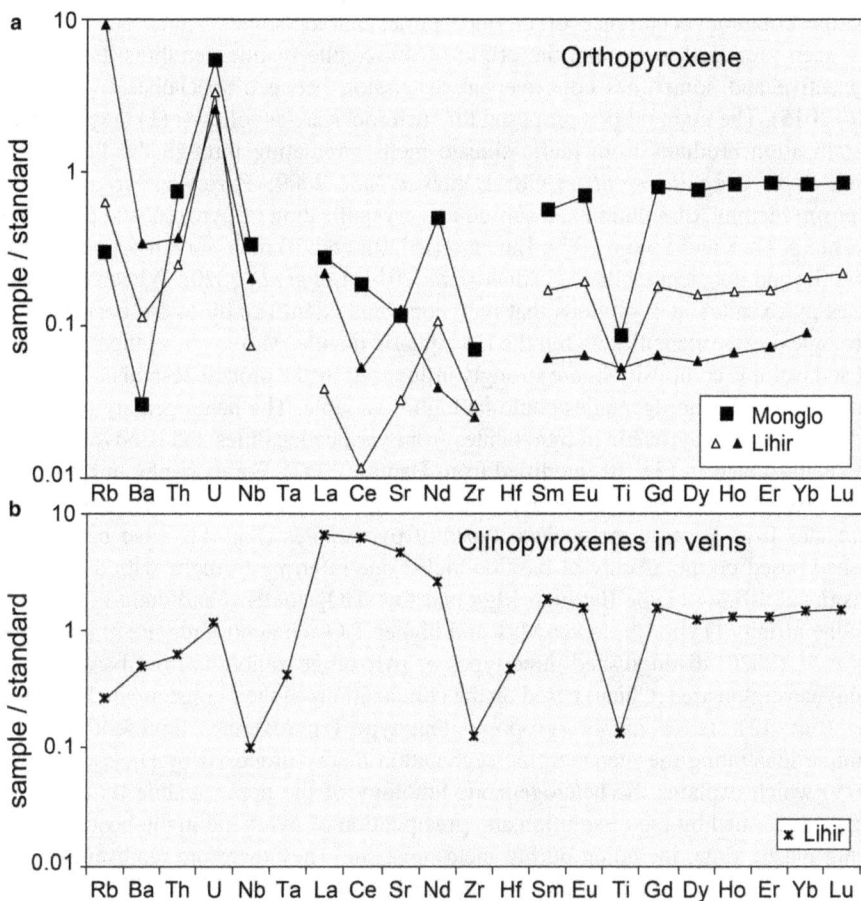

Fig. 9. Primitive mantle-normalized (McDonough and Sun, 1995) incompatible trace element patterns for fibrous metasomatic orthopyroxene (OPX2) occurring in mantle xenoliths from Monglo and Lihir and for secondary clinopyroxene associated with the fibrous orthopyroxene (OPX2) of the Lihir mantle xenoliths analysed by LA-ICP-MS (modified after Grégoire *et al.*, 2001, for Lihir, and after Grégoire *et al.*, 2008, for Monglo).

ratio of such metasomatized peridotites is commonly large (≫1), as in the metasomatic pyroxenes.

The main petrological differences between the two main types of Phanerozoic mantle metasomatism described above are summarized in Table 1.

4. Mantle pyroxenites: rare, important and diverse

Pyroxenite xenoliths are subordinate to peridotite xenoliths at almost all localities, but they represent a petrologically significant mantle rock type (*e.g.* France *et al.*, 2015). This rock type ranges from orthopyroxenite through websterite to clinopyroxenite, with the common occurrence of olivine, garnet and/or spinel. Numerous processes have been proposed to explain the origin of pyroxenite mantle xenoliths leading to a very active and sometimes controversial discussion between specialists (*e.g.* France *et al.*, 2015). The main origins proposed for such rocks are as follows: (1) they represent crystallization products from mafic silicate melts circulating through the lithospheric mantle (*e.g.* Grégoire *et al.*, 1998; Dantas *et al.*, 2009; Puziewicz *et al.*, 2011); (2) *in situ* melting/dissolution and subsequent crystallization of pyroxenes in pyroxenite layers (*e.g.* Dick and Sinton 1979; Liu *et al.*, 2020); and (3) melt–rock reaction between peridotite and transient melts (*e.g.* Chen *et al.*, 2014; Liu *et al.*, 2020). Whatever process causes pyroxenites, it is obvious that they contribute significantly to the occurrence of petrological heterogeneities within the lithospheric mantle. Moreover, as their geochemical and isotopic compositions are strongly influenced by the process responsible for their formation, these heterogeneities could be highly variable. The heterogeneity of compositions of the clinopyroxene in pyroxenites from various localities and geodynamic contexts is illustrated in Fig. 10 (modified from Dantas, 2007). For example, in the case of the pyroxenite xenoliths from the Kerguelen Islands, they are all considered to be deep segregates from basaltic melts. Two types of pyroxenites (Fig. 11) have been distinguished based on the affinity of basaltic melts; one referring to melts with a tholeiitic-transitional affinity (Type IIa; high Mg# and low TiO_2 content) and one to melts with alkaline affinity (Type IIb; lower Mg# and higher TiO_2 content; Grégoire *et al.*, 1998). Liu *et al.* (2020) distinguished three types of pyroxenite xenoliths from basalts of the Yangyuan craton area (China) based on the compositions of their constituent clinopyroxenes (Fig. 12). Those authors proposed that type I pyroxenites represent a natural example illustrating the metamorphic segregation theory proposed by Dick and Sinton (1979) which explains the heterogeneous lithology of the upper mantle by pyroxene-rich veins formed by the dissolution and precipitation of pyroxene in the host peridotite during plastic flow, including during metasomatism. They therefore represent original peridotites with pyroxene enrichments at the local centimeter scale. They also show that type II pyroxenites formed during reactions between an asthenosphere or juvenile lithospheric mantle-derived melt and the host mantle peridotite. Finally, the type III pyroxenites were suggested to originate from fractional crystallization of various parental magmas (Liu *et al.*, 2020) and therefore would resemble the deep mantle segregates from the Kerguelen Islands (Figure 3b).

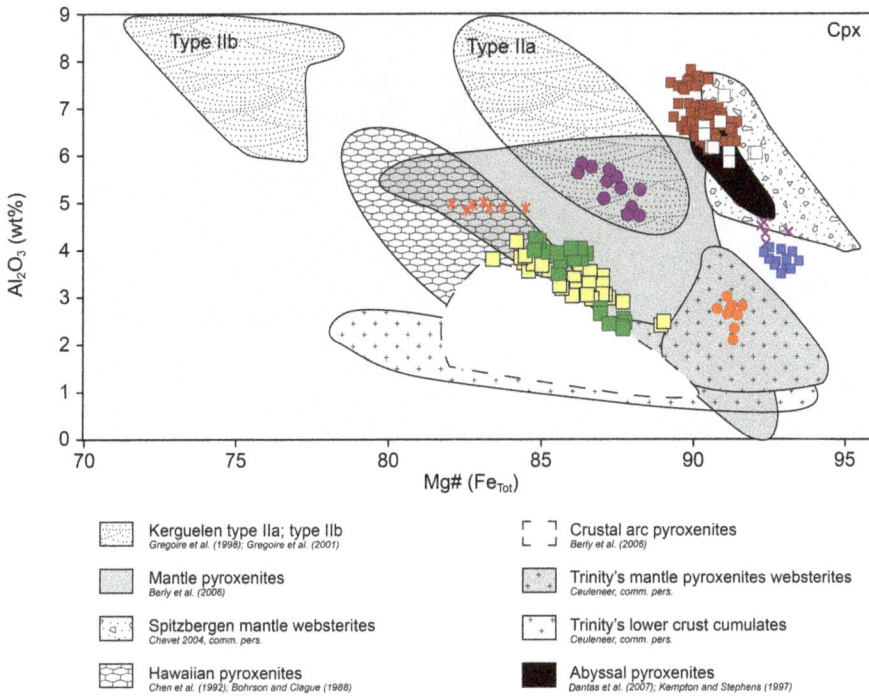

Fig. 10. Al_2O_3 *vs.* Mg# diagram for clinopyroxenes from Patagonian mantle pyroxenite xenoliths, in comparison with pyroxenites from several geodynamic contexts (abyssal, ophiolite, arc and oceanic intraplate) worldwide (modified from Dantas, 2007). The different types of symbol represent the diversity of the pyroxenite xenoliths from Patagonia (see Dantas, 2007 for details).

5. Implications

It is now well recognized that a link does exist between the main petrogenetic processes affecting the mantle peridotites, the formation of mantle pyroxenites and the physical properties of the upper mantle. Indeed, all these processes imply changes in the petrological, mineralogical and chemical characteristics of the upper mantle and therefore changes in its physical properties (density, porosity, seismic properties...). For example, partial melting processes in the mantle decrease its density, increase its seismic velocity and decrease the heat production of the mantle residue generated by extracting heat-producing elements (U, Th, K) and therefore affect the rheology of the residual mantle. Metasomatic processes usually have the opposite effects to those of partial melting. The lithospheric mantle has recorded multiple partial melting, magmatic and metasomatic events since its formation. These events have repeatedly overprinted primary mantle rocks leading to a complex heterogeneous lithospheric mantle. Finally, when pieces of the heterogeneous lithospheric upper mantle are recycled within the convective mantle they will imply changes in the composition of the latter and will participate in its heterogeneity.

Fig. 11. Photomicrographs of pyroxenite mantle xenoliths from the Kerguelen Archipelago (scale bar: 1 mm, from the collection of M. Grégoire). (*a*) Cumulative olivine-bearing websterite of tholeiitic-transitional affinity; and (*b*) garnet-bearing clinopyroxenite of alkaline affinity. Cpx: clinopyroxene; Grt: garnet; Ol: olivine; Opx: orthopyroxene.

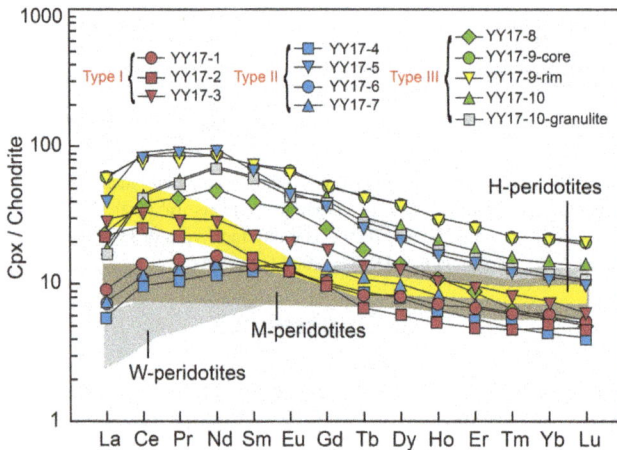

Fig. 12. Chondrite-normalized REE patterns for the clinopyroxene from the Yangyuan pyroxenite xenoliths (modified after Liu *et al.*, 2020). Yangyuan peridotites: H-peridotites: highly metasomatized peridotites; M-peridotites: moderately metasomatized peridotites and W-peridotites: weakly metasomatized peridotites.

Acknowledgments

The present authors thank all of their friends, students and colleagues with whom they have worked on this subject for >20 years. There are too many to name individually here but they are all listed in the references below. A special thanks to Anne-Marie Cousin, graphic designer, who reworked all the figures in order to make them as homogeneous, as legible and as aesthetically pleasing as possible.

References

Baker, M.B. and Stolper, E.M. (1994) Determining the composition of high-pressure mantle melts using diamond aggregates. *Geochimica et Cosmochimica Acta*, **58**, 2811–2827.

Bernstein, S., Kelemen, P.B. and Hanghoj, K. (2007) Consistent olivine Mg# in cratonic mantle reflects Archean mantle melting to the exhaustion of orthopyroxene. *Geology*, **35**, 459–462.

Chen, M.-M., Tian, W., Suzuki, K., Tejada, M.-L-G., Liu, F.-L., Senda, R., Wei, C.-J. , Chen, B. and Chu, Z.-Y. (2014) Peridotite and pyroxenite xenoliths from Tarim, NW China: Evidences for melt depletion and mantle refertilization in the mantle source region of the Tarim flood basalt. *Lithos*, **204**, 97–111.

Collerson, K.D., Hapugoda, S., Kamber, B.S. and Williams, Q. (2000) Rocks from the mantle transition zone: Majorite-bearing xenoliths from Malaita, Southwest Pacific. *Science*, **288**, 1215–1223.

Coltorti, M., Bonadiman, C., Faccini, B., Grégoire, M., O'Reilly S.Y. and Powell, W. (2007) Amphiboles from suprasubduction and intraplate lithospheric mantle. *Lithos*, **99**, 68–84.

Créon, L., Rouchon, V., Youssef, S., Rosenberg, E., Delpech, G., Szabó, C., Remusat, L., Mostefaoui, S., Asimow, P.D., Antoshechkina, P.M., Ghiorso, M.S., Boller, E. and Guyot, F. (2017) Highly CO_2-super-saturated melts in the Pannonian lithospheric mantle – A transient carbon reservoir? *Lithos*, **7**, 519–533.

Dantas, C. (2007) *Caractérisation du manteau supérieur patagonien: les enclaves ultramafiques et mafiques dans les laves alcalines*. Géochimie. Université Paul Sabatier – Toulouse III. Français. fftel-00163376f.

Dantas, C., Grégoire, M., Koester, E., Conceicao, R.V. and Rieck Jr., N. (2009) The lherzolite-websterite xeno-lith suite from Northern Patagonia (Argentina): evidence of mantle-melt reaction processes. *Lithos*, **107**, 107–120.

Dawson, L.B. (1984) Contrasting types of upper mantle metasomatism? In: *Kimberlites II. The mantle and crust-mantle relationships* (J. Kornprobst, editor). Elsevier, Amsterdam, pp. 289–294.

Delpech, G., Grégoire, M., O'Reilly, S.Y., Cottin, J.Y., Moine, B. and Michon, G. (2004) Feldspar from car-bonate-rich metasomatism in the oceanic mantle under Kerguelen Islands (South Indian Ocean). *Lithos*, **75**, 209–237.

Dick, H.J. and Sinton, J.M. (1979) Compositional layering in alpine peridotites: evidence for pressure solution creep in the mantle. *The Journal of Geology*, **87**, 403–416.

Drury, M.R. and van Roermund, H.L.M. (1988) Metasomatic origin for Fe-Ti-rich multiphase inclusions in olivine from kimberlite xenoliths. *Geology*, **16**, 1035–1038.

Farnetani, C.Z. (2024) Dynamics of mantle plumes carrying compositional and rheological heterogeneities. Pp. 19–38 in: Chemical Geodynamics of the Earth's Mantle: New Paradigms (C. Bonadiman and E. Rampone, editors). EMU Notes in Mineralogy, **21**. European Mineralogical Union and the Mineralogical Society of the United Kingdom and Ireland.

France, L., Chazot, G., Kornprobst, J., Dallai, L., Vannucci, R., Grégoire, M., Bertrand, H. and Boivin, P. (2015) Mantle refertilization and magmatism in old orogenic regions: The role of late-orogenic pyrox-enites. *Lithos*, **232**, 49–75.

Gasparik, T. (1984) Two-pyroxene thermobarometry with new experimental data in the system CaO-MgO-Al_2O_3-SiO_2. *Contributions to Mineralogy and Petrology*, **87**, 87–97.

Green, D.H. and Wallace, M.E. (1988) Mantle metasomatism by ephemeral carbonatite melts. *Nature*, **336**, 459–462.

Grégoire, M. (1994) *Les Enclaves Basiques et Ultrabasiques des Iles Kerguelen*. PhD thesis, Editions Univer-sitaires Européennes, 305 pp.

Grégoire, M., Cottin, J.Y., Mattielli, N., Giret, A. and Weis, D. (1998) The metaigneous xenoliths from Ker-guelen archipelago: Evidence of a continent nucleation in an oceanic setting. *Contributions to Mineralogy and Petrology*, **133**, 259–283.

Grégoire, M., Moine, B.N., O'Reilly, S.Y., Cottin, J.Y. and Giret, A. (2000a) Trace element residence and par-titioning in mantle xenoliths metasomatised by high alkaline silicate and carbonate-rich melts (Kerguelen Islands, Indian Ocean). *Journal of Petrology*, **41**, 477–509.

Grégoire, M., Lorand, J.-P., O'Reilly, S.-Y. and Cottin, J.-Y. (2000b) Armalcolite-bearing, Ti-rich metasomatic assemblages in harzburgitic xenoliths from the Kerguelen Islands: Implications for the oceanic mantle budget of high-field strength elements. *Geochimica et Cosmochimica Acta*, **64**, 673–694.

Grégoire, M., McInnes, B.I.A. and O'Reilly, S.Y. (2001) Hydrous metasomatism of oceanic sub-arc mantle, Lihir, Papua New Guinea. Part 2: Trace element characteristics of slab-derived fluids. *Lithos*, **59**, 91–108.

Grégoire, M., Bell, D.R and Le Roex, A.P. (2003) Garnet lherzolites from the Kaapvaal craton (South Africa): trace element evidence for a metasomatic history. *Journal of Petrology*, **44**, 629–657.

Grégoire, M., Jégo, S., Maury, R.C., Polvé, M., Payot, B., Tamayo, Jr. R.A. and Yumul, Jr., G.P. (2008) Metasomatic interactions between slab-derived melts and depleted mantle: Insights from xenoliths within Monglo adakite (Luzon arc, Philippines). *Lithos*, **103**, 415–430.

Grégoire, M., Langlade, J.A., Delpech, G., Dantas, C. and Ceuleneer, G. (2009) Nature and evolution of the lithospheric mantle beneath the passive margin of East Oman: evidence from mantle xenoliths sampled by Cenozoic alkaline lavas. *Lithos*, **112**, 203–216.

Grégoire, M., Chevet, J. and Maaloe, S. (2010) Composite xenoliths from Spitsbergen: evidence of the circulation of MORB-related melts within the upper mantle. Pp. 71–86 in: *Petrological Evolution of the European Lithospheric Mantle* (M. Coltorti, H. Downes, M. Grégoire and S.Y. O'Reilly, editors). Special Publications, **337**, Geological Society, London.

Harte, B. (1983) Mantle peridotites and process – the kimberlite sample. Pp. 46–91 in: *Continental Basalts and their Mantle Xenoliths* (C.J. Hawkesworth and M.J. Norry, editors). Shiva Publishing Limited, Nantwitch, Cheshire, UK.

Harte, B. and Harris, J.W. (1994) Lower mantle mineral associations preserved in diamonds. *Mineralogical Magazine*, **58A**, 384–385.

Hervig, R.L., Smith, J.V. and Dawson, J.B. (1986) Lherzolite xenoliths in kimberlites and basalts: Petrogenetic and crystallochemical significance of some minor and trace elements in olivine, pyroxenes, garnets and spinel. *Royal Society of Edinburgh Transactions, Earth Science*, **77**, 181–201.

Hidas, K., Guzmics, T., Szabó, C., Kovács, I., Bodnar, R.J., Zajacz, Z., Nédli, Z., Vaccari, L. and Perucchi, A. (2010) Coexisting silicate melt inclusions and H_2O-bearing, CO_2-rich fluid inclusions in mantle peridotite xenoliths from the Carpathian–Pannonian region (central Hungary). *Chemical Geology*, **274**, 1–18.

Hirschmann, M.M., Ghiorso, M.S., Wasylenki, L.E., Asimow, P.D. and Stolper, E.M. (1998) Calculation of peridotite partial melting from thermodynamic models of minerals and melts. I. Review of methods and comparison with experiments. *Journal of Petrology*, **3**, 1091–1115.

Ionov, D. A., Grégoire, M. and Ashchepkov, I.V. (1999) Feldspar-Ti oxide metasomatism in off-cratonic continental and oceanic upper mantle. *Earth Planetary Science Letters*, **165**, 37–44.

Ishimaru, S., Arai, S., Ishida, Y., Shirasaka, M. and Okrugin, V.M. (2007) Melting and multi-stage metasomatism in the mantle wedge beneath a frontal arc inferred from highly depleted peridotite xenoliths from the Avacha Volcano, southern Kamchatka. *Journal of Petrology*, **48**, 395–433.

Kelemen, P.B. (1990) Reaction between ultramafic rock and fractionating basaltic magma I. Phase relations, the origin of calc-alkaline magma series, and the formation of discordant dunite. *Journal of Petrology*, **31**, 51–98.

Klemme, S. (2004) The Influence of Cr on the garnet-spinel transition in the Earth's Mantle: Experiments in the system $MgO-Cr_2O_3-SiO_2$ and thermodynamic modeling. *Lithos*, **77**, 639–646.

Liu, Y.-D., Ying, J.-F., Li, J., Sun, Y. and Teng, F.-Z. (2020) Diverse origins of pyroxenite xenoliths from Yangyuan, North China Craton: implications for the modification of lithosphere by magma underplating and melt–rock interactions. *Lithos*, **372–373**, 105680.

Lu, J.,Tilhac, R., Griffin, W.L., Zheng, J., Xiong, Q., Oliveira, B. and O'Reilly, S.-Y. (2020) Lithospheric memory of subduction in mantle pyroxenite xenoliths from rift-related basalts. *Earth and Planetary Science Letters*, **544**, 116365.

McDonough, W.F. and Sun, S.-S. (1995) The composition of the Earth. *Chemical Geology*, **120**, 223–253.

McInnes, B.I.A, Grégoire, M., Binns, R.A., Herzing, P.M. and Hannington, M.D. (2001) Hydrous metasomatism of oceanic sub-arc mantle, Lihir, Papua New Guinea. Part 1: Petrology and geochemistry of fluid-metasomatised mantle wedge xenoliths. *Earth and Planetary Sciences Letters*, **188**, 169–183.

Moine, B.N., Grégoire, M., O'Reilly, S.Y., Sheppard, S.M.F. and Cottin, J.Y. (2001) High field strength element (HFSE) fractionation in the upper mantle: evidence from amphibole-rich composite mantle xenoliths from the Kerguelen Islands (Indian Ocean). *Journal of Petrology*, **42**, 2145–2167.

Moine, B.N., Grégoire, M., O'Reilly, S.Y., Delpech, G., Sheppard, S.M.F., Lorand, J.P., Renac, C., Giret A. and Cottin, J.Y. (2004) Carbonatite melt in oceanic upper mantle beneath the Kerguelen Archipelago. *Lithos*, **75**, 239–252.

Niu, Y. (1997) Mantle melting and melt extraction processes beneath ocean ridges: evidence from abyssal peridotites. *Journal of Petrology*, **38**, 1047–1074.

Norman, M.D. (1998) Melting and metasomatism in the continental lithosphere: laser ablation ICPMS analysis of minerals in spinel lherzolites from eastern Australia. *Contributions to Mineralogy and Petrology*, **130**, 240–255.

O'Reilly, S.Y. and Griffin, W.L. (1988) Mantle metasomatism beneath western Victoria, Australia: I. Metasomatic processes in Cr-diopside lherzolites. *Geochimica et Cosmochimica Acta*, **52**, 433–457.

O'Reilly, S.Y. and Griffin, W.L. (2013) Mantle metasomatism. Pp. 471–533 in: *Metasomatism and the Chemical Transformation of Rock: the Role of Fluids in Terrestrial and Extraterrestrial Processes* (D. E. Harlov and H. Austrheim, editors). Lecture Notes in Earth System Sciences, Springer Nature.

Pearson, D.G., Brenker F.E., Nestola, F., McNeill, J., Nasdala, L., Hutchison, M.T., Matveev, S., Mather K., Silversmit, G., Schmitz, S., Vekemans, B. and Vincze, L. (2014) Hydrous mantle transition zone indicated by ringwoodite included within diamond. *Nature*, **507**, 221–224.

Python, M. and Ceuleneer, G. (2003) Nature and distribution of dykes and related melt migration structures in the mantle section of the Oman ophiolite. *Geochemistry Geophysics Geosystems*, 4–7. doi:10.1029/2002GC000354.

Puziewicz, J., Koepke, J., Grégoire, M., Ntaflos, T. and Matusiak-Malek, M. (2011) Lithospheric mantle modification during Cenozoic rifting in Central Europe: Evidence from the Ksieginki Nephelinite (SW Poland) Xenolith Suite. *Journal of Petrology*, **52**, 2107–2145.

Schiano, P., Clocchiatti, R., Shimizu, N., Weis, D. and Mattielli, N. (1994) Cogenetic silica-rich and carbonate-rich melts trapped in mantle minerals in Kerguelen ultramafic xenoliths: implications for metasomatism in the oceanic upper mantle. *Earth and Planetary Science Letters*, **123**, 167–178.

Seitz, H.-M., Brey, G.P., Harris, J.W., Durali-Müller, S., Ludwig, T. and Höfer, H.E. (2018) Ferropericlase inclusions in ultradeep diamonds from Sao Luiz (Brazil): high Li abundances and diverse Li-isotope and trace element compositions suggest an origin from a subduction mélange. *Mineralogy and Petrology*, **112**, 291–300.

Simon, N.S.C., Neumann, E.-R., Bonadiman, C., Coltorti, M., Delpech, G., Grégoire, M. and Widom, E. (2008) Ultra-depleted domains in the oceanic mantle lithosphere: evidence from major element and modal relationships in mantle xenoliths from ocean islands. *Journal of Petrology*, **49**, 1223–1251.

Streckeisen, A.L. (1976) Classification and nomenclature of igneous rocks. *Neues Jahrbuch fur Mineraogie Abhandlungen*, **107**, 144–240.

Teitchou, M.I., Grégoire, M., Temdjim, R., Ghogomu, R.T., Ngwa, C. and Aka, F.T. (2011) Mineralogical and geochemical fingerprints of mantle metasomatism beneath Nyos volcano (Cameroon volcanic line). *Geological Society of America Special Paper*, **478**, 193–210.

Tilhac, R., Grégoire, M., O'Reilly, S.Y., Griffin, W.L., Henry, H. and Ceuleneer, G. (2017) Sources and timing of pyroxenite formation in the sub-arc mantle: Case study of the Cabo Ortegal Complex, Spain. *Earth and Planetary Science Letters*, **474**, 490–502.

Walter, M.J. (1998) Melting of garnet peridotite and the origin of kKomatiite and depleted lithosphere. *Journal of Petrology*, **39**, 29–60.

Whitney, D.L. and Evans, B.W. (2010) Abbreviations for names of rock-forming minerals. *American Mineralogist*, **95**, 185–187.

Wyllie, P.J. (1981) Plate tectonics and magma genesis. *Geologische Rundschau*, **70**, 128–153.

Xu, X., O'Reilly, S.-Y., Griffin, W.L. and Zhou, X. (2000). Genesis of young lithospheric mantle in southeastern China: a LAM–ICPMS trace element study. *Journal of Petrology*, **41**, 111–148.

EMU Notes in Mineralogy, Vol. 21 (2024), Chapter 4, 57–110

Simple models for trace element fractionation during decompression melting of a two-lithology mantle

YAN LIANG

Department of Earth, Environmental and Planetary Sciences, Brown University, Providence, Rhode Island 02912, USA e-mail: yan_liang@brown.edu

Batch melting, fractional melting, continuous melting and two-porosity melting models have been used widely in geochemical studies of trace element fractionation during mantle melting. These simple melting models were developed for melting an homogeneous mantle source. Here we revisit and further develop these melting models in the context of decompression melting of a two-lithology mantle. Each lithology has its own source composition and melting parameters. During decompression melting, melt and solid flow vertically in the melting column. Part of the melt produced in one lithology is transferred to the other lithology at a prescribed rate. We use a set of conservation equations to solve for melt and solid mass fluxes, extent of melting and concentrations of a trace element in interstitial melt and aggregated melt in each lithology and mixed-column melt between the two lithologies. We uncover conditions under which batch melting, fractional melting, continuous melting and two-porosity melting models are realized during decompression melting through four case studies. We show that porosity in the continuous melting model varies along the melting column during decompression melting, contrary to what was assumed in its original development. We unify the batch melting, fractional melting, continuous melting and two-porosity melting models through a two-lithology melting model for decompression melting in a two-lithology mantle column. We discuss basic features of the two-lithology melting model through worked examples. We show that it is possible to produce partial and well-mixed melts with a range of REE patterns, from LREE depleted to LREE enriched, similar to those observed in mid-ocean ridge basalts by decompression melting of a two-lithology mantle.

1. Introduction

Models for trace element fractionation during mantle melting are essential to interpretation of basalts and peridotites. Simple models that have been used widely in geochemical studies of mantle melting include batch melting, fractional melting and continuous melting models (*e.g.* Shaw, 2006; Zou, 2007 and references therein). The continuous melting model is also referred to as the dynamic melting model in the literature (*e.g.* McKenzie, 1985; Albarède, 1995; Zou, 1998; Shaw, 2000). These simple melting models were originally developed by considering mass balance for a mantle parcel, irrespective of flows of partial melt and residual solid in the melting region (*e.g.* Gast, 1968; Shaw, 1970, Langmuir *et al.*, 1977; McKenzie, 1985; Albarède, 1995; Zou, 1998; Shaw, 2000). It has been shown in subsequent studies that these simple melting models can also be derived by considering flows of melt and solid in an upwelling steady-state melting column (*e.g.* Ribe, 1985; Spiegelman and Elliott, 1993; Iwamori, 1994; Asimow and

DOI: 10.1180/EMU-notes.21.4

Stolper, 1999; Lundstrom, 2000; Liang, 2008; Liang and Peng, 2010). During batch melting, both partial melt and residual solid flow upwards in the melting column and no melt is extracted to melt conduits or high-porosity channels along the melting column (*e.g.* Ribe, 1985; Spiegelman and Elliott, 1993; Asimow and Stolper, 1999). During fractional melting, melt generated at any point in the melting column is completely and instantaneously removed from the residual solid. The melt fraction or porosity is zero in the residuum. During continuous or dynamic melting, a small but constant fraction of melt (typically <2%) is retained in the residuum after an initial stage of batch melting (Langmuir *et al.*, 1977; McKenzie, 1985; Albarède, 1995; Zou, 1998; Shaw, 2000). In a physically more realistic setting, only a fraction of melt is extracted through high-porosity channels and the remaining melts percolate through the melting column. Consequently, the porosity increases upwards in the melting column (*e.g.* Ribe, 1985; Hewitt and Fowler, 2008; Liang and Liu, 2018). This leads to the two-porosity melting models (*e.g.* Iwamori, 1994; Lundstrom, 2000; Ozawa, 2001; Jull *et al.*, 2002; Liang and Parmentier, 2010; Liang and Peng, 2010). In this chapter, we will take a closer look at these simple melting models from a standpoint of mass conservation of the melt and residual solid in an upwelling steady-state melting column.

The batch, fractional, continuous and two-porosity melting models were originally developed for modelling trace element variations during partial melting of a homogeneous mantle source. There is growing evidence that the source region for mantle-derived magmas is chemically (depleted *vs.* enriched) and lithologically (peridotite *vs.* pyroxenite) heterogeneous. The heterogeneities are long lived and probably produced by tectonic processes involving crustal formation, crust and mantle recycling and core-mantle interaction (*e.g.* Zindler and Hart, 1986; Hofmann, 1997; Stracke, 2012; White, 2015). There are several types of heterogeneities in the mantle. In terms of radiogenic isotope ratios, the mantle source is identified as the depleted mantle and the enriched mantle of various types (*e.g.* EM I, EM II, HIMU, FOZO, Zindler and Hart, 1986; Hofmann, 2003). The lithology of the depleted mantle is generally ascribed to lherzolite and its composition is relatively well constrained (*e.g.* Salters and Stracke, 2004; Workman and Hart, 2005). Compositions of enriched mantle components are model dependent as they are estimated by mixing the depleted mantle with various proportions of the recycled or subducted oceanic crust, ancient pelagic or terrigenous sediments and/or lower continental crust (*e.g.* Weaver, 1991; Stracke *et al.*, 2003; Willbold and Stracke, 2006; Turner *et al.*, 2017). The lithology of the enriched mantle could be pyroxenite, eclogite, or peridotite. One important petrological observation is that the solidi of garnet pyroxenite, eclogite and carbonated peridotite are lower than the solidus of anhydrous lherzolite (*e.g.* Yasuda *et al.*, 1994; Pertermann and Hirschmann, 2003; Kogiso *et al.*, 2004; Dasgupta *et al.*, 2006; Lambart *et al.*, 2016). During decompression melting of a veined mantle, garnet pyroxenite and eclogite melt at greater depth than their surrounding peridotites. Melts derived from the garnet pyroxenite and eclogite react with their surrounding peridotites, producing secondary pyroxenites that are enriched in orthopyroxene (Opx) or clinopyroxene (Cpx) (*e.g.* Yaxley and Green, 1998; Lo Cascio, 2008; Lambart *et al.*, 2012; Mallik and Dasgupta, 2012; Wang *et al.*, 2013, 2020; Borghini *et al.*, 2017, 2020; Soderman *et al.*, 2022). One of the objectives of this study is to expand the existing melting models so we can include a second lithology in an upwelling melting column.

To model trace element fractionation during partial melting of a heterogeneous mantle source, the standard geochemical treatment is melting followed by mixing (*e.g.* Vollmer, 1976; Langmuir *et al.*, 1978). In this two-step approach, one first uses one of the simple melting models to calculate compositions of melt derived from the depleted mantle source and the enriched mantle source in two independent calculations. One then mixes the two melts in different proportions to obtain a set of mixed melt compositions and compares the modelling results with geochemical observations. This melting followed by mixing approach has been used widely in the interpretation of trace element and isotope data of oceanic basalts (*e.g.* Langmuir *et al.*, 1978; Zindler *et al.*, 1984; Niu *et al.*, 2002; Ito and Mahoney, 2005; Stracke and Bourdon, 2009; Rudge *et al.*, 2013; Shimizu *et al.*, 2016; Shorttle *et al.*, 2016). As the two melting calculations are independent of each other, there is no mass transfer between the enriched and depleted mantle in the melting region. Mixing takes place at the top of the melting region.

The purpose of this chapter is to present the batch, fractional, continuous and two-porosity melting models in the context of decompression melting of an upwelling two-lithology melting column in which melts produced in one lithology flow into and interact with the melt and residual solid in another lithology. The presence of the second lithology also allows us to model the formation of high-porosity channels, a process that has not been considered in the two-porosity melting models (*e.g.* Iwamori, 1994; Lundstrom, 2000; Ozawa, 2001; Jull *et al.*, 2002; Liang and Peng, 2010; but see Liang and Parmentier, 2010). Here we consider a set of general problems in which the mantle source consisted of two lithologies: A and B. Figure 1 presents a simplified treatment in which the background lithology A has a larger volume fraction in the melting region. To obtain steady-state solutions, we assume that the shape of lithology B is in the form of long strings (Fig. 1a). We subdivide the melting region into vertical columns; each contains a pair of A-B. For a given melting column (Fig. 1b), the two lithologies have their own volume fractions (ψ_A, ψ_B), source compositions (C_A^0, C_B^0), melting rates (Γ_A, Γ_B), degrees of melting (F_A, F_B), porosities (ϕ_f^A, ϕ_f^B), melting reactions and bulk solid-melt partition coefficients for the trace element of interest (k_A, k_B). In the lowest part of the melting column ($F_A \leq F_A^d$), melt fractions are small. There is no lateral melt flow across the lithological boundary and batch melting prevails in the two lithologies. As there is no chemical interaction between the two lithologies, the models presented in this study are also applicable to cases when the depth for the onset of melting for one lithology is different from another lithology. In the upper part of the melting column ($F_A > F_A^d$), part of the melt generated in lithology A at a given location in the melting column (*e.g.* z_1 in Fig. 1b) is transferred laterally into lithology B where it mixes locally with the melt produced in lithology B. The mixed melt percolates and re-equilibrates with the residual solid in lithology B in the overlying melting column. The amount of melt flowing from lithology A to lithology B per unit volume of the two lithologies per unit time is called the melt suction rate ($\dot{S}$), a key parameter in the models presented in this study. At the top of the melting column, melts from lithology A and lithology B mix with each other, forming the mixed-column melt for the two-lithology melting column.

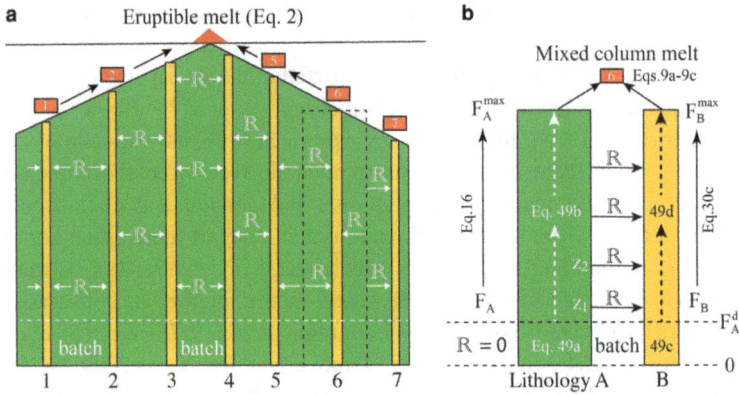

Fig. 1. (*a*) Schematic diagram showing an upwelling melting region that consists of two lithologies, A (green) and B (orange). Lithology B has a smaller volume fraction and is in the form of long vertical strings (labelled 1, 2, ..., 7). (*b*) Mass transfer in an idealized two-lithology melting column (equivalent to dashed box in panel a). The lower part of the melting column ($F_A \leq F_A^d$) experiences batch melting during which melts generated in the two lithologies do not interact with each other. The melt suction rate is zero. Part of the melt generated in lithology A in the upper part of the melting column ($F_A > F_A^d$) is transferred to lithology B at a prescribed melt suction rate of $\mathbb{R}$ (Eq. 17c). Mixing of melts from lithology A and lithology B at top of the melting column produces the mixed-column melt (boxes 1 to 7). The eruptible melt is formed by pooling mixed-column melts from columns 1, 2, ..., 7 in prescribed proportions. Key equations for concentrations of a trace element in the interstitial melt in the two lithologies, mixed-column melt at top of the melting column, and eruptible melt over the melting region are identified (Eqs. 9a–9c, 49a–49d and 2).

Finally, the mixed-column melts from selected melting columns aggregate and mix with each other across the melting region (*e.g.* columns 5, 6 and 7 in Fig. 1a), forming the eruptible melt. Although highly simplified, the scenario outlined here is broadly consistent with the process of high-porosity harzburgite and dunite channel formation in an upwelling mantle column (*e.g.* Liang *et al.*, 2010; Schiemenz *et al.*, 2011) and melt transport along decompaction channels at top of the melting region (*e.g.* Sparks and Parmentier, 1991). The flux of percolating melt must be high enough to form high-porosity channels in the upwelling column. This is achieved by the initial batch melting in the lowest part of the melting region and by the higher melting or/and dissolution rate of the enriched lithology.

Our main objective is to obtain simple mathematical expressions for concentrations of a trace element in interstitial melts and residual solids in lithology A and lithology B in the melting column, concentration of the trace element in the mixed-column melt collected at top of the melting column and concentration of the eruptible melt pooled across the melting region. In an upwelling melting column where melt and solid flow at different velocities, it is necessary to keep track of spatial variations of the melt and solid in the melting column. This is done through applications of mass conservation equations. Here we show how the conservation equations can be used to construct simple melting models and to understand their physical meanings. Appendices A and B present two sets of conservation equations for modelling trace element fractionation in lithology A and lithology B during concurrent melting and melt migration in a one-dimensional (1D) melting column. To connect to the batch, fractional, continuous and two-porosity melting models, we consider four steady-state problems in which concentrations, velocities and volume fractions of the melt and solid vary as a function of location in the melting column but are independent of time. For simplicity, we assume that the melting rate and volume fraction of each lithology are constant and uniform in the melting column and that the melt suction rate takes on one or two constant values. Following previous treatment, we assume that residual solid and interstitial melt are in local chemical equilibrium in each of the two lithologies. These simplifications allow us to obtain simple analytical solutions to the two-lithology melting problems and to gain new insights into the geochemical consequence of decompression melting of a two-lithology mantle.

Here, detailed step-by-step derivations of the four melting models are presented in sections 3–5. For the convenience of geochemical modelling, the main equations for each of the melting models in Appendix E are summarized. To set up the study in sections 3–5, simple derivations for the composition of mixed-column melt at top of the two-lithology melting column (C_f^{mix}) and the composition of eruptible melt pooled from several melting columns in the melting region ($C_f^{eruptible}$) are presented first. Mathematical expressions for these mixed-melt compositions are general and independent of assumptions of the underlining melting models (Appendix C). Figure 2 is a roadmap that outlines the relationships among the melting models presented in this study. For modelling trace element fractionation during decompression melting of a heterogeneous mantle, the more general two-lithology melting model which reduces to the batch melting, fractional melting, continuous melting and two-porosity melting models under special conditions is recommended.

2. Compositions of mixed-column melt and eruptible melt

2.1. Compositions of the eruptible melt and the overall average melt

If we pool all the melts collected at the top of each melting column, we obtain the average melt for the entire melting region. Let $C_{f,j}^{mix}$ be the concentration of a trace element of interest in the mixed-column melt derived from lithology A and lithology B at the top of column j and $\bar{F}_j$ be the average degree of melting for the two-lithology melting column j. The overall average melt composition (C_f^{avg}) is the weighted mean of mixed-column melts from all the melting columns, *i.e.*

$$C_f^{avg} = \frac{\bar{F}_1 C_{f,1}^{mix} + \bar{F}_2 C_{f,2}^{mix} + \ldots + \bar{F}_N C_{f,N}^{mix}}{\bar{F}_1 + \bar{F}_2 + \ldots + \bar{F}_N} \tag{1a}$$

$$\bar{F}_j = \psi_{A,j} F_{A,j} + \psi_{B,j} F_{B,j} \tag{1b}$$

$$\psi_{A,j} + \psi_{B,j} = 1 \tag{1c}$$

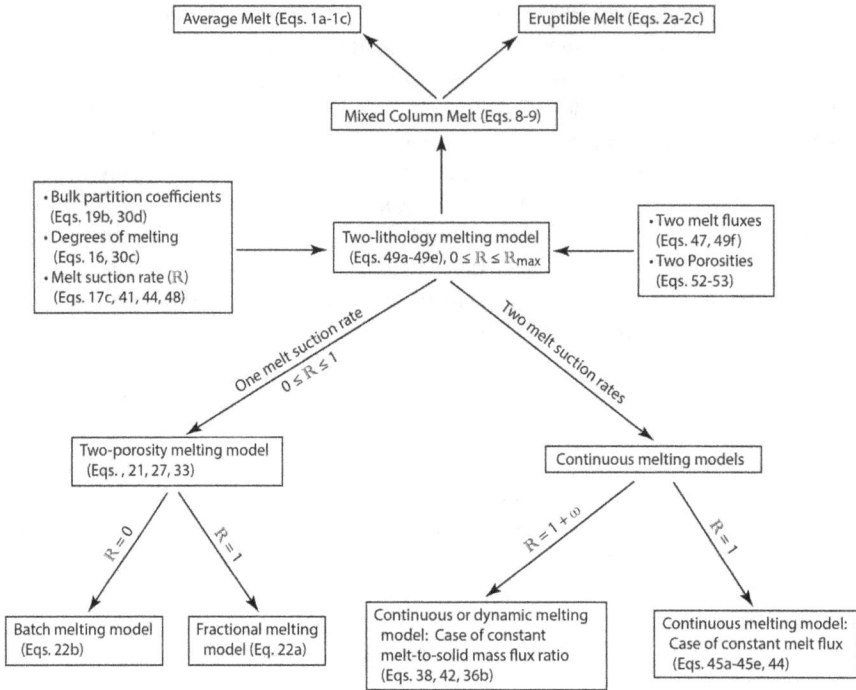

Fig. 2. Roadmap illustrating relationships among the various melting models presented here. The two-lithology melting model is the most general. It has three sets of melting parameters, two melt fluxes and two porosities. Depending on the choice of melt suction rate, the two-lithology melting model is reduced to the batch melting, fractional melting, continuous melting, and two-porosity melting models under specified limits. The mixed column melt is a mixture of melts derived from lithology A and lithology B at top of the melting column. The eruptible melt is formed by pooling mixed-column melts from different melting columns. The average melt is the weighted mean of all the melts collected at top of each melting column. Key equations are identified in the figure and summarized in Appendix E.

where N is the number of two-lithology melting columns in the melting region ($N = 7$ in Fig. 1a); and subscript j (=1, 2, ..., N) refers to properties of lithology A or lithology B of column j. For example, $\psi_{A,j}$ is the volume fraction of lithology A in column j. The volume fractions of A and B in one melting column can be different from those in another melting column. Key symbols used in this note are listed in Table 1.

If mixing of melts from different melting columns is incomplete, the composition of mixed melt would depend on the amount of melt contributed from each melting column. A collection of incompletely mixed melts from different melting columns is referred to as the eruptible melt, the composition of which is given by the general expression:

$$C_f^{eruptible} = \alpha_1 C_{f,1}^{mix} + \alpha_2 C_{f,2}^{mix} + \ldots + \alpha_N C_{f,N}^{mix} \qquad (2a)$$

Table 1. List of key symbols.

Symbol	Description
C_A^0, C_B^0	Concentration of the bulk solid A or B at the onset of melting
C_f^A, C_f^B	Concentration of a trace element in interstitial melt in lithology A or B
C_f^{avg}	Concentration of the average melt for the entire melting region
C_f^{mix}, C_{fj}^{mix}	Concentration of the mixed-column melt for column j
$\bar{C}_f^A$	Average concentration of melts sucked from lithology A to lithology B
C_s^A, C_s^B	Concentration of a trace element in residual solid in lithology A or B
d_A, d_B	Mean grain size of lithology A or lithology B
F_A, F_B	Degree of melting experienced by lithology A or lithology B
F_A^{max}, F_B^{max}	Maximum extent of melting at top of the melting column
F_d^A, F_d^B	Maximum extent of batch melting in the lower part of the melting column
$\bar{F}_j$	Average degree of melting for the two-lithology melting column j
k_A^0, k_B^0	Bulk partition coefficient at the onset of melting for lithology A or B
k_A^p, k_B^p	Bulk partition coefficient according to the melting reaction of A or B
k_A, k_B	Bulk solid-melt partition coefficient for lithology A or lithology B
n	Porosity exponent in the permeability model
p_j^A, p_j^B	Modal proportion of mineral j in the melting reaction of lithology A or B
$\mathbb{R}$	Dimensionless melt suction rate
$\dot{S}$	The rate of melt extracted from lithology A to lithology B
V_f^A, V_f^B	Velocity of the interstitial melt in lithology A or lithology B
V_s^A, V_s^B	Velocity of the solid in lithology A or lithology B in the melting column
V_s^0	Solid upwelling rate at the onset of melting
w_j^A, w_j^B	Modal proportion of mineral j in lithology A or lithology B
z	Vertical coordinate with origin at the bottom of the melting column
α_j	Fraction of contribution to the eruptible melt from column j
ϕ_f^A, ϕ_f^B	Porosity of lithology A or lithology B
$\phi_{ref}^A, \phi_{ref}^B$	Reference porosity of lithology A or lithology B
$\kappa_\phi^A, \kappa_\phi^B$	Permeability of lithology A or lithology B
ω	Ratio of melt-to-solid mass flux for the continuous melting model
Γ_A, Γ_B	Melting rate of lithology A or lithology B
ρ_f, ρ_s	Density of the melt or solid
ψ_A, ψ_B	Volume fraction of lithology A or lithology B

$$\alpha_1 + \alpha_2 + \ldots + \alpha_N = 1 \quad \text{and} \quad 0 \le \alpha_j \le 1 \tag{2b}$$

where the weighing factor α_j is model dependent. The weighing factor is zero for columns that do not contribute to the eruptible melt. Equation 2a is reduced to Eq. 1a when the weighing factor is proportional to the volume or volume flux of melt produced from each of the two-lithology melting columns, *i.e.*

$$\alpha_j = \frac{\bar{F}_j}{\bar{F}_1 + \bar{F}_2 + \ldots + \bar{F}_N} \qquad (2c)$$

The mixing proportion (α_j) is determined by a mixing mechanism which is not well constrained. Rudge *et al.* (2013) proposed a statistical model for calculating the weighing factor in Eq. 2a for the eruptible melt produced by fractional melting of a two-lithology mantle. Shimizu *et al.* (2016) and Shorttle *et al.* (2016) used this model to study variations in trace elements and Nd isotope ratio in MORB samples. Liu and Liang (2020) studied the effect of incomplete mixing on Sr-Nd-Hf-Pb isotope ratios and incompatible trace element abundances in pooled melts collected at top of a triangular melting region by setting the weighing factor in Eq. 2a to zero for selected melting columns. Liang (2022) used Eq. 1a-1c and a mixing scheme to model along ridge variations in Sr-Nd-Hf isotope ratios in pooled melts produced by fractional melting of a two-lithology mantle. These studies have demonstrated the importance of incomplete mixing in producing large variations in radiogenic isotope ratios and highly incompatible trace element abundances in oceanic basalts. However, a process-based mixing model still awaits future developmemt, a topic that is beyond the scope of this chapter.

In the next section, general expressions for the concentration of a trace element in the well mixed-column melt for the two-lithology melting column j are presented. To simplify notations, we drop the subscript j for all the variables for column j in the equations presented in the remainder of this chapter.

2.2. Composition of the mixed-column melt

Let's start with the classic problem of batch melting in a closed box. As no mass has entered or left the box, concentrations of a trace element in the melt (C_f) and residual solid (C_s) are related to each other through the mass balance equation:

$$FC_f + (1 - F)C_s = C_s^0 \qquad (3)$$

where F is the fraction of melt in the system and C_s^0 is the solid concentration before melting. If the melt and residual solid are in chemical equilibrium, we obtain the batch melting model:

$$C_f = \frac{C_s^0}{k + (1 - k)F} \qquad (4)$$

where k is the bulk solid-melt partition coefficient. Let us now expand this exercise by considering batch melting of lithology A and lithology B in the same box. The two lithologies are isolated from each other except at top of the melting column where melts derived from the two lithologies mix completely with each other, forming the well mixed-column melt. Applying Eq. 3 to lithology A and lithology B, respectively, we have:

$$F_A C_f^A + (1 - F_A) C_s^A = C_A^0 \tag{5a}$$

$$F_B C_f^B + (1 - F_B) C_s^B = C_B^0 \tag{5b}$$

where superscripts A and B refer to properties of lithology A and lithology B, respectively. Taking a weighted sum of Eqs 5a and 5b, we have:

$$\psi_A \left[F_A C_f^A + (1 - F_A) C_s^A \right] + \psi_B \left[F_B C_f^B + (1 - F_B) C_s^B \right] = \psi_A C_A^0 + \psi_B C_B^0 \tag{6}$$

where ψ_A and ψ_B are volume fractions of A and B in the melting column. Rearranging Eq. 6, we have:

$$\psi_A F_A C_f^A + \psi_B F_B C_f^B = \psi_A \left[C_A^0 - (1 - F_A) C_s^A \right] + \psi_B \left[C_B^0 - (1 - F_B) C_s^B \right] \tag{7}$$

The mixed-column melt concentration is the weighted average of melts from lithologies A and B, *i.e.*

$$C_f^{mix} = \frac{\psi_A F_A C_f^A + \psi_B F_B C_f^B}{\psi_A F_A + \psi_B F_B} \tag{8}$$

The denominator in Eq. 8 is the average degree of melting of the two-lithology melting column. Substituting Eq. 7 into Eq. 8, we have an alternative expression for the mixed-column melt concentration:

$$C_f^{mix} = \frac{\psi_A \left[C_A^0 - (1 - F_A) C_s^A \right] + \psi_B \left[C_B^0 - (1 - F_B) C_s^B \right]}{\psi_A F_A + \psi_B F_B} \tag{9a}$$

which can also be written in terms of concentrations of interstitial melts in lithologies A and B:

$$C_f^{mix} = \frac{\psi_A \left[C_A^0 - (1 - F_A) k_A C_f^A \right] + \psi_B \left[C_B^0 - (1 - F_B) k_B C_f^B \right]}{\psi_A F_A + \psi_B F_B} \tag{9b}$$

For non-modal melting during which the bulk partition coefficient varies as a function of the degree of melting, we have:

$$C_f^{mix} = \frac{\psi_A\left[C_A^0 - (k_A^0 - k_A^p F_A)C_f^A\right] + \psi_B\left[C_B^0 - (k_B^0 - k_B^p F_B)C_f^B\right]}{\psi_A F_A + \psi_B F_B} \tag{9c}$$

where k^0 is the bulk solid-melt partition coefficient at the onset of melting; and k^p is the bulk solid-melt partition coefficient according to melting reaction for lithology A or lithology B (see Eqs 18–19 below). The differences in the two square brackets on the right-hand-side of Eqs 9a–9c are the amount of melt produced by melting of lithology A and lithology B, respectively. In Appendix C, we show that Eqs 8 and 9 are general expressions for the concentration of mixed-column melt produced by decompression melting of a two-lithology mantle, irrespective of how the two lithologies interact in the melting column. These equations highlight the importance of composition and volume proportion of the mantle source and extent of melting in geochemical mixing calculations.

For highly incompatible trace elements, their concentrations in residual solids become negligible after a small to moderate extent of melting. Equation 9a can be simplified as:

$$C_f^{mix} \approx \frac{\psi_A C_A^0 + \psi_B C_B^0}{\psi_A F_A + \psi_B F_B} \tag{10a}$$

Hence mixing of melts from lithology A and lithology B at top of the melting column is equivalent to mixing of the two sources weighted by the fraction of melts produced in the two-lithology column. As the demonimator in Eq. 10a is generally less than one, the concentration of the mixed melt is greater than the concentration of the mixed-mantle source. The demonimator in Eq. 10a is cancelled out when taking a ratio of two highly incompatible trace elements or two isotopes of the same element. For $^{87}Sr/^{86}Sr$, we have

$$\frac{^{87}C_f^{mix}}{^{86}C_f^{mix}} \approx \frac{\psi_A\,^{87}C_A^0 + \psi_B\,^{87}C_B^0}{\psi_A\,^{86}C_A^0 + \psi_B\,^{86}C_B^0} \tag{10b}$$

In terms of isotope or element ratios of highly incompatible trace elements, mixing of melts from lithology A and lithology B at top of the melting column is equivalent to mixing of the two sources. Hence one cannot distinguish source mixing from magma mixing using highly incompatible trace elements unless the extent of melting is very small. This statement is independent of melting models.

For moderately incompatible and compatible trace elements, the mixed-column melt composition depends on melt or residual solid compositions at top of the melting column, hence the melting models. In the next three sections, we present two classes of melting models, one has a constant and uniform melt suction rate (section 3) and the other has two melt suction rates (sections 4–5). We show how these melting models are related to the batch, fractional, continuous and two-porosity melting models. In section 6, we take a closer look at the various melting models by comparing porosities derived from the melting models. Finally in section 7, we present examples of calculated melt compsitions using the two-lithology melting model.

3. Case of constant and uniform melt suction rate

First, a simple case in which the melt suction rate is constant and uniform in the melting column is considered. This case was first examined by Iwamori (1994) for mantle melting with diffuse and channelized porous flow. The melting region is treated as two overlapping continua consisting of low-porosity matrix (lithology A) and interconnected high-porosity channels (lithology B). However, the process of high-porosity channel formation was not considered in this class of two-porosity melting models (*e.g.* Iwamori, 1994; Lundstrom, 2000; Ozawa, 2001; Jull *et al.*, 2002; Liang, 2008; Liang and Peng, 2010). Here we complete this model by explicitly modelling high-porosity channel formation in lithology B. Section E1 in Appendix E summarizes the main results. Below we provide a detailed derivation.

3.1. Trace element concentrations in lithology A

We start with the steady-state mass conservation equation for concentration of a trace element in interstitial melt in lithology A in the 1D melting column:

$$\left[\rho_f \phi_f^A V_f^A + \rho_s \left(1 - \phi_f^A \right) V_s^A k_A \right] \frac{dC_f^A}{dz} = \left(k_A^p - 1 \right) C_f^A \Gamma_A \tag{11a}$$

where ρ_f and ρ_s are densities of the melt and solid, respectively; ϕ_f^A and V_f^A are the porosity and velocity of interstitial melt in the melting column; V_s^A is the residual solid velocity; k_A^p is the bulk solid-melt partition coefficient according to melting reaction (Eq. A9 in Appendix A); and z is the vertical coordinate, positive upwards. For convenience, we set the origin of our coordinate system ($z = 0$) to the solidus of lithology A (Fig. 1b). At the solidus, concentration of the melt is related to concentration of the mantle source of A (C_A^0) via equilibrium partitioning:

$$C_f^A(0) = \frac{C_A^0}{k_A^0} \tag{11b}$$

where k_A^0 is the bulk solid-melt partition coefficient at the onset of melting.

To solve the melt concentration from Eqs 11a and 11b, we need to know how mass fluxes of the melt, $\rho_f \phi_f^A V_f^A$ and solid, $\rho_s \left(1 - \phi_f^A \right) V_s^A$ vary spatially in the melting column. This is an important feature of steady-state melting: concentration of a trace element in the melt depends on the product of volume fraction and velocity of the melt and solid, respectively, not their individual values. The mass fluxes can be calculated using the steady-state mass conservation equations for the interstitial melt and residual solid in lithology A. From Eqs A1 and A2, we have:

$$\frac{d\rho_f \phi_f^A V_f^A \psi_A}{dz} = \psi_A \Gamma_A - \dot{S} \tag{12a}$$

$$\frac{d\rho_s \left(1 - \phi_f^A\right) V_s^A \psi_A}{dz} = -\psi_A \Gamma_A \tag{12b}$$

At the solidus ($z = 0$), melt fraction is zero and solid velocity equals to the upwelling velocity of the mantle source (V_s^0). Integrating Eqs 12a and 12b for constant melting rate and melt suction rate, we have:

$$\rho_f \phi_f^A V_f^A = \left(\Gamma_A - \frac{\dot{S}}{\psi_A}\right) z \tag{13a}$$

$$\rho_s \left(1 - \phi_f^A\right) V_s^A = \rho_s V_s^0 - \Gamma_A z \tag{13b}$$

In response to melting and melt extraction, the mass flux of the melt increases upwards in the melting column. This is an important feature of the constant melt suction rate model.

In geochemical studies, it is common practice to express melt and solid concentrations as a function of the extent of melting experienced by residual solid. The evolution equation relating the degree of melting (F_A) to melting rate (Γ_A) is Eq. A6 in Appendix A. At steady state, we have:

$$V_s^A \frac{dF_A}{dz} = \frac{(1 - F_A)\Gamma_A}{\rho_s \left(1 - \phi_f^A\right)} \tag{14}$$

Substituting Eq. 13b into Eq. 14, we have

$$\frac{dF_A}{dz} = \frac{(1 - F_A)\Gamma_A}{\rho_s V_s^0 - \Gamma_A z} \tag{15}$$

Integrating Eq. 15 from solidus ($z = 0$, $F_A = 0$), we obtain the degree of melting for lithology A:

$$F_A = \frac{1}{\rho_s V_s^0} \int_0^z \Gamma_A dz = \frac{\Gamma_A z}{\rho_s V_s^0} \tag{16}$$

Hence for constant melting rate, the degree of melting increases linearly as a function of z in the melting column. For variable melting rate, we use the integral version of Eq. 16 to calculate the degree of melting. In terms of the degree of melting, the mass fluxes of the melt and solid in Eqs 13a and 13b take on the expressions (Iwamori, 1994; Lundstrom, 2000; Jull *et al.*, 2002; Liang and Peng, 2010):

$$\rho_f \phi_f^A V_f^A = \rho_s V_s^0 (1 - \mathbb{R}) F_A \tag{17a}$$

$$\rho_s \left(1 - \phi_f^A\right) V_s^A = \rho_s V_s^0 (1 - F_A) \tag{17b}$$

where $\mathbb{R}$ is the dimensionless melt suction rate, defined as

$$\mathbb{R} = \frac{\dot{S}}{\psi_A \Gamma_A} \tag{17c}$$

For the problem considered here $0 \le \mathbb{R} \le 1$.

Finally, during mantle melting, mineral modes in the residuum change according to the melting reaction. The bulk solid-melt partition coefficient varies as a function of extent of melting experienced by the solid. We can use the following steady-state equation to calculate spatial variations of the bulk partition coefficient in the melting column (see also Eq. A7):

$$V_s^A \frac{dk_A}{dz} = \frac{\left(k_A - k_A^p\right) \Gamma_A}{\rho_s \left(1 - \phi_f^A\right)} \tag{18a}$$

Substituting the solid mass flux (Eq. 13b) into Eq. 18a, we have:

$$\frac{dk_A}{dz} = \frac{\left(k_A - k_A^p\right) \Gamma_A}{\rho_s V_s^0 - \Gamma_A z} \tag{18b}$$

For constant k_A^p, we have:

$$\frac{k_A - k_A^p}{k_A^0 - k_A^p} = \frac{\rho_s V_s^0 - \Gamma_A z}{\rho_s V_s^0} = 1 - \frac{\Gamma_A z}{\rho_s V_s^0} \tag{19a}$$

which can also be written in terms of the degree of melting experienced by lithology A:

$$k_A = \frac{k_A^0 - k_A^p F_A}{1 - F_A} \tag{19b}$$

Equation 19b is the familiar expression relating bulk partition coefficient to degree of melting and partition coefficients for the melting reaction and at the onset of melting (*e.g.* Shaw, 1970).

We can now calculate the interstitial melt composition by substituting Eqs 16, 17 and 19b into Eq. 11a which takes on the form:

$$\frac{dC_f^A}{dF_A} = \frac{\left(k_A^p - 1\right)C_f^A}{k_A^0 + \left(1 - k_A^p - \mathbb{R}\right)F_A} \tag{20}$$

When $1 - k_A^p - \mathbb{R} \neq 0$, we have an expression for the interstitial melt composition:

$$C_f^A = \frac{C_A^0}{k_A^0}\left[\frac{k_A^0 + \left(1 - k_A^p - \mathbb{R}\right)F_A}{k_A^0}\right]^{\frac{k_A^p - 1}{1 - k_A^p - \mathbb{R}}} \tag{21a}$$

When $1 - k_A^p - \mathbb{R} = 0$, we have:

$$C_f^A = \frac{C_A^0}{k_A^0}\exp\left(\frac{k_A^p - 1}{k_A^0}F_A\right) \tag{21b}$$

Equation 21a was first obtained by Iwamori (1994). Equation 21b is a special case that arises from integration of Eq. 20. There are two physical parameters in this model: degree of melting experienced by lithology A (F_A) and dimensionless melt suction rate ($\mathbb{R}$). When $\mathbb{R} = 1$, all the melt produced in lithology A is sucked into lithology B along the melting column. The vertical melt flux, hence porosity, reduces to zero in lithology A (Eq. 17a). Equation 21a reduces to the non-modal perfect fractional melting model of Shaw (1970), *i.e.*

$$C_f^A = \frac{C_A^0}{k_A^0}\left(1 - \frac{k_A^p}{k_A^0}F_A\right)^{\frac{1 - k_A^p}{k_A^p}} \tag{22a}$$

When $\mathbb{R} = 0$, no melt produced in lithology A is transferred into lithology B. Equation 21a reduces to the non-modal batch melting model, i.e.

$$C_f^A = \frac{C_A^0}{k_A^0 + \left(1 - k_A^p\right)F_A} \tag{22b}$$

Hence batch and perfect fractional melting models are special cases of the steady-state melting model with a constant and uniform melt suction rate. Given the melt composition, concentration of the trace element in the residual solid or minerals can be calculated using solid-melt or mineral-melt partition coefficients. For the residual solid, we have:

$$C_s^A = k_A C_f^A \tag{23}$$

3.2. Average concentration of a trace element in lithology A

The average concentration of a trace element in melts sucked from lithology A to lithology B is a collection of transferred melts along the melting column, *i.e.*

$$\bar{C}_f^A = \frac{1}{\mathbb{R}F_A} \int\limits_0^{F_A} \mathbb{R}C_f^A dF_A$$

$$= \frac{C_A^0}{\mathbb{R}F_A} \left\{ 1 - \left[\frac{k_A^0 + \left(1 - k_m^p - \mathbb{R}\right)F_A}{k_A^0} \right]^{\frac{-\mathbb{R}}{1 - k_m^p - \mathbb{R}}} \right\} \tag{24}$$

Equation 24 is valid for $\mathbb{R} > 0$. When $\mathbb{R} = 0$, the average melt is undefined as no melt is transferred from lithology A to lithology B. To gain additional insights into the average melt composition, we seek an alternative expression for the average melt composition using the conserved form of the mass conservation equation for lithology A (Eq. A3). The steady-state version of Eq. A3 is:

$$\frac{d\left[\rho_f \phi_f^A V_f^A C_f^A + \rho_s \left(1 - \phi_f^A\right) V_s^A C_s^A \right]}{dz} = -\frac{\dot{S}}{\psi_A} C_f^A \tag{25}$$

Replacing the melt and solid mass fluxes and spatial coordinate in Eq. 25 by the degree of melting via Eqs 16–17, we have an ordinary differential equation with variable F_A:

$$\frac{d\left[(1 - \mathbb{R})F_A C_f^A + (1 - F_A)C_s^A\right]}{dF_A} = -\mathbb{R}C_f^A \tag{26}$$

Integrating both sides of Eq. 26 from the solidus, we have:

$$\bar{C}_f^A = \frac{C_A^0 - \left[(1 - \mathbb{R})F_A C_f^A + (1 - F_A)C_s^A\right]}{\mathbb{R}F_A} \tag{27a}$$

where values of C_f^A and C_s^A correspond to the extent of melting F_A. When the solid and melt are in local chemcial equilibrium, we have:

$$\bar{C}_f^A = \frac{C_A^0 - [(1 - \mathbb{R})F_A + (1 - F_A)k_A]C_f^A}{\mathbb{R}F_A} \tag{27b}$$

The physical meaning of Eq. 27 becomes clear after the rearrangment:

$$C_A^0 = (1 - \mathbb{R})F_A C_f^A + (1 - F_A)C_s^A + \mathbb{R}F_A \bar{C}_f^A \tag{28a}$$

Multiplying both sides of Eq. 28a by the total mass flux of lithology A, we have an overall mass flux balance equation for the trace element in lithology A:

$$\rho_s V_s^0 C_A^0 = \rho_s V_s^0 (1 - \mathbb{R}) F_A C_f^A + \rho_s V_s^0 (1 - F_A) C_s^A + \rho_s V_s^0 \mathbb{R} F_A \bar{C}_f^A \quad (28b)$$

In terms of melt production, Eq. 28b can also be written as:

$$\rho_s V_s^0 (1 - \mathbb{R}) F_A C_f^A + \rho_s V_s^0 \mathbb{R} F_A \bar{C}_f^A = \rho_s V_s^0 C_A^0 - \rho_s V_s^0 (1 - F_A) C_s^A \quad (28c)$$

Equation 28b states that the total mass flux of the trace element in lithology A feeding into the melting column from below equals to the sum of mass fluxes of the trace element in interstitial melt and residual solid at a given location or F_A in the melting column (first two terms on the right hand side) and the total mass flux of the trace element in the melt transferred from lithology A to lithology B (the last term). Equation 28c states that the the total melt flux produced by decompression melting of lithology A is the difference in solid mass flux feeding into the melting column from below and solid mass flux at the given location in the melting column. These are general statements of mass flux balance for a non-radioactive chemical species in lithology A in the upwelling steady-state melting column, irrespective of the melting process and whether the partial melt and residual solid are in local equilibrium in the melting column.

3.3. Trace element concentrations in lithology B

Here the conserved form of mass conservation equation for lithology B (Eq. B3) is used to calculate interstitial melt composition. The steady-state version of Eq. B3 takes on the form:

$$\frac{d\left[\rho_f \phi_f^B V_f^B C_f^B + \rho_s \left(1 - \phi_f^B \right) V_s^B C_s^B \right] \psi_B}{dz} = \dot{S} C_f^A \quad (29)$$

Similar to the case of lithology A, we replace the melt and solid mass fluxes in Eq. 29 by the degree of melting experienced by lithology B using solutions from the steady-state version of Eqs B1, B2, B6 and B7. The results are as follows:

$$\rho_f \phi_f^B V_f^B = \rho_s V_s^0 F_B + \rho_s V_s^0 \mathbb{R} F_A \frac{\psi_A}{\psi_B} \quad (30a)$$

$$\rho_s \left(1 - \phi_f^B \right) V_s^B = \rho_s V_s^0 (1 - F_B) \quad (30b)$$

$$F_B = \frac{\Gamma_B z}{\rho_s V_s^0}, \quad k_B = \frac{k_B^0 - k_B^p F_B}{1 - F_B} \quad (30c, 30d)$$

The degree of melting experienced by lithology B is related to the degree of melting experienced by lithology A through their melting rates, *i.e.*

$$F_B = \frac{\Gamma_B}{\Gamma_A} F_A \tag{30e}$$

Substituting Eqs 30a–30b into Eq. 29 and replacing z by F_A via Eqs 30c and 30e, we have:

$$\frac{d\left[(\psi_B F_B + \psi_A \mathbb{R} F_A)C_f^B + \psi_B(1 - F_B)C_s^B\right]}{dF_A} = \mathbb{R}\psi_A C_f^A \tag{31}$$

Equation 31 can be integrated along the melting column, starting from the solidus of lithology B,

$$(\psi_B F_B + \psi_A \mathbb{R} F_A)C_f^B + \psi_B(1 - F_B)C_s^B = \psi_B C_B^0 + \mathbb{R}\psi_A \int_0^{F_A} C_f^A dF_A \tag{32a}$$

Recalling the definition of average melt composition for lithology A (Eq. 24), we have:

$$(\psi_B F_B + \psi_A \mathbb{R} F_A)C_f^B + \psi_B(1 - F_B)C_s^B = \psi_B C_B^0 + \mathbb{R}\psi_A F_A \bar{C}_f^A \tag{32b}$$

Equation 32b states that the sum of mass flux of the element of interest in the interstitial melt and residual solid in lithology B (left hand side) is balanced by the mass flux of the element in the mantle source of lithology B feeding into the melting column from below and the melt flux flowing from lithology A into lithology B in the melting column (right hand side). When the melt and residual solid are in local chemical equilibrium, we obtain the following expressions for concentrations of the trace element in the melt and residual solid in lithology B:

$$C_f^B = \frac{\psi_B C_B^0 + \psi_A \mathbb{R} F_A \bar{C}_f^A}{\psi_B\left[k_B^0 + (1 - k_B^p)F_B\right] + \psi_A \mathbb{R} F_A} \tag{33a}$$

$$C_s^B = k_B C_f^B \tag{33b}$$

When $\mathbb{R} = 0$, no melt is transferred from lithology A to lithology B. Equation 33a is reduced to the non-modal batch melting model for lithology B. Hence Eq. 33a is a more general model for batch melting in which an external melt source contributes to the interstitial melt composition. For this reason, we refer to Eq. 33a as the fluxed batch melting model. Interestingly, we can obtain a similar expression for the interstitial melt in lithology A. Rearranging Eq. 28a, we have:

$$C_f^A = \frac{C_A^0 - \mathbb{R}F_A\bar{C}_f^A}{k_A^0 + \left(1 - k_A^p\right)F_A - \mathbb{R}F_A} \tag{33c}$$

Equation 33c can be interpreted as a 'defluxed' batch melting model, as part of the melt prouced has been removed from lithology A.

4. Two continuous melting models

The 'continuous melting model', also referred to in the literature as the 'dynamic melting model', has been used widely in geochemical studies of mantle-derived rocks. In the continuous melting model, a constant and uniform fraction of melt is retained in the residuum after an initial stage of batch melting (e.g. Langmuir et al., 1977; McKenzie, 1985; Albarède, 1995; Zou, 1998; Shaw, 2000). Any additional melt produced by melting is instantaneously removed from residual solid through an unspecified mechanism. Flows of the melt and solid are not considered explicitly in this class of melting models. Here we use steady-state mass conservation equations to derive the continuous melting model. Following the spirit of continuous melting, we assume that melting in the lower part of the melting column is characterized by batch melting (i.e. $\mathbb{R} = 0$). In the upper part of the melting column, a fraction of melt produced in lithology A at a given location is sucked into lithology B at a constant rate (Fig. 1b). We consider two choices of the relative melt suction rate. We show that the continuous or dynamic melting model discussed in the literature is equivalent to the steady-state melting model for a specific choice of melt suction rate (section 4.1) and that porosity in the melting column is not constant and uniform (section 5). We seek an alternative model for continuous melting in which porosity in lithology A in the upper part of the melting column is constant and uniform (section 4.2). Sections E2 and E3 in Appendix E summarize the main results.

4.1. Case of constant melt-to-solid mass flux ratio: The continuous melting model

For readability and convenience of derivation, the steady-state conservation equations for lithology A are relisted below:

$$\left[\rho_f \phi_f^A V_f^A + \rho_s\left(1 - \phi_f^A\right)V_s^A k_A\right]\frac{dC_f^A}{dz} = \left(k_A^p - 1\right)C_f^A \Gamma_A \tag{34a}$$

$$\frac{d\rho_f \phi_f^A V_f^A \psi_A}{dz} = \psi_A \Gamma_A - \dot{S} \tag{34b}$$

$$\frac{d\rho_s\left(1 - \phi_f^A\right)V_s^A\psi_A}{dz} = -\psi_A\Gamma_A \tag{34c}$$

$$V_s^A\frac{dF_A}{dz} = \frac{(1 - F_A)\Gamma_A}{\rho_s\left(1 - \phi_f^A\right)} \tag{34d}$$

$$V_s^A\frac{dk_A}{dz} = \frac{(k_A - k_A^p)\Gamma_A}{\rho_s\left(1 - \phi_f^A\right)} \tag{34e}$$

The boundary conditions at the solidus are:

$$C_f^A(0) = \frac{C_A^0}{k_A^0}, \ F_A = \phi_f^A = 0, \ V_s^A = V_s^0, \ k_A = k_A^0 \tag{34f}$$

For constant and uniform melting rate, solutions for the solid mass flux, degree of melting and bulk partition coefficient are the same as before, *i.e.*

$$\rho_s\left(1 - \phi_f^A\right)V_s^A = \rho_s V_s^0(1 - F_A), \ F_A = \frac{\Gamma_A z}{\rho_s V_s^0}, \ k_A = \frac{k_A^0 - k_A^p F_A}{1 - F_A} \tag{35}$$

Equation 34a can be integrated given the melt flux in lithology A. Following the spirit of continuous melting, we assume that the melt suction rate in the lower part of the melting column ($F_A < F_A^d$) is zero. Instead of solving the melt flux from Eq. 34b for a given melt suction rate, here we consider a special case in which the ratio between the melt mass flux and the solid mass flux in lithology A (designated as ω) is constant and uniform in the upper part of the melting column (Liang, 2008). To ensure mass conservation, we require that the melt flux at the boundary that divides the upper and lower melting column is continuous. Hence the melt flux in lithology A in the melting column takes on the form:

$$\rho_f \phi_f^A V_f^A = \begin{cases} \rho_s V_s^0 F_A & \text{for } F_A < F_A^d \\ \omega[\rho_s V_s^0(1 - F_A)] & \text{for } F_A \geq F_A^d \end{cases} \tag{36a}$$

$$F_A^d = \frac{\omega}{1 + \omega} \tag{36b}$$

where F_A^d is the degree of melting at which the melt fluxes from the two expressions in Eq. 36a are equal. We will discuss the physical meaning of ω at the end of this derivation. Substituting Eq. 36a in Eq. 34a, we obtain a set of ordinary differential equations for the melt composition. In terms of degree of melting experienced by lithology A, we have:

$$[F_A + (1 - F_A)k_A]\frac{dC_f^A}{dF_A} = (k_A^p - 1)C_f^A, \quad \text{for } F_A < F_A^d \tag{37a}$$

$$(1 - F_A)(\omega + k_A)\frac{dC_f^A}{dF_A} = (k_A^p - 1)C_f^A, \quad \text{for } F_A \geq F_A^d \tag{37b}$$

Equations 37a–37b can be integrated sequentially, starting from the solidus of lithology A. We have:

$$C_f^A = \frac{C_A^0}{k_A^0 + (1 - k_A^p)F_A}, \quad \text{for } F_A < F_A^d \tag{38a}$$

$$C_f^A = \left[\frac{C_A^0}{k_A^0 + (1 - k_A^p)F_A^d}\right]\left[\frac{k_A^0 + \omega - (k_A^p + \omega)F_A}{k_A^0 + \omega - (k_A^p + \omega)F_A^d}\right]^{\frac{1 - k_A^p}{k_A^p + \omega}} \quad \text{for } F_A \geq F_A^d \tag{38b}$$

In mathematical forms, Eqs 38a–38b are identical to the continuous or dynamic melting model discussed in the literature (*e.g.* McKenzie, 1985; Albarède, 1995; Zou, 1998; Shaw, 2000). In the previous treatment of continuous melting, ω in Eq. 38b is replaced by a melting parameter α which is the ratio between mass density of the melt and mass density of the solid in a unit volume of the partially molten system (*e.g.* Shaw, 2000),

$$\alpha = \frac{\rho_f \phi_f^A}{\rho_s\left(1 - \phi_f^A\right)} \quad \text{and} \quad F_A^d = \frac{\alpha}{1 + \alpha} = \frac{\rho_f \phi_f^A}{\rho_f \phi_f^A + \rho_s\left(1 - \phi_f^A\right)} \tag{39}$$

In the model presented above, ω is the ratio between the mass flux of the melt and the mass flux of the solid in lithology A, *i.e.*

$$\omega = \frac{\rho_f \phi_f^A V_f^A}{\rho_s\left(1 - \phi_f^A\right)V_s^A} \quad \text{and} \quad F_A^d = \frac{\rho_f \phi_f^A V_f^A}{\rho_f \phi_f^A V_f^A + \rho_s\left(1 - \phi_f^A\right)V_s^A} \tag{40}$$

The melt velocity is generally larger than the solid velocity in the melting column (*i.e.* $V_f^A > V_s^A$). Hence one would overestimate the melt fraction in residual solid in an upwelling melting column using the continuous melting model with melting parameters defined by Eq. 39.

In the steady-state melting model described in section 3.1, the dimensionless melt suction rate takes on a constant value in the entire melting column ($0 \leq \mathbb{R} \leq 1$). The melt flux increases linearly upwards, independent of the solid mass flux. In the continuous or dynamic melting model, the dimensionless melt suction rate takes on two constant values in the melting column:

$$\mathbb{R} = \frac{\dot{S}}{\psi_A \Gamma_A} = \begin{cases} 0 & \text{for } F_A < F_A^d \\ 1 + \omega & \text{for } F_A \geq F_A^d \end{cases} \tag{41}$$

Equation 41 is obtained from the mass conservation equation for interstitial melt (Eq. 34b) using the prescribed melt flux of Eq. 36a. Hence in the context of decompression melting, the continuous melting model is characterized by batch melting in the lower part of the melting column ($F_A < F_A^d$) and steady-state melting with a constant melt suction rate of $1 + \omega$ in the upper part of the melting column ($F_A \geq F_A^d$) where the mass flux ratio between the melt and solid takes on a constant value of ω. The greater than unit dimensionless melt suction rate ($1 + \omega$) in the upper part of the melting column arises from the influx of batch melt at $F_A = F_A^d$.

Finally, given the interstitial melt composition (Eq. 38a–38b) and melt suction rate (Eq. 41) for lithology A, we can use the conserved form of mass conservation equation to calculate concentrations of the trace element in the melt and residual solid in lithology B. Integrating Eqs 25 and 29, we have:

$$C_f^B = \frac{C_B^0}{k_B^0 + \left(1 - k_B^p\right)F_B}, \quad \text{for } F_A < F_A^d \tag{42a}$$

$$C_f^B = \frac{\psi_B C_B^0 + \psi_A(1 + \omega)\left(F_A - F_A^d\right)\bar{C}_f^A}{\psi_B\left[k_B^0 + \left(1 - k_B^p\right)F_B\right] + \psi_A(1 + \omega)\left(F_A - F_A^d\right)}, \quad \text{for } F_A \geq F_A^d \tag{42b}$$

where the average concentration of melts transferred from lithology A to lithology B is defined as:

$$\bar{C}_f^A = \frac{1}{F_A - F_A^d} \int_{F_A^d}^{F_A} C_f^A dF_A = \frac{C_A^0 - \left[k_A^0 + \omega - \left(k_A^p + \omega\right)F_A\right]C_f^A}{(1 + \omega)\left(F_A - F_A^d\right)}, \quad \text{for } F_A \geq F_A^d \tag{42c}$$

Note that in Eq. 42c batch melts are excluded from the integration because they are not extracted to lithology B in the lower part of the melting column.

4.2. Case of constant melt mass flux: The constant porosity model

The assumption of constant and uniform melt-to-solid mass flux ratio in the upper part of the melting column in lithology A is somewhat arbitrary. It is used for the convenience of model derivation and comparison with the continuous melting model. A different set of melting models can be obtained by assuming, for example, a constant melt flux in the upper part of the melting column in lithology A, *i.e.*

$$\rho_f \phi_f^A V_f^A = \begin{cases} \rho_s V_s^0 F_A & \text{for } F_A < F_A^d \\ \rho_s V_s^0 F_A^d & \text{for } F_A \geq F_A^d \end{cases} \tag{43}$$

This condition can be realized when the dimensionless melt suction rate takes on the following values in melting column:

$$\mathbb{R} = \frac{\dot{S}}{\psi_A \Gamma_A} = \begin{cases} 0 & \text{for } F_A < F_A^d \\ 1 & \text{for } F_A \geq F_A^d \end{cases} \tag{44}$$

The concentration of the trace element in the melt in lithology A takes on the expressions

$$C_f^A = \frac{C_A^0}{k_A^0 + (1 - k_A^p)F_A}, \quad \text{for } F_A < F_A^d \tag{45a}$$

$$C_f^A = \left[\frac{C_A^0}{k_A^0 + (1 - k_A^p)F_A^d} \right] \left[\frac{k_A^0 + F_A^d - k_A^p F_A}{k_A^0 + (1 - k_A^p)F_A^d} \right]^{\frac{1 - k_A^p}{k_A^p}}, \quad \text{for } F_A \geq F_A^d \tag{45b}$$

The concentration of the trace element in the melt in lithology B is given by:

$$C_f^B = \frac{C_B^0}{k_B^0 + (1 - k_B^p)F_B}, \quad \text{for } F_A < F_A^d \tag{45c}$$

$$C_f^B = \frac{\psi_B C_B^0 + \psi_A (F_A - F_A^d)\bar{C}_f^A}{\psi_B [k_B^0 + (1 - k_B^p)F_B] + \psi_A (F_A - F_A^d)}, \quad \text{for } F_A \geq F_A^d \tag{45d}$$

where the average concentration of melts transferred from lithologies A to B is defined as:

$$\bar{C}_f^A = \frac{1}{F_A - F_A^d} \int_{F_A^d}^{F_A} C_f^A dF_A = \frac{C_A^0 - (k_A^0 + F_A^d - k_A^p F_A)C_f^A}{F_A - F_A^d}, \quad \text{for } F_A \geq F_A^d \tag{45e}$$

Equation 45b is slightly different from Eq. 38b for the continuous melting model. We will compare these two models in sections 5 and 6.

5. A general two-lithology melting model

The melt suction rate in the upper part of the melting column takes on the special value of 1 or $1 + \omega$ in the two cases presented in section 4. In a more general case, the constant melt suction rate in the upper part of the melting column can take on a range of values, from <1 to >1 and the degree of melting or depth that divides the two melting regimes (F_A^d) does not have to be related to the melting parameter ω. In this section, this more general case of steady-state melting is considered. We start with mass conservation for the interstitial melt in lithology A which takes on the form:

$$\frac{d\rho_f \phi_f^A V_f^A \psi_A}{dz} = \psi_A \Gamma_A, \quad \text{for } F_A < F_A^d \tag{46a}$$

$$\frac{d\rho_f \phi_f^A V_f^A \psi_A}{dz} = \psi_A \Gamma_A - \dot{S}, \quad \text{for } F_A \geq F_A^d \tag{46b}$$

where the melt suction rate in Eq. 46b is a constant. Integrating Eqs 46a–46b from the solidus, we have the melt flux in lithology A:

$$\rho_f \phi_f^A V_f^A = \begin{cases} \rho_s V_s^0 F_A & \text{for } F_A < F_A^d \\ \rho_s V_s^0 (1 - \mathbb{R}) F_A + \rho_s V_s^0 F_A^d \mathbb{R} & \text{for } F_A \geq F_A^d \end{cases} \tag{47}$$

The second term on the right-hand-side of Eq. 47 is the influx of excess melt from the lower part of the melting column $(F_A < F_A^d)$ where the melt suction rate is zero. The dimensionless melt suction rate (Eq. 17c) is a constant that can take on values of $0 \leq \mathbb{R} \leq \mathbb{R}_{\max}$ in the upper part of the melting column. The upper bound for the dimensionless melt suction rate $(\mathbb{R}_{\max})$ is constrained by the condition when the melt flux in lithology A becomes zero. From Eq. 47, we have the upper bound:

$$\mathbb{R}_{\max} = \frac{F_A^{\max}}{F_A^{\max} - F_A^d} \tag{48}$$

where $F_A^{\max}$ is the maxium extent of melting experienced by lithology A at top of the melting column. For $F_A^{\max} = 20\%$ and $F_A^d = 5\%$, we have $\mathbb{R}_{\max} = 1.33$.

Expressions for the mass flux of solid and degree of melting are the same as the two cases discussed in section 4 (Eq. 35). Substituting the melt and solid mass fluxes into the mass conservation equation for trace elements (Eq. 34a) and upon integration, we obtain the following expressions for the concentration of a trace element in the interstitial melt in lithology A:

$$C_f^A = \frac{C_A^0}{k_A^0 + (1 - k_A^p)F_A}, \quad \text{for } F_A < F_A^d \tag{49a}$$

$$C_f^A = \left[\frac{C_A^0}{k_A^0 + (1 - k_A^p)F_A^d}\right]\left[\frac{k_A^0 + F_A^d\mathbb{R} + (1 - k_A^p - \mathbb{R})F_A}{k_A^0 + (1 - k_A^p)F_A^d}\right]^{\frac{k_A^p - 1}{1 - k_A^p - \mathbb{R}}},$$

$$\text{for } F_A \geq F_A^d. \tag{49b}$$

Solutions for the concentration of the trace element in the interstitial melt in lithology B are

$$C_f^B = \frac{C_B^0}{k_B^0 + (1 - k_B^p)F_B}, \quad \text{for } F_A < F_A^d \tag{49c}$$

$$C_f^B = \frac{\psi_B C_B^0 + \psi_A \mathbb{R}(F_A - F_A^d)\bar{C}_f^A}{\psi_B[k_B^0 + (1 - k_B^p)F_B] + \psi_A \mathbb{R}(F_A - F_A^d)}, \quad \text{for } F_A \geq F_A^d \tag{49d}$$

where the average concentration of melts transferred from lithologies A to B is defined as:

$$\bar{C}_f^A = \frac{1}{F_A - F_A^d}\int_{F_A^d}^{F_A} C_f^A dF_A = \frac{C_A^0 - [k_A^0 + \mathbb{R}F_A^d + (1 - k_A^p - \mathbb{R})F_A]C_f^A}{\mathbb{R}(F_A - F_A^d)},$$

$$\text{for } F_A \geq F_A^d \tag{49e}$$

Finally, the melt and solid mass fluxes in lithology B are

$$\rho_f \phi_f^B V_f^B \psi_B = \begin{cases} \rho_s V_s^0 F_B \psi_B & \text{for } F_A < F_A^d \\ \rho_s V_s^0[F_B\psi_B + \mathbb{R}(F_A - F_A^d)\psi_A] & \text{for } F_A \geq F_A^d \end{cases} \tag{49f}$$

$$\rho_s\left(1 - \phi_f^B\right)V_s^B = \rho_s V_s^0(1 - F_B) \tag{49g}$$

Equations 49a–49b unify the simple models for trace element fractionation during decompression melting in an upwelling melting column in which the melt suction rate takes on one or two constant values. In addition to partition coefficients, there are three physical parameters in this melting model: degree of melting experienced by the residuum (F_A), the depth or degree of melting above which interstitial melt flows from lithology A to lithology B (F_A^d) and the dimensionless melt suction rate ($\mathbb{R}$). Different choices of these melting parameters lead to different melting models presented in the preceding sections. Specifically, Eqs 49a–49b are reduced to:

(a) the batch melting model when $\mathbb{R} = 0$;
(b) the perfect fractional melting model when $\mathbb{R} = 1$ and $F_A^d = 0$;
(c) the two-porosity melting model with constant and uniform melt suction rate (Eq. 21a) when $0 < \mathbb{R} < 1$ and $F_A^d = 0$;
(d) the continuous melting model with constant melt flux or porosity when $\mathbb{R} = 1$ in the upper part of the melting column; and
(e) the continuous or dynamic melting model discussed in the literature when $\mathbb{R} = 1 + \omega$ and $F_A^d = \omega/(1 + \omega)$.

Figure 2 summarizes these results. When the volume fraction of lithology B is reduced to zero ($\psi_B = 0$), the melt composition of lithology B (Eq. 49d) equals to the average melt composition of lithology A, *i.e.* $C_f^B = \bar{C}_f^A$. The average melt composition from lithology A has been used widely in modelling trace elements and isotope ratios in pooled melt derived from partial melting of a single lithology source.

Another case of special interest is when lithology B is dunite in the upper part of the melting column. The melting rate of dunite is neglegible. Melt composition in dunite channels varies along the melting column according to the fraction of melt transferred from lithology A to lithology B. Let F_B^{dunite} be the extent of melting at which pyroxene-free dunite is formed from lithology B. Melt composition in the dunite channel is given by the expression:

$$C_f^B = \frac{\psi_B C_B^0 + \psi_A \mathbb{R}(F_A - F_A^d)\bar{C}_f^A}{\psi_B[k_B^0 + (1 - k_B^p)F_B^{dunite}] + \psi_A \mathbb{R}(F_A - F_A^d)} \tag{50}$$

where the first terms in the numerator and the denominitor account for dunite formation. Hence, melt composition in the dunite channel is a mixture of melts transferred from lithology A to lithology B and the melt produced by converting lithology B to dunite (*via* partial melting and reactive dissolution). For transport of incompatible trace elements in dunite channels, we can further simplify Eq. 50 by setting partition coefficients in lithology B to zero, *i.e.*

$$C_f^B = \frac{\psi_B C_B^0 + \psi_A \mathbb{R}(F_A - F_A^d)\bar{C}_f^A}{\psi_B F_B^{dunite} + \psi_A \mathbb{R}(F_A - F_A^d)} \tag{51}$$

Equation 51 indicates that formation of dunite channels in the melting column results in a dilution of incompatible trace element concentrations in the channel melt, *i.e.* $C_f^B < \bar{C}_f^A$.

This effect, which has not been considered in most two-porosity melting models in the literature, may provide a simple mechanism for producing highly depleted melts, such as those observed in some olivine-hosted melt inclusions (*e.g.* Sobelev and Shimizu, 1994; Shimizu, 1998).

In summary, Eqs 49a–49g are a set of more general solutions for trace element fractionation during decompression melting in an upwelling two-lithology melting column. This new model features batch melting in the lower part of the melting column for both lithology A and lithology B, steady-state melting with a constant melt suction rate in the upper part of the melting column for lithology A and fluxed batch melting in the upper part of the melting column for lithology B. It recovers the batch melting, fractional melting, continuous melting and two-porosity melting models under specific limits. To minimize confusion with the previous melting models, we call this more general steady-state melting model (Eqs 49a–49g, 8–9) the two-lithology melting model. In the next section, we show that this two-lithology melting model also has two porosities.

6. Porosity in the two-lithology melting column

Porosity or melt fraction in an upwelling melting column depends primarily on the melt flux. Hence, porosities in the two lithologies vary along the melting column and are different for different choices of the melt suction rate. To further differentiate the various melting models considered in this study, we use Darcy's law to calculate porosities in the upwelling melting column for a given set of melt suction rates. Appendix D provides a derivation of porosities for the two-lithology melting model presented in section 5. Approximate expressions for the porosity in lithology A are:

$$\phi_f^A \approx \phi_{ref}^A \left(\frac{F_A}{F_A^{max}} \right)^{\frac{1}{3}}, \quad \text{for } F_A < F_A^d \tag{52a}$$

$$\phi_f^A \approx \phi_{ref}^A \left[\frac{(1 - \mathbb{R})F_A + \mathbb{R}F_A^d}{F_A^{max}} \right]^{\frac{1}{3}}, \quad \text{for } F_A \geq F_A^d \tag{52b}$$

Similarly, for lithology B, we have:

$$\phi_f^B \approx \phi_{ref}^B \left(\frac{F_B}{F_B^{max}} \right)^{\frac{1}{3}}, \quad \text{for } F_A < F_A^d \tag{53a}$$

$$\phi_f^B \approx \phi_{ref}^B \left[\frac{\psi_B F_B + \psi_A \mathbb{R}\left(F_A - F_A^d\right)}{\psi_B F_B^{max}} \right]^{\frac{1}{3}}, \quad \text{for } F_A \geq F_A^d \tag{53b}$$

Here ϕ_{ref}^A and ϕ_{ref}^B are reference porosities for lithology A and lithology B defined in Eqs D5a–D5b in Appendix D. The two reference porosities are related to each other via the relationship:

$$\phi_{ref}^B = \left(\frac{F_B^{max} d_A^2}{F_A^{max} d_B^2}\right)^{\frac{1}{3}} \phi_{ref}^A = \left(\frac{\Gamma_B d_A^2}{\Gamma_A d_B^2}\right)^{\frac{1}{3}} \phi_{ref}^A \tag{54}$$

where d_A and d_B are average grain sizes of lithology A and lithology B, respectively. If $F_B^{max} = 2F_A^{max}$, $d_A = d_B$ and $\phi_{ref}^A = 2\%$, we have $\phi_{ref}^B = 2.5\%$. Hence porosities in the two lithologies are generally different for different choices of melting rate and grain size.

Given the reference porosities, we can calculate variations of porosity in the two lithologies in the melting column. Figure 3a displays porosity variations in lithology A (solid lines) and lithology B (dashed lines) for five choices of the melt suction rate. In response to melting, porosities increase upwards for a given melt suction rate in the two lithologies. With increasing melt suction rate, porosity in lithology A decreases, whereas porosity in lithology B increases at a given depth in the melting column. For perfect fractional

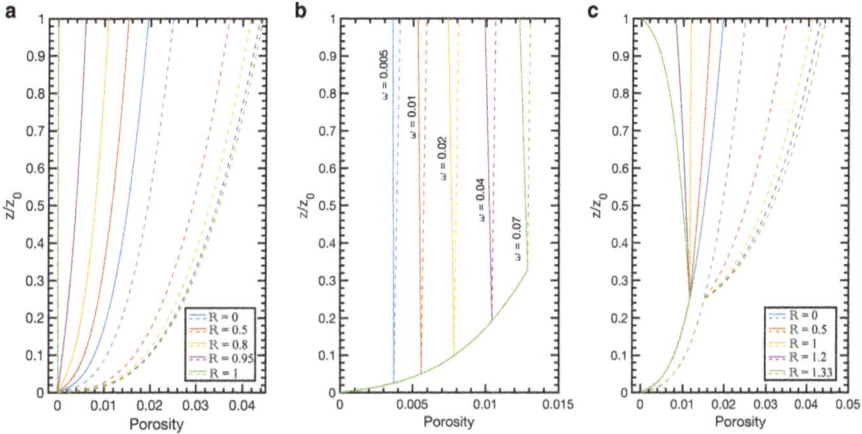

Fig. 3. (*a*) Spatial variations in porosity in lithology A (solid curves) and lithology B (dashed curves) in the melting column for decompression melting with a constant and uniform melt suction rate. (*b*) Variations in porosity in lithology A during continuous melting with constant melt-to-solid mass flux ratio (solid lines) or constant melt flux (dashed lines) in the upper part of the melting column. (*c*) Variations in porosity in lithology A (solid curves) and lithology B (dashed curves) in the melting column according to the two-lithology melting model in which batch melting takes place in the lower quarter of the melting column in the two lithologies. Five choices of melting parameters are provided in respective panels. Porosities in the three panels are solved numerically from Eqs D6a and D6b using the MATLAB function fzero. Porosity obtained from the approximate solutions (Eqs 52–55, not shown here) are slightly smaller than the 'exact' values (by <5%, relative). Note, there is a very small upwards increase in porosity for the case of constant melt flux (dashed lines in Fig. 2b). This is due to the upwards reduction in the solid volume flux, which is second order (the third term in the square bracket of Eq. D4a) and is ignored in the approximate solution (Eq. 52b).

melting, porosity in lithology A is zero in the melting column while porosity in lithology B reaches maximum values along the melting column (cf. green solid and dashed lines in Fig. 3a).

Figure 3b compares porosity distributions in lithology A for the case of constant melt-to-solid mass flux ratio (solid lines) and the case of constant melt flux (dashed lines) in the upper part of the melting column for five choices of the melting parameter ω ($\mathbb{R} = 1 + \omega$) or F_A^d. In terms of trace element fractionation, the case of constant melt-to-solid mass flux ratio is equivalent to the continuous or dynamic melting model used in the literature. As more melt is transferred from lithology A to lithology B than that produced in lithology A in this part of the melting column ($\mathbb{R} > 1$), the porosity in lithology A decreases upwards in the melting column in the continuous melting model (solid lines in Fig. 3b). This is in contrast with the assumption of constant porosity used in the derivation of the continuous melting model in the literature (*e.g.* McKenzie, 1985; Albarède, 1995; Zou, 1998; Shaw, 2000).

The porosity in lithology A is effectively constant and uniform in the upper region of the melting column when the amount of melt produced in lithology A by melting is balanced by the amount of melt transferred from lithology A to lithology B. This is the case when $\mathbb{R} = 1$. From Eq. 52b, we have an approximate expression for the porosity in the upper part of lithology A:

$$\phi_f^A \approx \phi_{ref}^A \left(\frac{F_A^d}{F_A^{max}} \right)^{\frac{1}{3}}, \quad \text{for } F_A \geq F_A^d \tag{55}$$

If $F_A^d = 5\%$, $F_A^{max} = 20\%$ and $\phi_{ref}^A = 2\%$, we have $\phi_f^A = 1.26\%$. The case of constant melt flux is shown as vertical dashed lines in Fig. 3b. In general, porosities derived from the constant melt-to-solid mass flux ratio model are slightly smaller than porosities derived from the constant melt flux model for the same amount of batch melting in the lower part of the melting column (cf. solid *vs.* dashed lines of the same colour in Fig. 3b).

Figure 3c displays porosity distributions in lithology A (solid lines) and lithology B (dashed lines) for the two-lithology melting model. In this example, batch melting takes place in the lower quarter of the melting column ($\mathbb{R} = 0$) and the melt suction rate in the upper part of the melting column takes on a wide range of values. When $0 \leq \mathbb{R} < 1$ in the upper three quarters of the melting column, only part of the melt generated in lithology A is transferred to lithology B. The remaining melt percolates upwards in lithology A, resulting in an upwards increase in porosity in lithology A. When $\mathbb{R} = 1$, the porosity and melt flux in lithology A remain constant and uniform in the upper three quarters of the melting column. When $1 < \mathbb{R} \leq \mathbb{R}_{max}$, more melt is transferred from lithology A to lithology B than that produced in lithology A in the upper three quarters of the melting column. Consequently, the porosity in lithology A decreases upwards. Regardless of the melt suction rate, the porosity in lithology B increases upwards in the melting column. The larger the melt suction rate, the larger the porosity is in lithology B.

In summary, porosities in lithologies A and B in the two-lithology melting model take on a range of values, depending on the melting parameters. They recover the special cases displayed in Figs 3a-3b under stated limits. Furthermore, porosities in lithology A and lithology B are continuous along the melting column. This is in contrast with previous two-porosity melting models in which porosity of the channel is discontinuous in the melting column (*e.g.* Jull *et al.*, 2002). The discontinuous channel porosity results from the assumption of instantaneous channel formation at a prescribed depth in the melting column. In our models, high-porosity channels are formed gradually in lithology B. The discontinuity in porosity gradient in the models shown in Figs 3b and 3c is due to the assumption of constant melt suction rate in the upper part of the melting column. This discontinuity can be eliminated by allowing the melt suction rate to vary continuously in the melting column, starting from zero in the lower part of the melting column. Finally, we note that given the melting parameters, Eqs 52–53 can be used to infer porosities in the melting column. Liang and Peng (2010) provided an example in which they used the extent of melting and melt suction rate derived from REE patterns in clinopyroxene in residual abyssal peridotites to calculate the porosity and permeability in the meting region beneath mid-ocean ridge spreading centers.

7. Variations in melt composition

In this section we use the two-lithology melting model (Eqs. 49a-49e) to illustrate how REE in interstitial melts in lithology A and lithology B, average melt in lithology A and mixed-column melt bewteen the two lithologies in the melting column vary as a function of the degree of melting experienced by lithology A (F_A) or lithology B (F_B). For a given melting rate, the degree of melting scales with the column height (Eqs. 16 or 30c). Melt compositions at different locations in a tall melting column (*e.g.* column 10 in the inset to Fig. 4c) can also be taken as melt compositions at top of shorter melting columns (columns 1 to 9). We consider two cases: lithology B is dunite in the upper part of the melting column (section 7.1); and (2) lithology B is a pyroxenite (section 7.2 and section 7.3).

7.1. Dilution effect due to dunite channel formation in the melting column

To highlight the dilution effect, we consider decompression melting of a homogeneous mantle (lherzolite = 15% Cpx + 28% Opx + 57% olivine) in which lithology A and lithology B have the same composition and lithology in the mantle source. A fraction of the residual mantle (identified as lithology B) is transformed into pyroxene-free dunite through a combination of fluxed melting and reactive dissolution. For the purpose of demonstration, we assume that the combined melting and dissolution rate of lithology B is four times the melting rate of lithology A and that the pyroxenite-free dunite forms in the upper part of lithology B when $F_B = F_{dunite} = 50\%$. Figure 4a shows that REE concentrations in the interstitial melt in lithology A are depleted progressively by a combination of 4% batch melting followed by 16% continuous melting

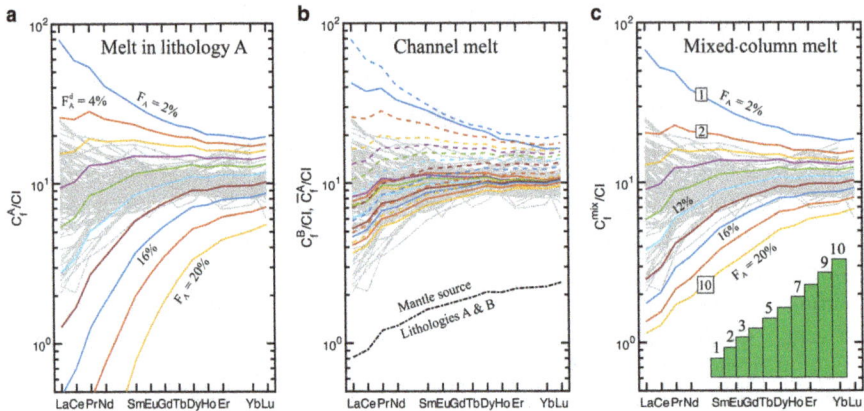

Fig. 4. Chondrite normalized REE patterns in interstitial melts in lithology A (*a*) and lithology B (*b*, solid lines), average melts transferred from lithology A to lithology B (dashed lines in *b*), and mixed-column melts at top of melting columns (*c*) produced by decompression melting of a homogeneous mantle with channelized melt migration in lithology B. Curves of the same colour correspond to a set of melt compositions produced by the same extent of melting (at 2% increment in F_A). The mixed-column melts 1, 2, …, 10 correspond to the numbered melting columns in the inset in panel (*c*). The mantle sources for lithology A and lithology B have the same lithology (lherzolite = 15% Cpx, 28% Opx and 57% olivine) and composition (DMM of Workman and Hart, 2005, dash-dotted line in panel *b*). For simplicity, we ignore spinel in the REE modelling. REE concentrations are calculated using Eqs 49a–49e, Eq. 50, and Eq. 8 with the following melting parameters: $\mathbb{R} = 1$ and $F_A^d = 4\%$. The maximum extent of melting for lithology A is 20% and the maximum extent of melting for lithology B is 50% at which pyroxene-free dunite forms. Mineral-melt REE partition coefficients are from Sun and Liang (2012, 2013) and Yao *et al.* (2012). For demonstration, we assume non-modal melting for the lherzolite and modal melting for lithology B. Melting reaction for the lherzolite is 0.79 Cpx + 0.41 Opx = 0.2 olivine + melt (modified from Walter, 2014). REE patterns in the dunite are not sensitive to the details of how pyroxenes are exhausted in the residual solid. Hence it is sufficient to use modal melting for lithology B in this example. The background gray lines are REE in olivine-hosted melt inclusions from the FAMOUS segment of the Mid-Atlantic Ridge (Shimizu, 1998; Laubier *et al.*, 2012).

($\mathbb{R} = 1$) of the starting lherzolite (solid curves with 2% increment in F_A). The LREE depleted patterns are typical of fractional or near fractional melting of lherzolite. During near fractional melting of lithology A (lherzolite), a large fraction of melt produced in lithology A is transferred to lithology B (high-porosity channels), resulting in strong depletion in incompatible trace elements in the residuum.

The interstitial melt in dunite channels is a mixture of melts transferred from lithology A to lithology B and melts produced by dissolution of pyroxene and precipitation of olivine in lithology B. Figure 4b shows that the interstitial melt in lithology B (solid lines, Eqs 49d and 50) is more depleted than the average melt from lithology A (dashed lines, Eq. 49e) at a given location in the melting column (lines of the same colour). The extent of dilution depends on the fraction of dunite in the melting column (ψ_B) and the 'degree of melting' at which dunite channel initiates (F_{dunite}): the larger the ψ_B and F_{dunite}, the stronger the dilution effect (cf. Eq. 51). However, the extent of dilution is not sensitive to the details of how pyroxenes are exhausted in the residual solid, *i.e.* whether it is by melting or dissolution. The average melt from lithology A has been used widely as the channel melt in the literature. Here we show that incompatible trace elements in the channel melt are diluted by the dunite formation reaction: pyroxene + melt$_1$ = olivine + melt$_2$, where melt$_1$ is the melt transferred from lithology A to lithology B; and melt$_2$ is the interstitial melt in the dunite channel. Unlike melt$_2$, melt$_1$ cannot be sampled in the melting column or at its top.

Finally, for the depleted starting mantle composition, LREE abundances in the mixed-column melt and interstitial melts are depleted progressively by melting (Fig. 4a–4c). Figure 4c shows that the LREE-depleted patterns in olivine-hosted melt inclusions from the FAMOUS segment of the Mid-Atlantic Ridge (Shimizu, 1998; Laubier *et al.*, 2012) can be explained by the high degree of mixed-column melts ($8\% < F_A < 16\%$) or melts from incomple mixing of mixed-column melts from short and tall melting columns (not shown). These LREE depleted patterns are broadly similar to those observed in D-MORB and N-MORB. However, the flat and LREE-enriched patterns may require an enriched mantle source which we will explore below.

7.2. Melting a two-lithology mantle: Two incompatible trace elements

The presence of a second lithology changes the bulk partition coefficient and the extent of melting for lithology B, resulting in additional fractionation of the trace element of interest. To set up a stage for the more practical examples presented in section 7.3, we first compare two trace elements with constant bulk partition coefficients of (0.01, 0.03) and (0.1, 0.3) in the two lithologies (A, B). For the purpose of illustration, we assume that concentrations of the two incompatible trace elements in lithology B in the mantle source are 10 times those in lithology A and that the melting rate of lithology B is twice that of lithology A.

Figure 5a,c displays concentrations of the two trace elements in the interstitial melt (solid curves) and average melt (dashed lines) in lithology A for five choices of the melt suction rate. Three general observations can be made. First, incompatible trace element cocentrations in interstitial melts become more and more depleted with increasing melt suction rate and extent of melting. The depletion is due to removal of incompatible trace elements in residual solid by melting and in interstitial melt by transport to lithology B. Second, the difference in trace element concentration between the continuous melting model (case of $\mathbb{R} = 1.042$) and the constant melt flux model ($\mathbb{R} = 1$) is small for both the interstitial melt and average melt of lithology A. This is consistent with their small differences in porosity and melt flux as the melt suction rate between the two models is very similar. Finally, concentrations of the two incompatible trace elements in average melts at a given degree of melting are higher than those in interstitial melts produced by batch melting ($\mathbb{R} = 0$). This happens because progressively less melt is left behind in residual solid with increasing melt suction rate (Fig. 3a).

Figure 5b,d compares concentrations of the two incompatible elements in interstitial melt in lithology B (solid curves in the upper part of each panel), average melt from lithology A (solid curves in the lower part of each panel) and mixed-column melt

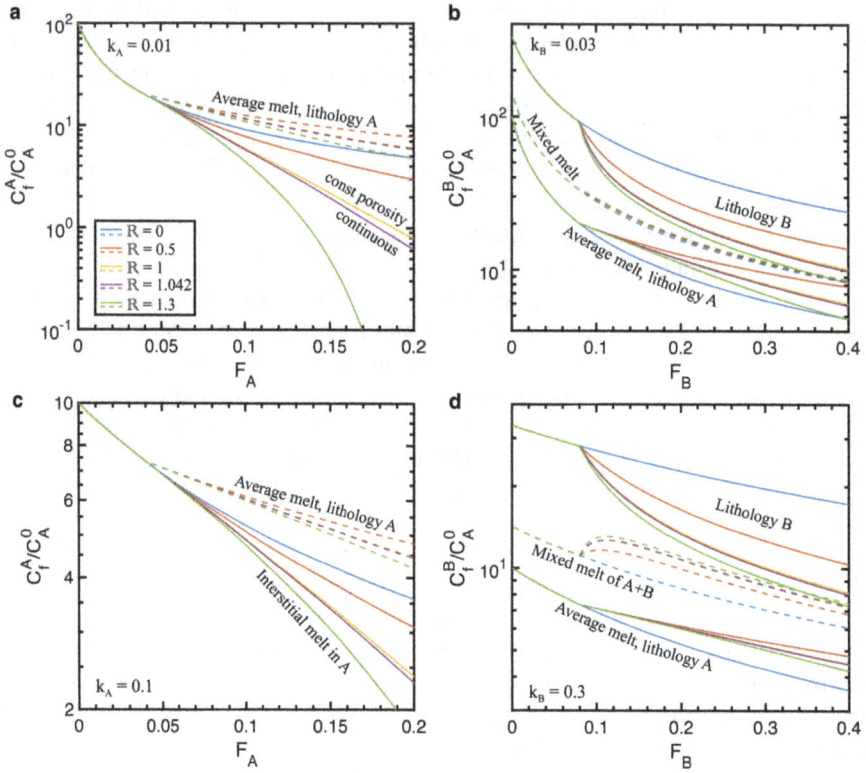

Fig. 5. Variations of two incompatible trace elements in interstitial melts (solid curves) and average melts (dashed curves) in lithology A (*a*, *c*), interstitial melts in lithology B (upper solid curves) and mixed-column melts (dashed curves) from lithology A and lithology B (*b*, *d*) as functions of the degree of melting experienced by the two lithologies. Trace element concentrations are calculated for five choices of the melt suction rate (colour-coded curves) using Eqs 49a–49e and Eq. 8 and bulk partition coefficients provided in each panel. For demonstration, we assume modal melting in the two lithologies and that batch melting takes place in the lower fifth of the melting column ($F_A^d = 4\%$). To facilitate comparison, concentrations of the two trace elements in melts are normalized by their respective concentrations in lithology A in the mantle source. Concentrations of the two trace elements in lithology B in the mantle source are ten times those in lithology A.

from the two lithologies (dashed curves). Five general observations can be made. First, incompatible trace element concentrations in interstitial melts in lithology B are less depleted with increasing extent of melting than those in lithology A for a given melt suction rate. This is due to a combination of fluxed batch melting in lithology B and its enriched mantle source composition. Upwards percoloation of lower-degree melts in the deeper part of the melting column to the overlying melting region increases incompatible trace element abundance in the melt, which alleviates the depletion of incompatible trace elements in the overlying melting column. This self-enrichment takes place when lower-degree melts percolate through a higher-degree melting region (*e.g.*

Richter, 1986; Liang, 2008; Liang and Peng, 2010). Second, the dilution effect discussed in section 7.1 and illustrated in Fig. 4b is obscured by the enriched mantle source composition in lithology B (by a factor of 10 relative to lithology A). If we lowered concentrations of the two trace elements in lithology B in mantle source by a factor of 10, the interstitial melts in lithology B would be plotted below the average melts of lithology A in Fig. 5b,d. Third, at a given location or extent of melting, trace element concentrations in the interstitial melt in lithology B decrease with the increase of melt suction rate. This is part of the dilution effect produced by the influx of more depleted melts from lithology A. Fourth, the mixed-column melt from lithology A and lithology B are plotted between the average melt from lithology A and the interstitial melt from lithology B. For highly incompatible trace elements (Fig. 5b), the mixed-column melt composition is practically independent of the melt suction rate or melting model, as residual solid is depleted extensively by melting (Eqs 9a and 10a). Hence, for highly incompatible trace elements ($k < 0.01$), mixing of melts from lithologies A and B are equivalent to mixing of the two sources, although their proportions are different from those in the mantle source. This can be seen from Eq. 10a in section 2.2. Finally, for moderately incompatible (and compatible) trace elements, mixing of melts from the two lithologies is sensitive to the melt suction rate or melting mechanism. Figure 5d presents one such example.

7.3. Melting a peridotite-pyroxenite mantle: Rare earth elements

Here we consider two examples in which the mantle source of lithology A is the same lherzolite as that used in the example shown in Fig. 4 (15% Cpx + 28% Opx + 57% olivine). The mantle source of lithology B is either a Cpx-rich pyroxenite (60% Cpx + 10% Opx + 30% olivine, Fig. 6a,d) or an Opx-rich pyroxenite (10% Cpx + 60% Opx + 30% olivine, Fig. 6b,e). These secondary pyroxenites can be formed by reaction between a peridotite and melts derived from garnet pyroxenite and eclogite in the deeper part of the mantle column, a process that is not modelled here. The melting rate of the secondary pyroxenite is probably greater than that of the peridotite, although the former is still not well characterized. For the examples presented below, we set the maximum extent of melting to 20% for lithology A and 40% for lithology B. Batch melting takes place in the lower fifth of the melting column. The melt suction rate in the upper four fifths of lithology A is 1, which is the case of constant porosity or constant melt flux melting model (section 4.2). REE abundances in the interstitial melt in lithology A are the same as those shown in Fig. 4a.

Figure 6a–c explores the effect of Opx to Cpx proportion in the pyroxenite on REE patterns in interstitial melts in lithology B and the mixed-column melt at top of the melting column. Here we fix the volume fractions of lherzolite (90%) and pyroxenite (10%) in the mantle source and vary the proportion of Cpx to Opx in lithology B (60:10 *vs.* 10:60, 30% olivine). For a small to moderate extent of melting of lithology B (*e.g.* short melting columns with $F_B < 16\%$), REE patterns in interstitial melts and mixed-column melts are sensitive to the fractions of Cpx and Opx in the pyroxenite. The greater the Opx to Cpx ratio in the pyroxenite, the smaller the bulk REE partition coefficients and the higher the REE concentrations in the melt will be. At a greater

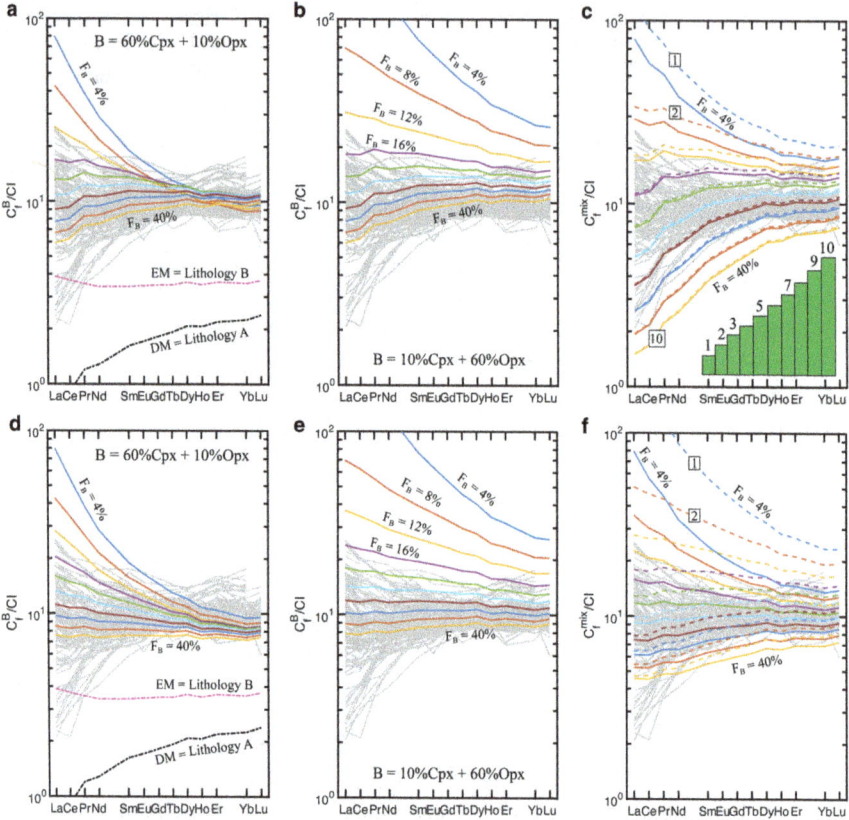

Fig. 6. Chondrite-normalized REE patterns in interstitial melts in lithology B (panels *a, b, d, e*) and mixed-column melts (panels *c, f*) produced by melting two peridotite-pyroxenite mantle sources. REE patterns in interstitial melts in lithology A are the same as those shown in Fig. 4a. Mantle source for lithology A is lherzolite (15% Cpx + 28% Opx + 57% olivine). Mantle source for lithology B is pyroxenite with either 60% Cpx + 10% Opx + 30% olivine (panels *a, d*, and solid lines in *c, f*) or 10% Cpx + 60% Opx + 30% Olivine (panels *b, e*, and dashed lines in *c, f*). Volume fractions of pyroxenite in the mantle source (ψ_B) are 10% for results presented in the first row (*a-c*) and 40% for results in the second row (*d-f*). The composition of lithology A is DMM of Workman and Hart (2005). The composition of lithology B is from the present author (dashed magenta lines in *a* and *d*). Melting parameters are $\mathbb{R} = 1$ and $F_A^d = 4\%$. The maximum extent of melting for lithology A is 20% and the maximum extent of melting for lithology B is 40%. Other parameters and legends are the same as those used to construct Fig. 4.

extent of melting ($F_B > 16\%$), almost all REE in lithology B are partitioned into the melt. Consequently, REE patterns in the interstitial melt and the mixed-column melt are practically independent of melting parameters ($\mathbb{R}$ and F_A^d) and the relative proportion of Opx and Cpx in the pyroxenite (cf. dashed and solid lines of the same colour in Fig. 6c). Figure 6c shows that the LREE depleted patterns of the mixed-column melt are very similar to those produced by decompression melting of the homogeneous lherzolite with channelized melt migration in the dunite (cf. Figs 4c and 6c). Hence it is difficult to resolve contributions of the enriched mantle component based on REE patterns in D-MORB and N-MORB when the volume fraction of the enriched mantle is small in the mantle source. However, we may be able to tell such contributions from Sr-Nd-Hf-Pb isotope ratios in the samples. Such samples would be depleted in LREE but variously enriched in radiogenic isotope ratios.

When the volume fraction of the enriched mantle is large in the mantle source, it is possible to produce flat to LREE-enriched patterns in high-degree melts. Figure 6d–f provides one such example in which the volume fraction of pyroxenite is 40% in the mantle source. Other parameters are the same as those for Fig. 6a–c. Figure 7a–c further expands this case by reversing melt flow direction in the upper part of the melting column, *i.e.* from the pyroxenite (lithology A) to the lherzolite (lithology B). The LREE-depleted pattern in small-degree melts in the lherzolite ($F_B \leq 4\%$ in Fig. 7b) is produced by batch melting in the lower fifth of the melting column where no melt flows from the pyroxenite to the lherzolite. The flat to enriched LREE patterns in the mixed-column melts (Fig. 7c) are produced by a combination of fluxed batch melting in the lherzolite and near fractional melting in the LREE-enriched pyroxenite where part of the enriched melts are sucked into the lherzolite.

The preceding examples (Figs 4, 6, 7) demonstrate that a range of REE patterns for the mixed-column melt can be produced by decompression melting of a two-lithology mantle. These REE patterns are broadly similar to those observed in MORB. Similarities in REE patterns between the mixed-column melt and MORB suggest that the eruptible melt sampled on the surface or seafloor is not well mixed across the melting region.

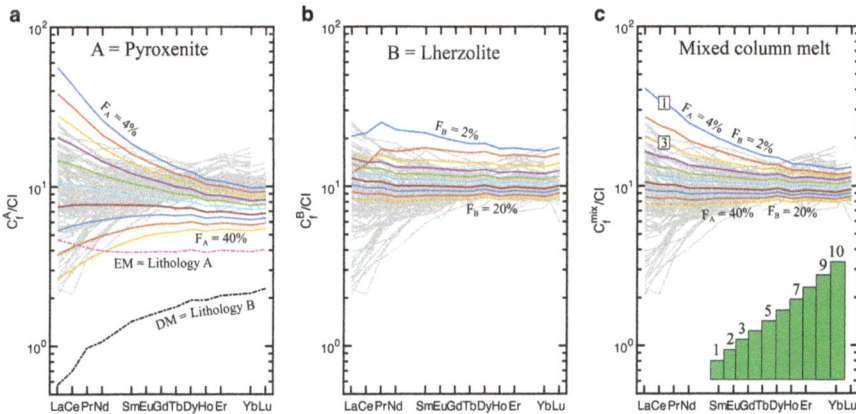

Fig. 7. Chondrite-normalized REE patterns in interstitial melts (***a***, ***b***) and mixed-column melts (***c***) produced by decompression melting of a two-lithology mantle in which lithology A is pyroxenite (60% Cpx + 10% Opx + 30% olivine, 40% by volume) and lithology B is lherzolite (15% Cpx + 28% Opx + 57% olivine, 60% by volume). Part of the melt produced in the pyroxenite is sucked into the lherzolite ($\mathbb{R} = 1$ and $F_A^d = 8\%$). The maximum extent of melting for lithology A is 40% and the maximum extent of melting for lithology B is 20%. Other parameters and legends are the same as those used to construct Fig. 6d–f.

Incomplete mixing of different column melts at the top of the melting region (Fig. 1a) is a complicated process and will be the subject of future investigation.

8. Summary and discussion

Decompression melting is a dynamic process in which melt and solid flow at different rates and different directions. Locations and flow rates of the melt and residual solid in the melting column can be tracked using conservation equations. In this chapter, we present a set of steady-state melting models for decompression melting of a two-lithology mantle in which the enriched mantle is in the form of long strings (Fig. 1a). We show that the widely used simple melting models can be obtained by solving a set of conservation equations under nearly identical setups (Fig. 1b). Melt and solid mass fluxes have played a central role in quantifying trace element concentrations. Different melting models have different melt fluxes. Batch melting, fractional melting, continuous melting and two-porosity melting models can be unified through the two-lithology melting model (Eqs 49a–g, 8–9, Fig. 2). The melt suction rate is a key parameter determining the style of melting in lithology A. It is shown here that melting in lithology B is characterized by fluxed batch melting (Eqs 49c–e) so long as melt flows from lithology A into lithology B. Melting in lithology A can be interpreted as a 'defluxed' batch melting (Eq. 33c) as long as melt flows out of lithology A. For modelling trace element fractionation during decompression melting of a heterogeneous mantle, we recommend the two-lithology melting model.

We demonstrate the usefulness of the conserved form of mass conservation equations in finding simple solutions for concentrations of the average melt and mixed-column melt at top of the melting column. The eruptible melt is formed by incomplete mixing of mixed-column melts from several melting columns across the melting region. However, the process of mixing is still not well understood and requires further study. Our formulations for the average melt (Eq. 1), eruptible melt (Eq. 2) and mixed-column melt compositions (Eqs 8–9) are general and can also be used for modelling major elements. For highly incompatible trace elements, we show that mixing of melts from the enriched and depleted lithologies at top of the melting column is equivalent to mixing of the two sources weighted by the fraction of melts produced in the two-lithology column (Eq. 10a). In terms of ratios of isotopes or highly incompatible trace elements, mixing of melts from the two lithologies at top of the melting column is equivalent to mixing of the two sources (Eq. 10b).

Through applications of mass conservation equations, we have gained a better understanding of the widely used continuous or dynamic melting model. Contrary to constant porosity assumption that underpins the original development of the continuous melting model, porosity decreases upwards in the upper part of the melting column during continuous melting in an upwelling column (Fig. 3b). This happens because the solid mass flux decreases upwards in the melting column. To maintain the original mathematical form for continuous melting, the melt-to-solid mass flux ratio must remain constant and uniform in the upper part of melting column. Constant and uniform porosity is established when the amount of melt produced is balanced by the amount of melt extracted in the melting column. This is the case of constant melt flux model which has a slightly different form than the continuous melting model. When the extent of melting experienced by lithology A is not large (<20%), the differences between the two cases are small in terms of concentration and porosity. Irrespective of the choice of melt flux in the continuous melting model, porosity in lithology B always increases upwards in the melting column (Eqs 53a and 53b).

The two-lithology melting model also has two porosities (Fig. 3c). We show that high-porosity channels form naturally during decompression melting in a two-lithology mantle when melt produced in one lithology flows into another lithology and the melting rates of the two lithologies are different. The formation of dunite channels has a dilution effect on incompatible trace element concentrations in the channel melt (Eqs 50–51), which may provide a simple mechanism for producing highly depleted melts, such as those found in some olivine-hosted melt inclusions. In the presence of an enriched mantle source, the dilution effect may be obscured. It is possible to produce melts with depleted incompatible trace elements patterns but isotopically enriched signals.

We show through simple examples that it is possible to produce partial and well-mixed melts with a range of REE patterns by decompression melting of a two-lithology mantle, from LREE depleted to LREE enriched, similar to those observed in MORB. We have reasonably good knowledge of the composition and melting parameters of the depleted mantle source (*i.e.* its lithology, melting reaction and mineral-melt trace element partition coefficients). However, such knowledge is incomplete for the enriched mantle and requires further studies.

Although the steady-state melting models presented in this study are derived under the simplification of constant melting rate, the solutions remain valid when the melting rate varies spatially in the melting column. The degree of melting is related to the variable melting rate through the integral in Eq. 16. The melt suction rate in the four models presented here takes on one or two constant values in the melting column. This allows us to obtain simple analytical solutions for the melting problems. In a more general case, the melt suction rate varies continuously along the melting column, starting from zero at the solidus. Melt may flow from lithology A to lithology B in one part of the melting column but reverse its direction in another part of the melting column. The latter may serve as a mechanism for mantle metasomatism. Part of lithology A may transform into lithology B or vice versa. Liang and Parmentier (2010) presented some numerical examples of these more general cases. Dygert *et al.* (2016) presented a field example of infiltration of dunite-hosted melts into the host harzburgite and lherzolite from the Trinity ophiolite.

Finally, the two-lithology models presented here are independent of time. They are developed under the assumption that the shape of lithology B is a long string, interconnected in the vertical direction (Fig. 1). When the size of lithology B is smaller than the height of the melting column, different parcels of lithology B enter the melting column at different times. The melting problem becomes time dependent. Liang (2008) and Liang and Liu (2018) presented time-dependent solutions for batch melting, fractional melting, continuous melting and two-porosity melting of a heterogeneous mantle column in which the enriched and depleted mantle sources have the same lithology and melting rate (*i.e.* a chemically heterogeneous mantle). Liang (2020, 2022) further expanded the perfect fractional melting model by allowing the depleted and enriched mantle sources to have different melting rate and partition coefficient (*i.e.* a lithologically heterogeneous mantle). However, explicit expressions for a more general time-dependent model in which melt percolates in the two lithologies still awaits further development (for a preview see supplementary Movie_4 in Liang, 2020).

Acknowledgements

I thank Romain Tilhac and Alessio Sanfilippo for constructive reviews that greatly improved the presentation of this manuscript. Nick Dygert, Noah Hooper, Boda Liu and Chenguang Sun read earlier versions of this manuscript and made many useful suggestions. I would also like to thank Boda Liu, Marc Parmentier and Chenguang Sun for fruitful collaborations and stimulating discussions on trace element partitioning, channelized melt migration and decompression mantle melting. Some of the materials presented here are based on lecture notes used in a Geochemical Modelling course taught by the author at Brown University. I thank participating students for their interest and contributions. Work on mantle melting and trace element fractionation at Brown University has been supported by NSF and NASA, mostly recently through grants OCE-1852088, EAR-2147598 and 80NSSC22K0975.

Appendix A. Conservation equations for lithology A

The one-dimensional mass-conservation equations for interstitial melt and residual solid in the upwelling melting column are:

$$\frac{\partial \rho_f \phi_f^A \psi_A}{\partial t} + \frac{\partial \rho_f \phi_f^A V_f^A \psi_A}{\partial z} = \psi_A \Gamma_A - \dot{S} \tag{A1}$$

$$\frac{\partial \rho_s \left(1 - \phi_f^A\right) \psi_A}{\partial t} + \frac{\partial \rho_s \left(1 - \phi_f^A\right) V_s^A \psi_A}{\partial z} = -\psi_A \Gamma_A \tag{A2}$$

where t is time; z is the vertical coordinate, positive upwards; ρ_f and ρ_s are densities of the melt and solid, respectively; ϕ_f^A and V_f^A are the porosity (volume fraction) and velocity of interstitial melt of lithology A in the melting column, respectively; V_s^A is the solid velocity; and $\dot{S}$ is the melt suction rate (amount of melt flowing from lithology A to lithology B per unit volume of the two lithology per unit time). The mass conservation equation for a trace element in the bulk lithology A (melt + solid) is:

$$\frac{\partial \left[\rho_f \phi_f^A C_f^A + \rho_s \left(1 - \phi_f^A\right) C_s^A\right] \psi_A}{\partial t} + \frac{\partial \left[\rho_f \phi_f^A V_f^A C_f^A + \rho_s \left(1 - \phi_f^A\right) V_s^A C_s^A\right] \psi_A}{\partial z}$$
$$= -\dot{S} C_f^A \tag{A3}$$

where C_f^A and C_s^A are concentrations of the trace element in the interstitial melt and residual solid in lithology A, respectively. Equation A3 states that the rate of change of total concentration of the trace element in lithology A in a representative element volume (REV) at location z in the melting column is due to flow of the melt and solid across the REV (second term on the left-hand-side) and transfer of the trace element in the melt from lithology A to lithology B (the term on the right-hand side). At steady state, these two terms balance with each other. When the solid and melt are in local chemical equilibrium, Eq. A3 can be simplified through the bulk solid-melt partition coefficient, *i.e.* $C_s^A = k_A C_f^A$. We have:

$$\frac{\partial\left[\rho_f \phi_f^A + \rho_s\left(1 - \phi_f^A\right)k_A\right]\psi_A C_f^A}{\partial t} + \frac{\partial\left[\rho_f \phi_f^A V_f^A + \rho_s\left(1 - \phi_f^A\right)V_s^A k_A\right]\psi_A C_f^A}{\partial z}$$
$$= -\dot{S}C_f^A \tag{A4}$$

With the help of Eqs A1, A2 and A7 below, Eq. A4 can be further simplified:

$$\frac{\partial\left[\rho_f \phi_f^A + \rho_s\left(1 - \phi_f^A\right)k_A\right]C_f^A}{\partial t} + \left[\rho_f \phi_f^A V_f^A + \rho_s\left(1 - \phi_f^A\right)V_s^A k_A\right]\frac{\partial C_f^A}{\partial z}$$
$$= \left(k_A^p - 1\right)C_f^A \Gamma_A \tag{A5}$$

We refer to Eqs A3 and A4 as the conserved form and to Eq. A5 as the non-conserved form. As shown in the main text, the conserved form is useful to calculate concentration of the aggregated or pooled melt at top of the melting column, whereas the non-conserved form is more convenient to calculate concentration of instantaneous melt in the melting column. To solve Eq. A4 or A5, we also need know the bulk partition coefficient and the extend of melting experienced by residual solid at a given time and location in the melting column. These melting parameters are defined with respect to the upwelling solid by the following evolution equations:

$$\frac{\partial F_A}{\partial t} + V_s^A \frac{\partial F_A}{\partial z} = \frac{(1 - F_A)\Gamma_A}{\rho_s\left(1 - \phi_f^A\right)} \tag{A6}$$

$$\frac{\partial k_A}{\partial t} + V_s^A \frac{\partial k_A}{\partial z} = \frac{\left(k_A - k_A^p\right)\Gamma_A}{\rho_s\left(1 - \phi_f^A\right)} \tag{A7}$$

The factor $1 - F_A$ in Eq. A6 accounts for the fact that the F_A fraction of fusible solid has already been converted to melt at location z in the melting column at time t. The difference in partition coefficient in Eq. A7 arises from non-modal melting. The left-hand side of Eqs A6 and A7 are material derivatives following the motion of solid.

The bulk solid-melt partition coefficient (k_A) and the partition coefficient according to melting reaction (k_A^p) for a trace element are defined in the usual way (Shaw, 1970):

$$k_A = \sum_{j=1}^{N} w_j^A k_j^A \tag{A8}$$

$$k_A^p = \sum_{j=1}^{N} p_j^A k_j^A \tag{A9}$$

where w_j^A is the weight fraction of mineral j in the solid; p_j^A is the fraction of mineral j participated in the melting reaction; and k_j^A is the mineral-melt partition coefficient for mineral j. The bulk solid-melt partition coefficient k_A depends on mineral mode, while k_A^p depends on the melting reaction. Hence both partition coefficients are lithology specific. Derivations of the conservation and evolution equations outline in Appendices A and B can be found in Liang and Parmentier (2010), Liang and Peng (2010) and Liang (2020).

Appendix B. Conservation equations for lithology B

Conservation and evolution equations for lithology B are similar to those for lithology A except we have to reverse the sign for the source term involving flow or suction of melt from lithology A to lithology B. These equations are:

$$\frac{\partial \rho_f \phi_f^B \psi_B}{\partial t} + \frac{\partial \rho_f \phi_f^B V_f^B \psi_B}{\partial z} = \psi_B \Gamma_B + \dot{S} \tag{B1}$$

$$\frac{\partial \rho_s \left(1 - \phi_f^B\right) \psi_B}{\partial t} + \frac{\partial \rho_s \left(1 - \phi_f^B\right) V_s^B \psi_B}{\partial z} = -\psi_B \Gamma_B \tag{B2}$$

$$\frac{\partial \left[\rho_f \phi_f^B C_f^B + \rho_s \left(1 - \phi_f^B\right) C_s^B\right] \psi_B}{\partial t} + \frac{\partial \left[\rho_f \phi_f^B V_f^B C_f^B + \rho_s \left(1 - \phi_f^B\right) V_s^B C_s^B\right] \psi_B}{\partial z}$$
$$= \dot{S} C_f^A \tag{B3}$$

Note the solid velocity for lithology B may not be the same as that for lithology A. When the solid and melt are in local chemical equilibrium, we have the conserved form of mass conservation equation for a trace element in lithology B:

$$\frac{\partial\left[\rho_f\phi_f^B + \rho_s\left(1-\phi_f^B\right)k_B\right]\psi_B C_f^B}{\partial t} + \frac{\partial\left[\rho_f\phi_f^B V_f^B + \rho_s\left(1-\phi_f^B\right)V_s^B k_B\right]\psi_B C_f^B}{\partial z}$$

$$= \dot{S}C_f^A \tag{B4}$$

The non-conserved form of Eq. B4 is:

$$\frac{\partial\left[\rho_f\phi_f^B + \rho_s\left(1-\phi_f^B\right)k_B\right]C_f^B}{\partial t} + \left[\rho_f\phi_f^B V_f^B + \rho_s\left(1-\phi_f^B\right)V_s^B k_B\right]\frac{\partial C_f^B}{\partial z}$$

$$= \left(k_B^p - 1\right)C_f^B\Gamma_B + \dot{S}\left(C_f^A - C_f^B\right) \tag{B5}$$

The last term on the right-hand side of Eq. B5 accounts for the flow of melt from lithology A to lithology B. Evolution equations for the extent of melting and bulk partition coefficient are:

$$\frac{\partial F_B}{\partial t} + V_s^B\frac{\partial F_B}{\partial z} = \frac{(1-F_B)\Gamma_B}{\rho_s\left(1-\phi_f^B\right)} \tag{B6}$$

$$\frac{\partial k_B}{\partial t} + V_s^B\frac{\partial k_B}{\partial z} = \frac{\Gamma_B\left(k_B - k_B^p\right)}{\rho_s\left(1-\phi_f^B\right)} \tag{B7}$$

The bulk partition coefficient (k_B) and the partition coefficient according to the melting reaction (k_B^p) for lithology B are defined in a similar way as those in Eqs A8 and A9, *i.e.*

$$k_B = \sum_{j=1}^{N} w_j^B k_j^B \tag{B8}$$

$$k_B^p = \sum_{j=1}^{N} p_j^B k_j^B \tag{B9}$$

Appendix C. Mixed-column melt composition

At the top of the melting column, melts derived from lithology A and lithology B mix in accordance with their mass fluxes. The composition of the well-mixed melt at top of the melting column (C_f^{mix}) is the weighted average:

$$C_f^{mix} = \frac{\rho_f \phi_f^A V_f^A \psi_A C_f^A + \rho_f \phi_f^B V_f^B \psi_B C_f^B}{\rho_f \phi_f^A V_f^A \psi_A + \rho_f \phi_f^B V_f^B \psi_B} \tag{C1}$$

The denominator in Eq. C1 is the total melt flux at top of the melting column which can also be written as:

$$\rho_f \phi_f^A V_f^A \psi_A + \rho_f \phi_f^B V_f^B \psi_B = \rho_s V_s^0 F_A \psi_A + \rho_s V_s^0 F_B \psi_B \tag{C2}$$

The numerator in Eq. C1 can be evaluated by summing the steady-state version of Eqs A3 and B3:

$$\frac{d\left[\rho_f \phi_f^A V_f^A C_f^A + \rho_s\left(1 - \phi_f^A\right)V_s^A C_s^A\right]\psi_A}{dz}$$
$$+ \frac{\partial\left[\rho_f \phi_f^B V_f^B C_f^B + \rho_s\left(1 - \phi_f^B\right)V_s^B C_s^B\right]\psi_B}{dz} = 0 \tag{C3}$$

Integrating Eq. C3 from the solidus to the top of the melting column, we have

$$\rho_f \phi_f^A V_f^A C_f^A \psi_A + \rho_f \phi_f^B V_f^B C_f^B \psi_B = \left[\rho_s V_s^0 C_A^0 - \rho_s\left(1 - \phi_f^A\right)V_s^A C_s^A\right]\psi_A$$
$$+ \left[\rho_s V_s^0 C_B^0 - \rho_s\left(1 - \phi_f^B\right)V_s^B C_s^B\right]\psi_B \tag{C4}$$

Equation C4 states that the total mass flux of the element of interest available for melt extraction (the left-hand side) is the net difference between the total mass flux of the mantle source materials feeding into the melting column from below and the total mass flux of the residual solid left behind at top of the melting column. This statement is independent of melting models and how lithology A and lithology B interact in the melting column. Hence Eq. C4 is a general statement of steady-state global mass balance in the two-lithology melting column. In terms of the degree of melting, Eq. C4 takes the form:

$$\rho_f \phi_f^A V_f^A C_f^A \psi_A + \rho_f \phi_f^B V_f^B C_f^B \psi_B = \rho_s V_s^0 \{[C_A^0 - (1 - F_A)C_s^A]\psi_A$$
$$+ [C_B^0 - (1 - F_B)C_s^B]\psi_B\} \tag{C5}$$

Substituting Eq. C5 into Eq. C1, we have a simple expression for the concentration of well-mixed melt at top of the two-lithology melting column:

$$C_f^{mix} = \frac{\psi_A[C_A^0 - (1 - F_A)C_s^A] + \psi_B[C_B^0 - (1 - F_B)C_s^B]}{\psi_A F_A + \psi_B F_B} \tag{C6}$$

In terms of instantaneous melt composition, Eq. C6 can be written as:

$$C_f^{mix} = \frac{\psi_A\left[C_A^0 - (1 - F_A)k_A C_f^A\right] + \psi_B\left[C_B^0 - (1 - F_B)k_B C_f^B\right]}{\psi_A F_A + \psi_B F_B} \qquad \text{(C7a)}$$

For non-modal melting, we have:

$$C_f^{mix} = \frac{\psi_A\left[C_A^0 - (k_A^0 - k_A^P F_A)C_f^A\right] + \psi_B\left[C_B^0 - (k_B^0 - k_B^P F_B)C_f^B\right]}{\psi_A F_A + \psi_B F_B} \qquad \text{(C7b)}$$

Equations C6–C7 are mixing models for modelling major and trace element variations during decompression melting of a steady-state two-lithology melting column. To calculate the concentration of the element of interest in the well-mixed melt at top of the melting column, we need to know volume proportions of the two lithologies in the mantle source, extents of melting experienced by the two lithologies at top of the melting column and concentrations of the melt or residual solid in the two lithologies at top of the melting column. These equations can be generalized to an N-lithology mantle source:

$$C_f^{mix} = \frac{\sum_{j=1}^{N} \psi_j\left[C_j^0 - (1 - F_j)C_s^j\right]}{\sum_{j=1}^{N} \psi_j F_j} \qquad \text{(C8a)}$$

$$C_f^{mix} = \frac{\sum_{j=1}^{N} \psi_j\left[C_j^0 - (1 - F_j)k_j C_f^j\right]}{\sum_{j=1}^{N} \psi_j F_j} \qquad \text{(C8b)}$$

where index j refers to properties of lithology j.

Appendix D. Porosity in the melting column

Porosity in the melting column depends on melt flux which in turn depends on the melt suction rate. To demonstrate the basic idea, here we ignore compaction in the melting column and use Darcy's law to estimate melt porosity in the two-lithology melting column. The procedure is the same as that outlined by Liang and Liu (2018) for decompression melting for a constant and uniform melt suction rate. Including solid upwelling,

Darcy's law for the two lithologies are (McKenzie, 1984):

$$\phi_f^A \left(V_f^A - V_s^A \right) = \frac{\kappa_\phi^A}{\eta_f} \left(1 - \phi_f^A \right) \Delta \rho g \tag{D1a}$$

$$\phi_f^B \left(V_f^B - V_s^B \right) = \frac{\kappa_\phi^B}{\eta_f} \left(1 - \phi_f^B \right) \Delta \rho g \tag{D1b}$$

where κ_ϕ^A and κ_ϕ^B are permeabilities of lithologies A and B in the melting column; η_f is the melt viscosity; $\Delta \rho = \rho_s - \rho_f$; and g is the acceleration due to gravity. With the help of Eqs 47 and 49f, we write the relative volume flux of the melt in Eq. D1 as

$$\phi_f^A \left(V_f^A - V_s^A \right) = V_s^0 \left[(1 - \mathbb{R}) F_A + \mathbb{R} F_A^d - \phi_f^A (1 - F_A) \right] \tag{D2a}$$

$$\phi_f^B \left(V_f^B - V_s^B \right) = V_s^0 \left[F_B + \mathbb{R} (F_A - F_A^d) \frac{\psi_A}{\psi_B} - \phi_f^B (1 - F_B) \right] \tag{D2a}$$

where we make use of the simplifications: $\phi_f^A \ll 1$, $\phi_f^B \ll 1$ and $\rho_s / \rho_f \to 1$. The permeability is related to porosity and mean grain size (d_A, d_B) through the power-law relationships:

$$\kappa_\phi^A = \frac{d_A^2}{b} \left(\phi_f^A \right)^n, \quad \kappa_\phi^B = \frac{d_B^2}{b} \left(\phi_f^B \right)^n \tag{D3a, D3b}$$

where n is the permeability exponent; and b is a constant. For the case presented in section 5, we set $n = 3$ based on the study of Wark and Watson (1998). Substituting Eqs D2 and D3 into Darcy's law, we have a set of algebraic equations for porosities in the two lithologies:

$$\left(\phi_f^A \right)^n = \frac{b \eta_f}{d_A^2 \Delta \rho g} V_s^0 \left[(1 - \mathbb{R}) F_A + \mathbb{R} F_A^d - \phi_f^A (1 - F_A) \right] \tag{D4a}$$

$$\left(\phi_f^B \right)^n = \frac{b \eta_f}{d_B^2 \Delta \rho g} V_s^0 \left[F_B + \mathbb{R} (F_A - F_A^d) \frac{\psi_A}{\psi_B} - \phi_f^B (1 - F_B) \right] \tag{D4b}$$

It is convenient to introduce two reference porosities in terms of the maximum extents of melting experienced by the two lithologies in the melting column:

$$\phi_{ref}^A = \left(\frac{b \eta_f}{d_A^2 \Delta \rho g} V_s^0 F_A^{max} \right)^{\frac{1}{n}}, \quad \phi_{ref}^B = \left(\frac{b \eta_f}{d_B^2 \Delta \rho g} V_s^0 F_B^{max} \right)^{\frac{1}{n}} \tag{D5a, D5b}$$

Equations D4a–D4b then take on the forms:

$$\left(\phi_f^A\right)^n = \frac{\left(\phi_{ref}^A\right)^n}{F_A^{max}}\left[(1-\mathbb{R})F_A + \mathbb{R}F_A^d - \phi_f^A(1-F_A)\right] \qquad \text{(D6a)}$$

$$\left(\phi_f^B\right)^n = \frac{\left(\phi_{ref}^B\right)^n}{\psi_B F_B^{max}}\left[\psi_B F_B + \psi_A \mathbb{R}\left(F_A - F_A^d\right) - \psi_B \phi_f^B(1-F_B)\right] \qquad \text{(D6b)}$$

Given the reference porosity and melting parameters, Eq. D6 can be solved 'exactly' using a numerical method. An approximate solution to Eqs D6a–D6b can be obtained by ignoring the third term in the square bracket on the right-hand side of these equations, *i.e.*

$$\phi_f^A \approx \phi_{ref}^A\left[\frac{(1-\mathbb{R})F_A + \mathbb{R}F_A^d}{F_A^{max}}\right]^{\frac{1}{n}} \qquad \text{(D7a)}$$

$$\phi_f^B \approx \phi_{ref}^B\left[\frac{\psi_B F_B + \psi_A \mathbb{R}\left(F_A - F_A^d\right)}{\psi_B F_B^{max}}\right]^{\frac{1}{n}} \qquad \text{(D7b)}$$

When $F_A^d = 0$, Eq. D7a is reduced to the expression for constant and uniform melt suction rate in the melting column (Liang and Liu, 2018), *i.e.*

$$\phi_f^A \approx \phi_{ref}^A\left[(1-\mathbb{R})\frac{F_A}{F_A^{max}}\right]^{\frac{1}{n}} \qquad \text{(D8)}$$

Porosity obtained from the approximate solutions (Eqs D7a and D7b) are slightly smaller than the 'exact' values given by Eq. D6a and D6b (by <5%, relative).

Appendix E. Summary of melting models

To facilitate geochemical modelling, here we list key equations for the four melting models presented in sections 3–5. These equations are identified by their original equation numbers so a reader can easily find relevant information in the main text. Common to the four models are the dimensionless melt suction rate, bulk partition coefficient, extent of melting, equilibrium partitioning between residual solid and interstitial melt and expressions for aggregated melts.

- Dimensionless melt suction rate

$$\mathbb{R} = \frac{\dot{S}}{\psi_A \Gamma_A} \tag{17c}$$

- Degrees of melting for the two litholgies

$$F_A = \frac{\Gamma_A z}{\rho_s V_s^0}, \quad F_B = \frac{\Gamma_B z}{\rho_s V_s^0} \tag{16, 30c}$$

- Bulk solid-melt partition coefficients

$$k_A = \frac{k_A^0 - k_A^p F_A}{1 - F_A}, \quad k_B = \frac{k_B^0 - k_B^p F_B}{1 - F_B} \tag{19b, 30d}$$

- Concentrations of residual solid in lithologies A and B

$$C_s^A = k_A C_f^A, \quad C_s^B = k_B C_f^B \tag{23, 33b}$$

- Mixed-column melt at the top of the two-lithology melting column

In terms of interstitial melt composition:

$$C_f^{mix} = \frac{\psi_A F_A C_f^A + \psi_B F_B C_f^B}{\psi_A F_A + \psi_B F_B} \tag{8}$$

In terms of residual solid composition:

$$C_f^{mix} = \frac{\psi_A \left[C_A^0 - (1 - F_A) C_s^A \right] + \psi_B \left[C_B^0 - (1 - F_B) C_s^B \right]}{\psi_A F_A + \psi_B F_B} \tag{9a}$$

- Eruptible melt composition

$$C_f^{eruptible} = \alpha_1 C_{f,1}^{mix} + \alpha_2 C_{f,2}^{mix} + \ldots + \alpha_N C_{f,N}^{mix} \tag{2a}$$

$$\alpha_1 + \alpha_2 + \ldots + \alpha_N = 1 \quad \text{and} \quad 0 \le \alpha_j \le 1 \tag{2b}$$

- Average melt composition over the entire melting region

$$C_f^{avg} = \frac{\bar{F}_1 C_{f,1}^{mix} + \bar{F}_2 C_{f,2}^{mix} + \ldots + \bar{F}_N C_{f,N}^{mix}}{\bar{F}_1 + \bar{F}_2 + \ldots + \bar{F}_N} \tag{1a}$$

$$\bar{F}_j = \psi_{A,j} F_{A,j} + \psi_{B,j} F_{B,j} \tag{1b}$$

$$\psi_{A,j} + \psi_{B,j} = 1 \tag{1c}$$

E1. The two-porosity melting model: Case of one melt suction rate (section 3)

- Concentration of instantaneous melt in lithology A

$$
C_f^A = \begin{cases}
\dfrac{C_A^0}{k_A^0}\left[\dfrac{k_A^0 + \left(1 - k_A^P - \mathbb{R}\right)F_A}{k_A^0}\right]^{\frac{k_A^P - 1}{1 - k_A^P - \mathbb{R}}} & \text{if } 1 - k_m^P - \mathbb{R} \neq 0 \\[4mm]
\dfrac{C_A^0}{k_A^0}\exp\left(\dfrac{k_A^P - 1}{k_A^0}F_A\right) & \text{if } 1 - k_m^P - \mathbb{R} = 0
\end{cases}
$$

$$(21a, 21b)$$

- Average melt composition for lithology A

$$
\bar{C}_f^A = \frac{C_A^0 - \left[(1 - \mathbb{R})F_A + (1 - F_A)k_A\right]C_f^A}{\mathbb{R}F_A}
$$

$$(27b)$$

- Concentration of instantaneous melt in lithology B

$$
C_f^B = \frac{\psi_B C_B^0 + \psi_A \mathbb{R}F_A \bar{C}_f^A}{\psi_B\left[k_B^0 + \left(1 - k_B^P\right)F_B\right] + \psi_A \mathbb{R}F_A}
$$

$$(33a)$$

- Equations 21a and 21b reduce to batch melting when $\mathbb{R} = 0$ and fractional melting when $\mathbb{R} = 1$.

E2. The continuous melting model: Case of constant melt-to-solid mass flux (section 4.1)

- Concentration of instantaneous melt in lithology A

$$
C_f^A = \frac{C_A^0}{k_A^0 + \left(1 - k_A^P\right)F_A} \quad \text{for } F_A < F_A^d
$$

$$(38a)$$

$$
C_f^A = \left[\frac{C_A^0}{k_A^0 + \left(1 - k_A^P\right)F_A^d}\right]\left[\frac{k_A^0 + \omega - \left(k_A^P + \omega\right)F_A}{k_A^0 + \omega - \left(k_A^P + \omega\right)F_A^d}\right]^{\frac{1 - k_A^P}{k_A^0 + \omega}}
$$
$$
\text{for } F_A \geq F_A^d
$$

$$(38b)$$

- Average melt composition for lithology A

$$\bar{C}_f^A = \frac{1}{F_A - F_A^d} \int_{F_A^d}^{F_A} C_f^A dF_A$$

$$= \frac{C_A^0 - \left[k_A^0 + \omega - \left(k_A^p + \omega\right)F_A\right]C_f^A}{(1 + \omega)\left(F_A - F_A^d\right)}$$

$$\text{for } F_A \geq F_A^d \tag{42c}$$

- Concentration of instantaneous melt in lithology B

$$C_f^B = \frac{C_B^0}{k_B^0 + \left(1 - k_B^p\right)F_B} \quad \text{for } F_A < F_A^d \tag{42a}$$

$$C_f^B = \frac{\psi_B C_B^0 + \psi_A(1 + \omega)\left(F_A - F_A^d\right)\bar{C}_f^A}{\psi_B\left[k_B^0 + \left(1 - k_B^p\right)F_B\right] + \psi_A(1 + \omega)\left(F_A - F_A^d\right)}$$

$$\text{for } F_A \geq F_A^d \tag{42b}$$

- Key melting parameters

$$F_A^d = \frac{\omega}{1 + \omega} \tag{36b}$$

$$\omega = \frac{\rho_f \phi_f^A V_f^A}{\rho_s \left(1 - \phi_f^A\right)V_s^A} \quad \text{and}$$

$$F_A^d = \frac{\rho_f \phi_f^A V_f^A}{\rho_f \phi_f^A V_f^A + \rho_s \left(1 - \phi_f^A\right)V_s^A} \tag{40}$$

$$\mathbb{R} = \frac{\dot{S}}{\psi_A \Gamma_A} = \begin{cases} 0 & \text{for } F_A < F_A^d \\ 1 + \omega & \text{for } F_A \geq F_A^d \end{cases} \tag{41}$$

E3. The continuous melting model: Case of constant porosity in lithology A (section 4.2)

• Concentration of instantaneous melt in lithology A

$$C_f^A = \frac{C_A^0}{k_A^0 + (1 - k_A^p)F_A} \quad \text{for } F_A < F_A^d \tag{45a}$$

$$C_f^A = \left[\frac{C_A^0}{k_A^0 + (1 - k_A^p)F_A^d} \right] \left[\frac{k_A^0 + F_A^d - k_A^p F_A}{k_A^0 + (1 - k_A^p)F_A^d} \right]^{\frac{1 - k_A^p}{k_A^p}}$$

$$\text{for } F_A \geq F_A^d \tag{45b}$$

• Average melt composition for lithology A

$$\bar{C}_f^A = \frac{1}{F_A - F_A^d} \int_{F_A^d}^{F_A} C_f^A dF_A = \frac{C_A^0 - (k_A^0 + F_A^d - k_A^p F_A)C_f^A}{F_A - F_A^d}$$

$$\text{for } F_A \geq F_A^d \tag{45e}$$

• Concentration of instantaneous melt in lithology B

$$C_f^B = \frac{C_B^0}{k_B^0 + (1 - k_B^p)F_B} \quad \text{for } F_A < F_A^d \tag{45c}$$

$$C_f^B = \frac{\psi_B C_B^0 + \psi_A (F_A - F_A^d)\bar{C}_f^A}{\psi_B [k_B^0 + (1 - k_B^p)F_B] + \psi_A (F_A - F_A^d)} \quad \text{for } F_A \geq F_A^d \tag{45d}$$

• Key melting parameter

$$\mathbb{R} = \frac{\dot{S}}{\psi_A \Gamma_A} = \begin{cases} 0 & \text{for } F_A < F_A^d \\ 1 & \text{for } F_A \geq F_A^d \end{cases} \tag{44}$$

E4. The two-lithology melting model (section 5)

• Concentration of instantaneous melt in lithology A

$$C_f^A = \frac{C_A^0}{k_A^0 + (1 - k_A^p)F_A} \quad \text{for } F_A < F_A^d \tag{49a}$$

$$C_f^A = \left[\frac{C_A^0}{k_A^0 + \left(1 - k_A^p\right)F_A^d} \right] \left[\frac{k_A^0 + F_A^d \mathbb{R} + \left(1 - k_A^p - \mathbb{R}\right)F_A}{k_A^0 + \left(1 - k_A^p\right)F_A^d} \right]^{\frac{k_A^p - 1}{1 - k_A^p - \mathbb{R}}}$$

$$\text{for } F_A \geq F_A^d \tag{49b}$$

- Average melt composition for lithology A

$$\bar{C}_f^A = \frac{1}{F_A - F_A^d} \int\limits_{F_A^d}^{F_A} C_f^A dF_A = \frac{C_A^0 - \left[k_A^0 + \mathbb{R}F_A^d + \left(1 - k_A^p - \mathbb{R}\right)F_A\right]C_f^A}{\mathbb{R}\left(F_A - F_A^d\right)}$$

$$\text{for } F_A \geq F_A^d \tag{49e}$$

- Concentration of instantaneous melt in lithology B

$$C_f^B = \frac{C_B^0}{k_B^0 + \left(1 - k_B^p\right)F_B} \quad \text{for } F_A < F_A^d \tag{49c}$$

$$C_f^B = \frac{\psi_B C_B^0 + \psi_A \mathbb{R}\left(F_A - F_A^d\right)\bar{C}_f^A}{\psi_B\left[k_B^0 + \left(1 - k_B^p\right)F_B\right] + \psi_A \mathbb{R}\left(F_A - F_A^d\right)}$$

$$\text{for } F_A \geq F_A^d \tag{49d}$$

- Key melting parameter

$$0 \leq \mathbb{R} \leq \mathbb{R}_{\max}, \quad \mathbb{R}_{\max} = \frac{F_A^{\max}}{F_A^{\max} - F_A^d} \tag{48}$$

- Mass fluxes of the melt and solid in lithology A

$$\rho_f \phi_f^A V_f^A = \begin{cases} \rho_s V_s^0 F_A & \text{for } F_A < F_A^d \\ \rho_s V_s^0 (1 - \mathbb{R})F_A + \rho_s V_s^0 F_A^d \mathbb{R} & \text{for } F_A \geq F_A^d \end{cases} \tag{47}$$

$$\rho_s \left(1 - \phi_f^A\right)V_s^A = \rho_s V_s^0(1 - F_A) \tag{17b}$$

- Mass fluxes of the melt and solid in lithology B

$$\rho_f \phi_f^B V_f^B \psi_B = \begin{cases} \rho_s V_s^0 F_B \psi_B & \text{for } F_A < F_A^d \\ \rho_s V_s^0 \left[F_B \psi_B + \mathbb{R}\left(F_A - F_A^d \right) \psi_A \right] & \text{for } F_A \geq F_A^d \end{cases}$$

(49f)

$$\rho_s \left(1 - \phi_f^B \right) V_s^B = \rho_s V_s^0 (1 - F_B)$$

(30b)

References

Albarède, F. (1995) *Introduction to Geochemical Modelling*. Cambridge University Press, New York.

Asimow, P.D. and Stolper, E.M. (1999) Steady-steady mantle-melt interactions in one dimension: I. Equilibrium transport and melt focusing. *Journal of Petrology*, **40**, 475–494.

Borghini, G. and Fumagalli, P. (2020) Melting relations of anhydrous olivine-free pyroxenite Px1 at 2 GPa. *European Journal of Mineralogy*, **32**, 251–264.

Borghini, G., Fumagalli, P. and Rampone, E. (2017) Partial melting of secondary pyroxenite at 1 and 1.5 GPa and its role in upwelling heterogeneous mantle. *Contributions to Mineralogy and Petrology*, **172**, 70.

Dasgupta, R., Hirschmann, M.M. and Stalker, K. (2006) Immiscible transition from carbonate-rich to silicate-rich melts in the 3 GPa melting interval of eclogite + CO_2 and genesis of silica-undersaturated ocean island lavas. *Journal of Petrology*, **47**, 647–671.

Dygert, N., Liang, Y. and Kelemen, P. (2016) Formation of plagioclase lherzolite and associate dunite-harzburgite-lherzolite sequences by multiple episodes of melt percolation and melt-rock reaction: An example from the Trinity ophiolite, California, USA. *Journal of Petrology*, **57**, 815–838.

Gast, P.W. (1968) Trace element fractionations and the origin of tholeiitic and alkaline magma types. *Geochimica et Cosmochimica Acta*, **32**, 1057–1086.

Hofmann, A.W. (1997) Mantle geochemistry: the message from oceanic volcanism. *Nature*, **385**, 219–229.

Hofmann, A.W. (2003) Sampling mantle heterogeneity through oceanic basalts: Isotopes and trace elements. In: *Treatise on Geochemistry: The Mantle and Core* 2nd Edition (ed. R. W. Carlson). Elsevier, New York, pp. 61–101.

Hewitt, I.J. and Fowler, A.C. (2008) Partial melting in an upwelling mantle column. *Proceedings of the Royal Society of London A: Mathematical, Physical and Engineering Sciences*, **464**, 2467–2491.

Ito, G. and Mahoney, J.J. (2005) Flow and melting of a heterogeneous mantle: 1. Method and importance to the geochemistry of ocean island and mid-ocean ridge basalts. *Earth and Planetary Science Letters*, **230**, 29–46.

Iwamori, H. (1994) ^{238}U-^{230}Th-^{226}Ra and ^{235}U-^{231}Pa disequilibra produced by mantle melting and porous and channel flows. *Earth and Planetary Science Letters*, **125**, 1–16.

Jull, M., Kelemen, P.B. and Sims, K. (2002) Consequences of diffuse and channeled porous melt migration on Uranium series disequilibra. *Geochimica et Cosmochimica Acta*, **66**, 4133–4148.

Kogiso, T., Hirschmann, M.M. and Pertermann, M. (2004) High-pressure partial melting of mafic lithologies in the mantle. *Journal of Petrology*, **45**, 2407–2422.

Lambart, S., Laporte, D., Provost, A. and Schiano, P. (2012) Fate of pyroxenite-derived melts in the peridotite mantle: Thermodynamic and experimental constraints. *Journal of Petrology*, **53**, 451–476.

Lambart, S., Baker, M.B. and Stolper, E.M. (2016) The role of pyroxenite in basalt genesis: melt-PX, a melting parameterization for mantle pyroxenites between 0.9 and 5 GPa. *Journal of Geophysical Research*, **121**, 5708–5735.

Langmuir, C. H., Bender, J F., Bence, A. E., Hanson, G. N. and Taylor, S. R. (1977) Petrogenesis of basalts from the FAMOUS area: Mid-Atlantic Ridge. *Earth and Planetary Science Letters*, **36**, 133–156.

Laubier, M., Gale, A. and Langmuir, C.H. (2012) Melting and crustal processes at the FAMOUS segment (Mid-Atlantic Ridge): New insights from olivine-hosted melt inclusions from multiple samples. *Journal of Petrology*, **53**, 665–698.

Liang, Y. (2008) Simple models for dynamic melting in an upwelling heterogeneous mantle column: Analytical solutions. *Geochimica et Cosmochimica Acta*, **72**, 3804–3821.

Liang, Y. (2020) Trace element fractionation and isotope ratio variation during melting of a spatially distributed and lithologically heterogenous mantle. *Earth and Planetary Science Letters*, **552**, 116594.

Liang, Y. (2022) Mixing loops, mixing envelopes, and scatter correlations among trace elements and isotope ratios produced by mixing of melts derived from a spatially heterogenous mantle. *Journal of Petrology*, **63**, 1–16.

Liang, Y. and Liu, B. (2018) Stretching chemical heterogeneity by melt migration in an upwelling mantle: An analysis based on time-dependent batch melting and fractional melting models. *Earth and Planetary Science Letters*, **498**, 275–287.

Liang, Y. and Parmentier, E.M. (2010) A Two-Porosity Double Lithology Model for Partial Melting, Melt Transport and Melt–rock Reaction in the Mantle: Mass Conservation Equations and Trace Element Transport. *Journal of Petrology*, **51**, 125–152.

Liang, Y. and Peng, Q. (2010) Non-modal melting in an upwelling mantle column: Steady-state models with applications to REE depletion in abyssal peridotites. *Geochimica et Cosmochimica Acta*, **74**, 321–339.

Liang, Y., Schiemenz, A., Hesse, M., Parmentier, E.M. and Hesthaven, J.S. (2010) High-porosity channels for melt migration in the mantle: Top is the dunite and bottom is the harzburgite and lherzolite. *Geophysical Research Letters*, **37**, L15306.

Liu, B. and Liang, Y. (2020) Importance of the size and distribution of chemical heterogeneities in the mantle source to the variations of isotope ratios and trace element abundances in mid-ocean ridge basalts. *Geochimica et Cosmochimica Acta*, **268**, 383–404.

Lo Cascio, M. (2008) Kinetics of partial melting and melt-rock reaction in the Earth's mantle. Ph.D. Thesis, Brown University, https://doi.org/10.7301/Z0SF2TK0.

Lundstrom, C.C. (2000) Models of U-series disequilibria generation in MORB: The effect of two scales of melt porosity. *Physics of the Earth and Planetary Interiors*, **121**, 189–204.

Mallik, A. and Dasgupta, R. (2012) Reaction between MORB-eclogite derived melts and fertile peridotite and generation of ocean island basalts. *Earth and Planetary Science Letters*, **329-330**, 97–108.

McKenzie, D. (1984) The generation and compaction of partially molten rocks. *Journal of Petrology*, **25**, 713–765.

McKenzie, D. (1985) ^{230}Th-^{238}U disequilibrium and the melting processes beneath ridge axes. *Earth and Planetary Science Letters*, **72**, 149–157.

Niu, Y., Regelous, M., Wendt, I.J., Batiza, R. and O'Hara, M.J. (2002) Geochemistry of near-EPR seamounts: importance of source *vs.* process and the origin of enriched mantle component. *Earth and Planetary Science Letters*, **199**, 327–345.

Ozawa, K. (2001) Mass balance equations for open magmatic systems: Trace element behavior and its application to open system melting in the upper mantle. *Journal of Geophysical Research*, **106**, 13,407–13,434.

Pertermann, M. and Hirschmann, M.M. (2003) Partial melting experiments on a MORB-like pyroxenite between 2 and 3 GPa: Constraints on the presence of pyroxenite in basalt source regions from solidus location and melting rate. *Journal of Geophysical Research*, **108**, 2125, doi: 10.1029/2000JB000118.

Ribe, N.M. (1985) The generation and composition of partial melting in the earth's mantle. *Earth and Planetary Science Letters*, **73**, 361–376.

Richter, F.M. (1986) Simple models for trace element fractionation during melt segregation. *Earth and Planetary Science Letters*, **77**, 333–344.

Rudge, J.F., Maclennan, J. and Stracke, A. (2013) The geochemical consequences of mixing melts from a heterogeneous mantle. *Geochimica et Cosmochimica Acta*, **114**, 112–143.

Salters, V.J.M. and Stracke, A. (2004) Composition of the depleted mantle. *Geochemistry, Geophysics, and Geosystems*, **5**, Q05B07, doi: 10.1029/2003GC000597.

Schiemenz A., Liang Y. and Parmentier, E.M. (2011) A high-order numerical study of reactive dissolution in an upwelling heterogeneous mantle: I. Channelization, channel lithology, and channel geometry. *Geophysics Journal International*, **186**, 641–664.

Shaw, D.M. (1970) Trace element fractionation during anatexis. *Geochimica et Cosmochimica Acta*, **34**, 237–243.

Shaw, D.M. (2000) Continuous (dynamic) melting theory revisited. *The Canadian Mineralogist*, **38**, 1041–1063.

Shaw, D.M. (2006) *Trace Elements in Magmas*. Cambridge University Press, New York.

Shimizu, K., Saal, A., Myers, C.E., Nagle, A.N., Hauri, E.H., Forsyth, D.W., Kemenetsky, V.S. and Niu, Y. (2016) Two-component mantle-mixing model for the generation of mid-ocean ridge basalts: Implications for the volatile content of the Pacific upper mantle. *Geochimica et Cosmochimica Acta*, **176**, 44–80.

Shimizu, N. (1998) The geochemistry of olivine-hosted melt inclusions in a FAMOUS basalt ALV519-4-1. *Physics of the Earth and Planetary Interiors*, **107**, 183–201.

Shorttle, O., Rudge, J.F., Maclennan, J. and Rubin, K.H. (2016) A statistical description of concurrent mixing and crystallization during MORB differentiation: Implications for trace element enrichment. *Journal of Petrology*, **57**, 2127–2162.

Sobolev, A.V. and Shimizu, N. (1993) Ultra-depleted primary melt included in an olivine from the Mid-Atlantic Ridge. *Nature*, **363**, 151–154.

Soderman, C.R., Shorttle, O., Matthews, S. and Williams, H.M. (2022) Global trends in novel stable isotopes in basalts: Theory and observations. *Geochimica et Cosmochimica Acta*, **318**, 388–414.

Sparks, D.W. and Parmentier, E.M. (1991) Melt extraction from the mantle beneath spreading centers. *Earth and Planetary Science Letters*, **105**, 368–377.

Spiegelman, M. and Elliot, T. (1993) Consequences of melt transport for Uranium series disequilibrium. *Earth and Planetary Science Letters*, **118**, 1–20.

Stracke, A. (2012) Earth's heterogeneous mantle: A product of convection-driven interaction between crust and mantle. *Chemical Geology*, **330-331**, 274–299.

Stracke, A., Bizimis, M. and Salters, V.J.M. (2003) Recycling oceanic crust: Quantitative constraints. *Geochemistry, Geophysics, and Geosystems*, **4**, 8003, doi: 10.1029/2001GC00022.

Stracke, A. and Bourdon, B. (2009) The importance of melt extraction for tracing mantle heterogeneity. *Geochimica et Cosmochimica Acta*, **73**, 218–238.

Sun, C. and Liang, Y. (2012) The distribution REE between clinopyroxene and basaltic melt along a mantle adiabat: effect of major element composition, water, and temperature. *Contributions to Mineralogy and Petrology*, **163**, 807–823.

Sun, C. and Liang, Y. (2013) The importance of crystal chemistry on REE partitioning between mantle minerals (garnet, clinopyroxene, orthopyroxene, and olivine) and basaltic melts. *Chemical Geology*, **358**, 23–36.

Turner, S.J., Langmuir, C.H., Dungan, M.A. and Escrig, S. (2017) The importance of mantle wedge heterogeneity to subduction zone magmatism and the origin of EM1. *Earth and Planetary Science Letters*, **472**, 216–228.

Vollmer, R. (1976) Rb-Sr and U-Th-Pb systematics of alkaline rocks: the alkaline rocks from Italy. *Geochimica et Cosmochimica Acta*, **40**, 283–295.

Walter, M. (2014) Melt extraction and compositional variability in mantle lithosphere. In: *Treatise on Geochemistry: The Mantle and Core* (editor: R. W. Carlson), 2nd Edition, Elsevier, New York, pp. 393–419.

Wang, C., Liang, Y., Xu, W. and Dygert, N. (2013) Effect of melt composition on basalt and peridotite interaction: Laboratory dissolution experiments with applications to mineral compositional variations in mantle xenoliths from the North China Craton. *Contributions to Mineralogy and Petrology*, **166**, 1469–1488.

Wang, C., Lo, Cascio, M., Liang, Y. and Xu, W. (2020) An experimental study of peridotite dissolution in eclogite-derived melts: Implications for styles of melt-rock interaction in lithospheric mantle beneath the North China Craton. *Geochimica et Cosmochimica Acta*, **278**, 157–176.

Wark, D.A. and Watson, B.E. (1998) Grain-scale permeability of texturally equilibrated, monomineralic rocks. *Earth and Planetary Science Letters*, **164**, 591–605.

Weavor, B.L. (1991) The origin of ocean island basalt end-member compositions: trace element and isotopic constraints. *Earth and Planetary Science Letters*, **104**, 381–397.

White, W.M. (2015) Isotopes, DUPAL, LLSVPs, and Anekantavada. *Chemical Geology*, **419**, 10–28.

Willbold, M. and Stracke, A. (2006) Trace element composition of mantle end-members: Implications for recycling of oceanic and upper and lower continental crust. *Geochemistry, Geophysics, Geosystems,* **7**, Q04004, doi: 10.1029/2005GC001005.

Workman, R.K. and Hart, S.R. (2005). Major and trace element composition of the depleted MORB mantle (DMM). *Earth and Planetary Science Letters,* **231**, 53–72.

Yao, L., Sun, C. and Liang, Y. (2012) A parameterized model for REE distribution between low-Ca pyroxene and basaltic melts with applications to REE partitioning in low-Ca pyroxene along a mantle adiabat and during pyroxenite-derived melt and peridotite interaction. *Contributions to Mineralogy and Petrology,* **164**, 261–280.

Yasuda, A., Fujii, T. and Kurita, K. (1994) Melting phase relations of an anhydrous mid-ocean ridge basalt from 3 to 20 GPa: Implications for the behavior of subducted oceanic crust in the mantle. *Journal of Geophysical Research,* **99**, 9401–9414.

Yaxley, G.M. and Green, D.H. (1998) Reactions between eclogite and peridotite: mantle refertilisation by subduction of oceanic crust. *Schweizerisches Mineralogisches Petrographisches Mitteilungen,* **78**, 243–255.

Zindler, A. and Hart, S. (1986) Chemical geodynamics. *Annual Reveiews in Earth and Planetary Science,* **14**, 493–571.

Zindler, A., Staudigel, H. and Batiza, R. (1984) Isotope and trace element geochemistry of young Pacific seamounts: implications for the scale of upper mantle heterogeneity. *Earth and Planetary Science Letters,* **70**, 175–195.

Zou, H. (1998) Trace element fractionation during modal and nonmodal dynamic melting and open-system melting: A mathematical treatment. *Geochimica et Cosmochimica Acta,* **62**, 1937-1945.

Zou, H. (2007) *Quantitative Geochemistry.* World Scientific.

EMU Notes in Mineralogy, Vol. 21 (2024), Chapter 5, 111–154

The shallow mantle as a reactive filter: a hypothesis inspired and supported by field observations

Georges Ceuleneer[1], Mathieu Rospabé[2], Michel Grégoire[1] and Mathieu Benoit[1]

[1]*Géosciences Environnement Toulouse, Toulouse University, CNRS, IRD, 14 Av. E. Belin, 31400 Toulouse, France e-mail: georges.ceuleneer@get.omp.eu*
[2]*Geo-Ocean, Université Brest, CNRS, Ifremer, UMR6538, F-29280 Plouzané, France*

The footprints of mafic melts travelling from the depths to the surface are abundant in the mantle section of ophiolites. They constitute an important source of information about the melt migration mechanisms and related petrological processes in the shallowest part of the mantle beneath former oceanic spreading centres. In the field, these so-called 'melt migration structures' attract attention when they consist of mineral assemblages contrasting with that of their host peridotite. They therefore record a particular moment in the migration history: when the melt becomes out of equilibrium with the peridotite and causes a reaction impacting its modal composition, and/or when a temperature drop initiates the crystallization of the melt.

The existence of cryptic effects of migration revealed by geochemical data shows that melts do not always leave a trail visible in the field. Although incomplete and patchy, the melt migration structures preserved in ophiolites are witnesses of processes that do actually occur in nature, which constitutes an invaluable support to the interpretation of geophysical data and inescapable constraints for numerical simulations and models of chemical geodynamics.

Here we show how field observations and related petrological and geochemical studies allow us to propose answers to fundamental questions such as these: At which temperature is porous flow superseded by dyking? What are the factors governing melt trajectories? What is the nature of the 'universal solvent' initiating infiltration melting and making channelized porous flow the most common mode of transport of magmas through a peridotite matrix regardless the tectonic setting?

A fundamental message delivered by ophiolites is that the shallow mantle behaves as a particularly efficient reactive filter between the depths and the surface of the Earth. Unexpectedly, the reactions occurring there are enhanced by the hybridization between mafic melts and a hydrous component, whatever its origin (*i.e.* magmatic vs. hydrothermal). This hybridization triggers out of equilibrium reactions, leading to the formation of exotic lithologies, including metallic ores, and impacting the global geochemical cycle of a whole range of chemical elements.

1. Introduction

"Few questions in geology are more important than that of the capacity of magmas to dissolve the country rocks. (…) Most geologists (…) have either not considered the problem seriously or have refrained from publishing the product of their thought concerning it." Daly (1914, p. 209).

DOI: 10.1180/EMU-notes.21.5

Geologists trust in observations. As much as possible, they anchor their models in field evidence rather than in theoretical speculations. The history of geology is full of examples of paradigm shifts that resulted from the providential finding of key outcrops. The granite veins intrusive in the meta-sedimentary rocks of Glen Tilt, Scotland, brought to light in 1785 by James Hutton, played such a role at the dawn of igneous petrology: it was the field demonstration that granites do issue from molten silicates and are not necessarily the oldest rocks on Earth. As stressed by Charles Lyell (1830, p. 62), "By this important discovery (...) Hutton prepared the way for the greatest innovation on the systems of his predecessors."

Melt transport from the source to the location of solidification, and its impact on the composition of the melt itself, have always been recognized as important issues by igneous petrologists although we are far from a consensus on the topic. Melt migration involves processes initiating at the scale of a few millimeters or less, which is the size of mineral grains in the mantle source rock. It is far below the spatial resolution of current geophysical methods used to probe the inner Earth. To quote Bowen's thought about the obsessive question of the origin of magmas, "Geophysical data help very little. (...). The decision on this point would still depend principally upon geologic considerations" (Bowen, 1928, p. 319).

Geological data alone cannot solve all the issues related to melt migration in the mantle but, at least, can provide evidence of processes that do happen in nature. Here we illustrate with a few examples what has been learned about melt migration and related melt-rock reactions beneath former oceanic spreading ridges through the study of the Oman and Trinity ophiolites. We evoke the potential impact of these observations on the hypotheses subtending chemical geodynamics.

2. Field evidence of melt migration in the mantle section of ophiolites

2.1. Feeding dykes

As in the granite realm, the dykes observed in the mantle section of ophiolites (commonly referred to as 'mantle dykes') are witnesses of migration of a melt through the solid peridotite matrix, residue of one or more partial melting events. Dykes are former cracks filled with rocks of various compositions and textures (Fig. 1). They crystallized in response to decreasing temperature during the journey of their parent melt from the deep partial melting region to the surface. The contacts between these crystallization products and the host mantle, *i.e.* the dyke walls, are generally parallel to each other and knife sharp (Fig. 1a), which is the criterion to ascribe the name 'dyke' to these structures. Although dykes are emplaced at shallow depth and are not witnesses of the early stages of melt migration, they carry a wealth of precious information in their composition, mineral texture, orientation, and cross cutting relationships.

The orientation of the dykes at the outcrop or even massif scale is generally not random and is of great use to infer the stress orientation at the time of the crack generation (Nicolas and Jackson, 1982; Sleep, 1988) (Fig. 2). The cross-cutting relationships are

Fig. 1. Field aspect of mantle dykes in the Trinity ophiolite (panels b and i) and Oman ophiolite (other panels). (*a*) A common type of gabbro dyke. (*b*) Cross-cutting relationships between two generation of pyroxenite dykes. (*c*) Complex intersection and branching relationships between gabbro dykes. (*d*) Two overlapping gabbro dykes with dyke tips showing opposing directions of propagation. (*e*) Detail of the fine-grained texture of a gabbro dyke. (*f*) Detail on the coarse-grained texture of a gabbro dyke. (*g*) Pegmatitic dyke several metres thick. (*h*) Detail of the gabbro pegmatite. (*i*) Olivine-clinopyroxenite dyke with a crescumulate texture: the crystal growth stopped when the pyroxene crystals from facing margins met in the centre of the dykes. (*j*) Coarse-grained gabbro dyke partially affected by a mylonitic shear zone.

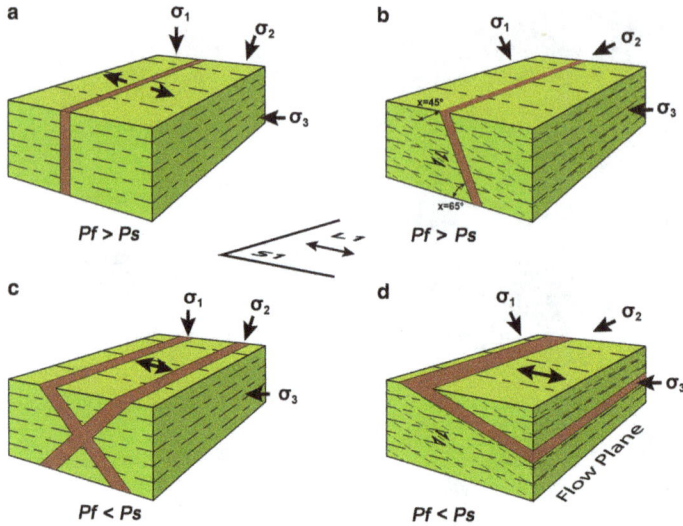

Fig. 2. Relationships between the orientation of the dykes and the stress field. *Pf* = melt pressure; *Ps* = pressure exerted by the surrounding solid; S1 and L1 are the plastic foliation and lineation related to solid-state mantle flow; σ1, σ2 and σ3 are the maximum, intermediate and minimum deviatoric stress components respectively. (*a*) Pure shear (extensional) regime with *Pf* > *Ps*; (*b*) simple shear regime with *Pf* > *Ps*; (*c*) pure shear regime with *Pf* < *Ps*; (*d*) simple shear regime with *Pf* < *Ps*. After Nicolas and Jackson (1982).

reliable indications to determine a relative chronology of injection events (Fig. 1b). Conversely, branching relationships can indicate that dykes of different orientations are coeval (Fig. 1c). The complex situation illustrated in Fig. 1c shows that the two kinds of relationships can occur on the same outcrop: the thin coarse-grained gabbro dykes (white) are cross cut almost orthogonally by thicker dykes with, locally, a coarse-grained margin of similar composition to the thin veins, and a fine-grained (darker) inner part. Cross-cutting could be interpreted in terms of change in stress direction between two injection events but the branching relationships show that the two directions of injections developed contemporaneously.

In well preserved ophiolites, the dyke orientation can be analysed in the ocean-spreading structural framework deduced from: (1) the mantle-crust boundary, proxy of the palaeo-horizontal; (2) the sheeted dyke complex, the average strike is equated with that of the former ridge axis; (3) the solid-state flow structures of the host peridotites, a record of the geometry of the asthenospheric currents; and (4) any other structural elements such as ductile shear zones and brittle faults. When observed, dyke tips indicate the direction of crack propagation (Fig. 1d) (Rubin, 1995).

The texture of the rocks filling the dyke reflects primarily the cooling rate which depends on many factors including the temperature difference between the melt and the host mantle. Apart from rare diabase dikes that were injected in a relatively cold country rock (<~400°C), most ophiolitic mantle dykes are filled with holocrystalline rocks, essentially gabbroic and pyroxenitic in composition, more rarely dunitic, dioritic, trondjhemitic or tonalitic. The crystal size ranges from fine- to coarse-grained (Fig. 1e, f) whereas texture ranges from cumulate to pegmatitic.

Dykes with a pegmatitic texture (Fig. 1g, h) are common. Pegmatites indicate rapid, disequilibrium crystal growth, often along a preferred lattice direction, in the context of a low degree of nucleation. Pegmatites are diagnostic of a significant temperature contrast between the melt and the host rock, although not high enough to induce sudden freezing of the melt as in the case of diabase. The presence of fluids is another factor favouring pegmatite formation, as suggested by the common occurrence of magmatic amphibole overgrowing pyroxenes.

Among the dykes presenting a cumulate texture, adcumulates are the most common. A continuum exists between dykes entirely filled with pegmatites and dykes entirely filled with cumulates as attested by the frequent occurrence of dykes showing coarse crystals at their walls only ('crescumulate') (Fig. 1i) and filled with rocks with an adcumulate texture in its inner part (Fig. 1c). Finally, cumulates filling the dykes can be affected by plastic deformation, which constitutes an interesting source of information concerning the interplay between igneous and tectonic activities (Fig. 1j).

The adcumulate texture points to efficient extraction of the interstitial melt and thus to fractional crystallization and mineral accumulation in an open system (Wager *et al.*, 1960). This was confirmed by Benoit *et al.* (1996) on the basis of major and trace element analyses and mass balance calculations. Briefly, an almost perfect fit exists between the trace element concentrations and patterns measured in the bulk rock and those calculated from mineral separate analyses and modal proportions, pointing to the virtual absence of trapped melt. It is thus reasonable to suppose that most of the melt that flowed within the cracks reached crustal magma chambers after a variable but generally low degree of fractional crystallization and crystal accumulation (Ceuleneer *et al.*, 1996). Most mantle dykes in ophiolites share common parent melts with the rocks making up the crustal section (*e.g.* Lippard *et al.*, 1986; Benoit *et al.*, 1996; Python and Ceuleneer, 2003), which is an argument supporting the view that they were the former 'feeding dykes' of the overlying magma chambers.

In the absence of trapped melt, the mineral proportions and chemical compositions are straightforward to interpret in terms of cotectic assemblages crystallizing along the liquid line of descent of common mafic melts, and are of great use to infer the nature of this melt (*e.g.* Ceuleneer *et al.*, 1996; Benoit *et al.*, 1996; Varfalvy *et al.*, 1996).

The hope of the geologists who analysed mantle dykes was, initially, to find in them the crystallization products from the primary melts that fed the crustal magma chambers (*e.g.* Malpas, 1978; Boudier and Coleman, 1981; Lippard *et al.*, 1986). However, the cumulate filling the dykes appeared to be issued from melts that were already and variably evolved. Although the quest for the primary melt seems an unreachable objective, the systematic petrographic and chemical studies of mantle dykes is, as in the case of the layered cumulates, a very efficient way to determine the magma series that contributed to the igneous evolution of an ophiolite (Python and Ceuleneer, 2003; Ceuleneer and le Sueur, 2008).

Most ophiolites are characterized by the occurrence of lavas, dykes and/or plutonic rocks belonging to two or more different magma series (*e.g.* Dilek and Furnes, 2011).

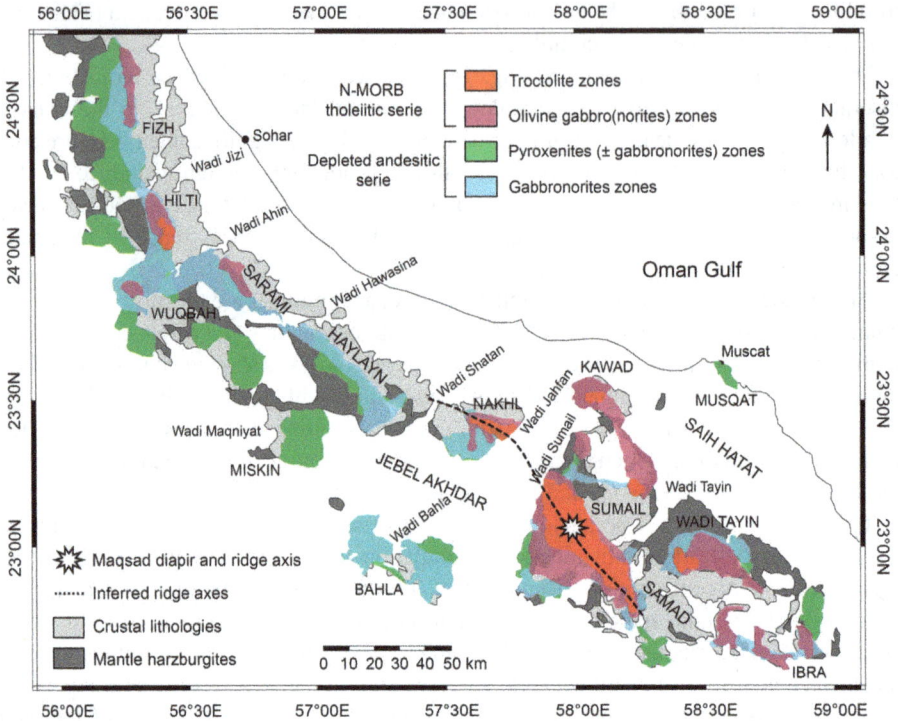

Fig. 3. Simplified map of the distribution of various lithological groups of mantle dykes and porous flow channels in the mantle section of ophiolites. Troctolites and olivine-gabbros (+rare evolved oxide-gabbronorites) are the primitive and differentiated cumulates from MORB-like olivine tholeiites, respectively; pyroxenites and opx-rich gabbronorites are the primitive and differentiated cumulates from depleted, Mg-rich andesites or high-Ca boninites, respectively. The main MORB area is an 80 km-long former ridge segment centred on the Maqsad diapir (white star). After Python and Ceuleneer (2003).

Mantle dykes appear to be a very convenient tool to map the spatial distribution of these magmatic series and to reconstruct the tectono-magmatic history of ophiolites. Figure 3 gives an example of such a distribution along the Oman ophiolite (Python and Ceuleneer, 2003). Bi-modal magmatism in Oman is generally interpreted in terms of successive episodes of melt generation and emplacement, typically a first-generation synchronous with the accretion of the ophiolite in a spreading context, mid-ocean ridge or arc-related basin, and a second one related to the emplacement events that preluded its final obduction on a continental margin (*e.g.* Belgrano *et al.*, 2019 and references therein).

However, the mapping of mantle dyke distribution revealed that the dykes of each petrological family do not crop out in the same areas (Benoit *et al.*, 1999; Python and Ceuleneer, 2003). Feeding dykes of MORB type are restricted to few districts corresponding to former ridge segments opening in a lithosphere hosting pre-existing dykes of boninitic-andesitic affinity that crystallized as pyroxenites and orthopyroxene-rich gabbronorites. These pyroxenite dykes and most of the gabbronorite dykes are restricted to the mantle section and do not invade the overlying crust, which should be the case if they represented a late volcanic episode related to subduction. So, the existence of two magmatic series could alternatively be attributed to the cyclicity and spatial variability in the spreading processes and modes of melt production than in a succession of distinct magmatic episodes (Benoit *et al.*, 1999; Python and Ceuleneer, 2003; Nonnotte *et al.*, 2005; Clénet *et al.*, 2010).

2.2. Porous flow structures

As mentioned above, mantle dykes represent an already evolved stage of melt migration, following significant pooling of incremental melt fractions. According to experiments and to the laws of thermodynamics and solutions, partial melting of a polymineralic assemblage initiates at the junctions between different mineral phases, where the melting temperature reaches a minimum value. In the case of mantle peridotites, it will take place at the contact between olivine, pyroxenes and aluminous phases. As these multiple junctions are evenly distributed in the peridotite, at least in a simplified frame not considering lithological heterogeneities, melt droplets produced by melting will likewise be spatially uniformly distributed (*e.g.* Zhu *et al.*, 2011). The early stages of melt pooling will thus involve migration at grain boundaries, currently referred to as 'porous flow' in the petrological literature.

To seek for field structures inherited from the initiation of melt migration at the depths of the partial melting region (several tens of kilometres or even deeper) would be illusive. However, even if they are more discrete and seem less common than mantle dykes, porous flow structures formed at shallower depths are present in the mantle section of ophiolites (Fig. 4). They are quite informative about the way a melt makes its way through the solid peridotitic matrix and it is reasonable to assume that it may be extrapolated, *mutatis mutandis,* to greater depths.

2.2.1. Macroscopic evidence of porous flow: no unique interpretation

Dykes are structures recognized as such by structural geologists. What we describe as porous flow 'structures' in the field refer actually to variations, more or less subtle, in the proportions and distribution of different minerals that catch our attention, and that we interpret as the footprint of migration of interstitial melt. In other words, the preservation of macroscopic evidence of former melt migration requires: (1) that the melt was in disequilibrium with at least one mineral of the solid matrix, inducing a dissolution reaction; (2) that it migrated through a temperature gradient leading to its partial crystallization; of (3) a combination of both processes with crystallization in response to dissolution, referred as "reactive crystallization" (Collier and Kelemen, 2010). So, the porous flow structures we can observe in the field give access either to the time-integrated distribution of a melt that was "aggressive" for its host peridotite or to a snapshot of the melt distribution at the time of crystallization, corresponding to the closure of the porosity and thus to the final stage of melt migration.

Strictly speaking, where the melt comes from and where it goes can generally not be deduced from field observations, although we currently interpret our data in the frame of

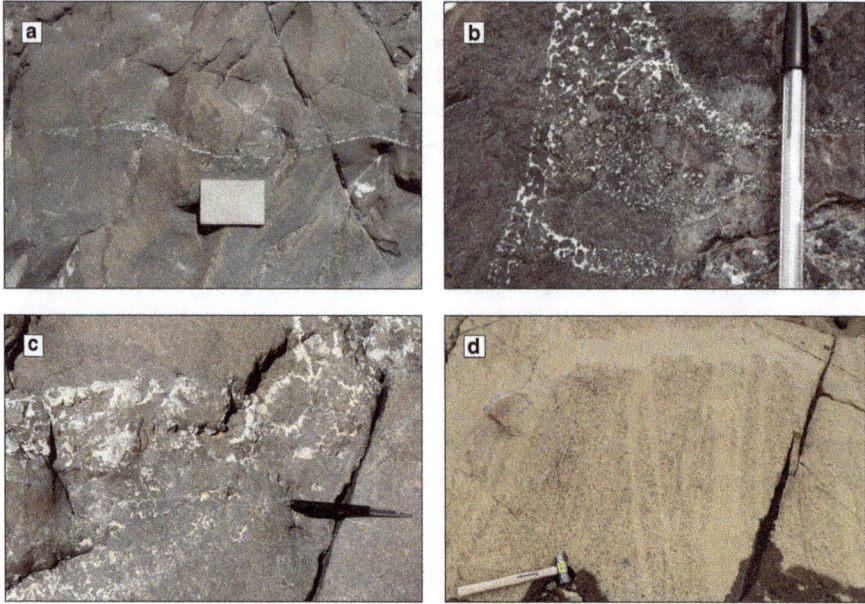

Fig. 4. Porous flow structures observed in the Oman (*a, b, c*) and Trinity (*d*) peridotites. In Oman, relics of the percolating melt are evidenced by whitish crystallization products, mostly plagioclase and, more rarely, clinopyroxene. In Trinity, former melt paths are shown by dunite (rock with a smooth surface) left after infiltration melting and dissolution of the pyroxenes from the lherzolite (rugged surface).

scenarios integrating melt trajectories and direction of flow. Field observations are objective but their interpretation is not, being anchored in assumptions (*e.g.* the melt being less dense than the country rock, it should migrate upward) and/or on petrogenetic and structural considerations. In the following, we illustrate with an example from the mantle section of the Trinity ophiolite the way how preconceived ideas may drive the interpretation.

Figure 5a shows a trail of clinopyroxene grains sandwiched symmetrically by dunite layers and then by the host lherzolite. The contacts between the different lithologies are contoured and follow the shape of the crystals, contrasting with the knife cut contacts shown by dykes. The lherzolite shows solid-state plastic deformation in high-temperature and low-stress conditions, as well as modal and chemical compositions typical of residual peridotite after a moderate degree of partial melting suggesting, thus, that this rock represents the ambient asthenospheric mantle. The clinopyroxene grains from the central trail are interstitial between olivine grains of the dunite and have a composition consistent with fractional crystallization from a primitive basaltic melt. Accordingly, this trail can reasonably be interpreted as the footprint of the melt that circulated in a percolation channel.

To determine the origin of the dunite is trickier. In theory, a dunite may be the residue of extreme degrees of decompression melting of a peridotite if the mantle potential

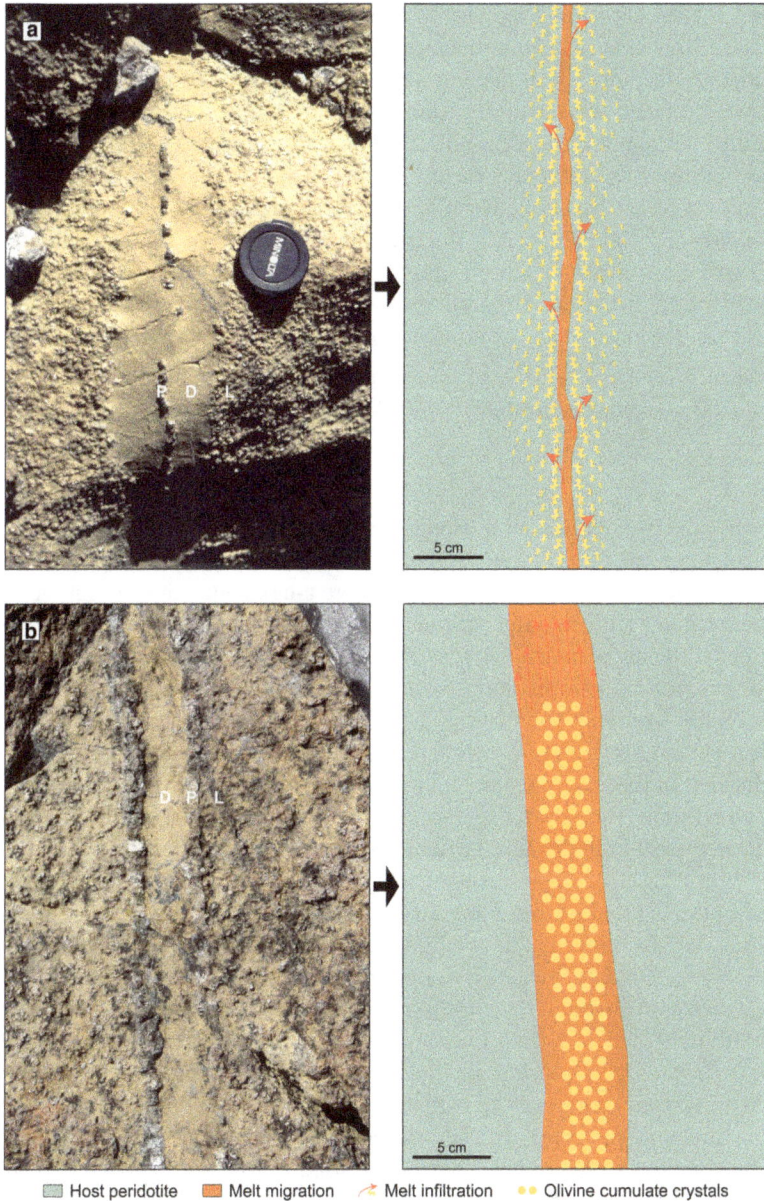

Fig. 5. Two different mutual relationships between lherzolite (L), dunite (D) and pyroxenite (P) in the Trinity ophiolite leading to different interpretations about melt migration.

temperature exceeds ~1500°C (*e.g.* Takahashi, 1986). However, this mechanism cannot account for the discrete occurrence of dunite in a host environment dominated by lherzolite, residue of decompression melting of a much colder mantle. The formation of both lithologies by decompression melting would imply unrealistic temperature gradients at small scale (a few centimetres) in the asthenosphere.

Considering decompression melting alone as implausible to account for dunite formation, we are faced with two physically and petrologically possible mechanisms for the origin of the dunite in Fig. 5a: (1) fractional crystallization of olivine from a Mg-rich melt followed by accumulation in the melt conduit (*e.g.* Elthon, 1979); and (2) the by-product of a reaction between the lherzolite and a melt in disequilibrium with the pyroxene (*e.g.* Kelemen *et al.*, 1992; Braun and Kelemen, 2002); in this case, the dunite would be a mixture between primary olivine grains from the lherzolite itself and secondary olivine issued from the incongruent melting of the pyroxenes according to the reaction (written for a simplified orthopyroxene (opx) composition):

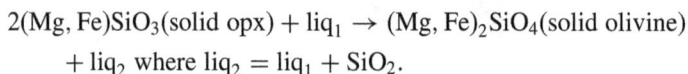

$$2(Mg, Fe)SiO_3(solid\ opx) + liq_1 \rightarrow (Mg, Fe)_2SiO_4(solid\ olivine)$$
$$+ liq_2\ where\ liq_2 = liq_1 + SiO_2.$$

To resolve the two options has a strong impact on our understanding and interpretation of physical processes occurring in the shallow mantle. The cumulate hypothesis implies that the dunite represents a former channel where a large amount of primitive, Mg-rich melt circulated and differentiated and the thickness of which may be approximated by the present distance between the two facing dunite/lherzolite contacts. The second melt-rock reaction hypothesis implies dunite is a relic of a system with a much lower melt/rock ratio with melt percolating interstitially between the grains of the peridotite, the central clinopyroxene trail showing us the path taken by the last melt that circulated in the channel. In this case the dunite/lherzolite contact would correspond to the lateral extent of a reaction front, *i.e.* the distance up to which the melt responsible for the reaction did percolate within the host lherzolite away from the central vein.

Cases where dunites are symmetrically distributed around a central vein are frequently observed in the field. However, other configurations exist and to see in all dunite veins the imprint left by "infiltration melting" might be a hazardous generalisation. For example, a few hundred meters away from the outcrop shown in Fig. 5a, we can observe a different situation: here, the dunite lies in the centre of the structure and clinopyroxenite trails locate at both interfaces between the dunite and the host lherzolite (Fig. 5b). Accordingly, considering field evidence alone, infiltration melting is not really the most convincing way to explain the origin of dunite in this specific case. In turn, this structure perfectly matches the expected distribution of the solid charge carried by a melt in a pipe flow situation. The solid charge would be the olivine grains that crystallized from the flowing melt and that were preferentially concentrated in the zones of weak velocity gradient, *i.e.* in the central part of the vein, while the clinopyroxenes would have crystallize from the more evolved melt that was expelled from the compacting central part and concentrated along the vein margins. This hypothesis was favoured by Maaloe (2005) for the dunite-pyroxenite association from the Leka ophiolite.

The reactional vs. cumulate origin of dunites formed during melt migration in the mantle is still a matter of debate. Geochemical criteria are frequently ambiguous when used alone. If interpreted with the support of other observations, mostly petrographic and structural, the answer has been shown to be non-unique, both origins can co-exist in a single section across tabular dunite (Abily and Ceuleneer, 2013). Nonetheless, geochemical and petrological studies tend to support the view that the reactional origin is the most common (Dick, 1977; Quick, 1981a,b; Kelemen *et al.*, 1992; Kelemen *et al.*, 1995; Suhr *et al.*, 2003, Dygert *et al.*, 2016). Dunites (or their olivine) showing no melt migration features observable in the field generally have slightly lower Mg# and Ni contents than in surrounding mantle peridotites. However, these values are far too high to consider such dunites as cumulates (*e.g.* Kelemen *et al.*, 1995; Godard *et al.*, 2000). They would imply a MgO-rich parent melt formed by high mantle partial melting rates that could not be reached in the modern Phanerozoic Earth. In addition, dunites share another common geochemical feature with mantle peridotites with enrichments in the most incompatible elements relative to expected partition coefficients in olivine, such as in light REE, defining U- or V-shaped REE and multi-element patterns (*e.g.* Prinzhofer and Allègre, 1985; Bodinier *et al.*, 1990; Godard *et al.*, 2000, 2008; Bouilhol *et al.*, 2009; Rospabé *et al.*, 2018) (Fig. 6). Such characteristics were either attributed to serpentinization (Gruau *et al.*, 1998) or to cryptic metasomatism, *i.e.* to the modification of the chemical composition of minerals from the peridotite under the action of migrating melts $\pm$ fluids (*e.g.* Navon and Stolper, 1987; Godard *et al.*, 1995; Vernières *et al.*, 1997).

2.2.2. Cryptic evidence for porous flow
The composition of mantle minerals is known to be deeply modified by the reactions leading to extreme modifications of the paragenetic assemblage. In the petrological

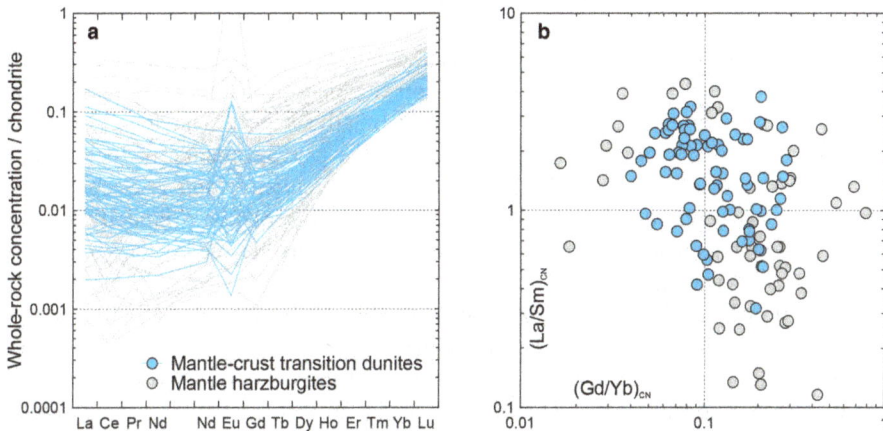

Fig. 6. Whole-rock REE patterns (*a*) and (La/Sm)$_{CN}$ *vs.* (Gd/Yb)$_{CN}$ ratios (*b*) CN for chondrite normalized) for DTZ dunites and mantle harzburgites from the Oman ophiolite. Data for DTZ dunites are from Godard *et al.* (2000), Rospabé *et al.* (2018, 2019a) and Kourim *et al.* (2022). Data for mantle harzburgites are from Godard *et al.* (2000), Gerbert-Gaillard (2002), Girardeau *et al.* (2002), Takazawa *et al.* (2003), Monnier *et al.* (2006), Hanghøj *et al.* (2010) and Kourim *et al.* (2022). Normalizing chondrite values are from Barrat *et al.* (2012).

literature devoted to melt migration, "cryptic" refers to evidence for melt-rock interaction that has no impact on the modal composition but is revealed through changes in the chemical composition of the different minerals.

We illustrate the reliability of this concept of "cryptic metasomatism" (Dawson, 1984) with an example from the Trinity ophiolite, where both modal and cryptic interactions coexist in the same outcrop. This is a different petrological context from that leading to dunite formation. In that case, the participants in the interaction were a mantle plagi-oclase-lherzolite and a vein of depleted, primitive, high-Ca boninitic melt that left an olivine-websterite cumulate after a small amount of fractional crystallization (Ceuleneer and le Sueur, 2008).

The sample (Fig. 7) can be described as a sandwich composed of a 16 mm-thick olivine-websterite vein surrounded by ~5 cm-thick walls of plagioclase-free lherzolite, surrounded, in turn, by plagioclase-bearing peridotite. Boundaries between these three lithologies are rather sharp, even if not knife-cut. No significant difference in the modal proportions of olivine and pyroxenes is observed between the plagioclase-lherzo-lite and the plagioclase-free lherzolite at the vein walls. Accordingly, the melt migrating in the vein and invading the country rock was apparently in equilibrium with olivine and both ortho- and clinopyroxenes but not with plagioclase, which is consistent with the Al-depleted nature of primitive boninite melts.

Therefore, plagioclase consumption is the only macroscopic evidence for melt-rock interaction in this specific case. To explore the consequences of melt-mantle interaction for the chemical compositions, we performed a detailed geochemical profile across this composite sample with the electron microprobe and using laser ablation inductively coupled plasma mass spectrometry (LA-ICP-MS). We show representative profiles with Ti, Zr, Y and Sr in clinopyroxene (cpx).

In the specific case of the Trinity ophiolite, there were fewer incompatible element in the clinopyroxene (cpx) from the lower crustal cumulates and in the vein-filling cumu-lates than in residual cpx in the mantle peridotite. This unusual situation results from the fact that the mantle peridotite from Trinity is relatively fertile, having experienced a moderate degree of partial melting and was not the parental mantle of the boninitic melt, which is issued from an ultra-depleted source (Ceuleneer and le Sueur, 2008). The Trinity massif is better described as a piece of mantle peridotite eventually intruded by boninitic plutons than as a typical ophiolite where the mantle and crustal sections are petrogenetically related.

The incompatible elements (Ti, Zr and Y) define V-shaped concentration profiles centred on the vein (Fig. 7). The cpx from the peridotite at the contact with the vein has concentrations in these trace elements that equal those in cpx from the vein. This content increases linearly and symmetrically away from the vein walls and reaches plateau values at distances of ~10 cm from them. Consequently, it happens that the thick-ness of the zone modified chemically by the interaction exceeds by a factor of ~2 the thickness of the 'macroscopic' reaction zone defined by plagioclase consumption (4–5 cm).

Clinopyroxene is the main reservoir of most incompatible trace elements in the peri-dotite (*i.e.* greater partition coefficients, D, in cpx than in opx and olivine for these

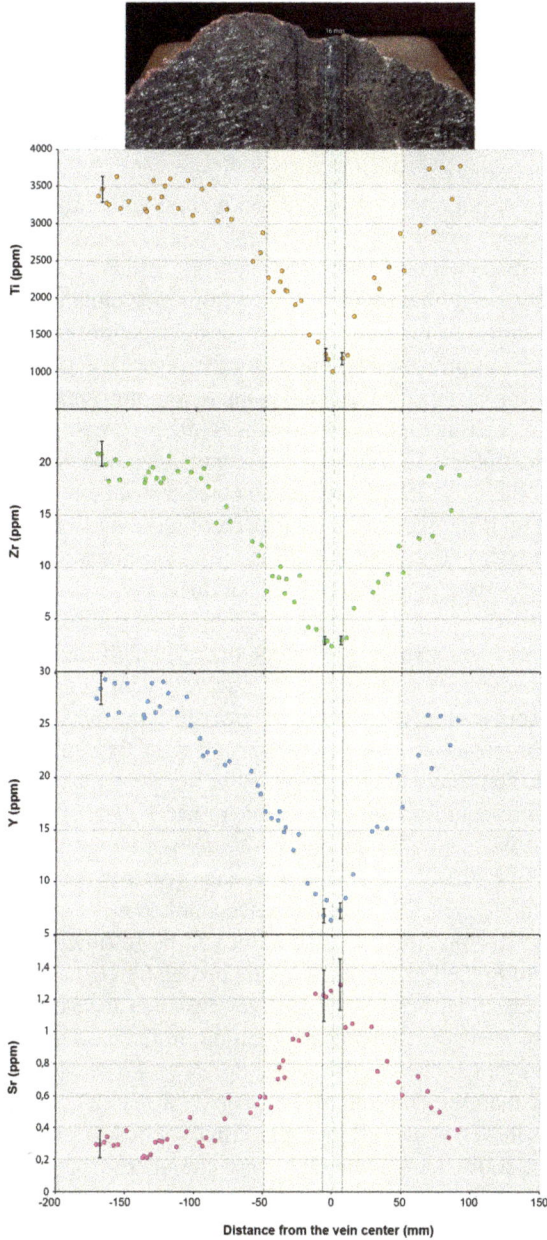

Fig. 7. Evidence for both modal and cryptic effects of melt–rock interaction in the Trinity plagioclase-lherzolite. A pyroxenite vein (pale green on the diagrams) is surrounded by lherzolite devoid of plagioclase. The cryptic impact of melt–rock interaction can be followed further away as revealed by the evolution in the content of incompatible trace elements in clinopyroxene (LA-ICP-MS unpublished data). The error bar is shown for a few analytical points.

chemical species). Accordingly, the dissolution of plagioclase had a minor effect on the concentration of these elements, and the profiles are scarcely disturbed if at all when crossing the plagioclase-out boundary. Sr shows a reverse evolution ('inverted V' profile, Fig. 7), which is consistent with the fact that its concentration in plagioclase is greater than in cpx (*i.e.* $^{plg/liquid}D_{Sr} \gg {}^{cpx/liquid}D_{Sr}$) and that cpx becomes the main host for Sr where plagioclase is dissolved (*i.e.* $^{cpx/liquid}D_{Sr} > {}^{opx/liquid}D_{Sr} > {}^{olivine/liquid}D_{Sr}$).

Cpx crystals present important grain size and texture variations from coarse porphyroclasts to minute grains in the recrystallized matrix and coarse cumulate grains in the vein. Nonetheless, these textural variations have no influence on the composition of cpx that appears to depend only on the distance from the vein and on the chemical concentration contrast between the vein and the far field peridotite. The fact that plastic deformation textures are preserved in the peridotite regardless of the distance from the vein indicates that the instantaneous melt volume was not great enough to erase this texture during the interaction process. Accordingly, the progressive decrease in incompatible element concentrations in the mantle pyroxenes approaching the vein, down to magmatic values at the vein walls, may be explained if mantle residual phases re-equilibrated with a melt with the melt/rock ratio decreasing linearly away from the vein. A chromatographic process controlled by diffusion in the interstitial liquid may also be invoked but would result in parabolic rather than linear concentration profiles (Bodinier *et al.*, 1990).

These simple data demonstrate the effect of cryptic metasomatism. The contrast between cryptic metasomatism recorded by trace element concentrations in cpx and modal metasomatism recorded by the occurrence/disappearance of plagioclase can be explained by the propagation of a percolation front in tiny inter-granular melt films away from the vein. The percolating melt, given the very low melt/rock ratio, is modified rapidly by chemical re-equilibration with mantle minerals. The composition of the modified interstitial melt is reset continuously by mixing with boninite melts flowing in the vein: the closer the peridotite is from the vein, the greater is the time-integrated contribution from the boninite melt, the more Al-undersaturated is the interstitial melt. Accordingly, the conditions for plagioclase dissolution will be fulfilled up to a given distance from the vein but not to the limit reached by the percolation front. It will result in a relatively sharp plagioclase-out boundary, half-way between the vein and the percolation front as revealed by the cryptic chemical evolution, in the case documented in Fig. 7.

To determine if cryptic metasomatism is negligible or if it may have a major impact on the composition of mantle peridotite at the scale of a massif is a fundamental matter. Regional variations in Ti content of cpx in the Trinity ophiolite are shown in Fig. 8 (Ceuleneer and le Sueur, 2008). Cpx in the ultramafic cumulates from the crustal section and in websterite veins have a small Ti content ($TiO_2 < 0.4\%$) and show a trend diagnostic of fractional crystallization from a depleted, boninitic melt with a slight increase in Ti as Mg# values decrease. Clinopyroxene in mantle lherzolite away from veins have variable but always large Ti contents ($0.3 < TiO_2 < 1.0\%$) with Mg# buffered around mantle values. Cpx in the peridotites sampled close to vein walls but showing no macroscopic evidence of melt-rock reaction (Fig. 8) have TiO_2 contents intermediate between those in veins and those in far field peridotite.

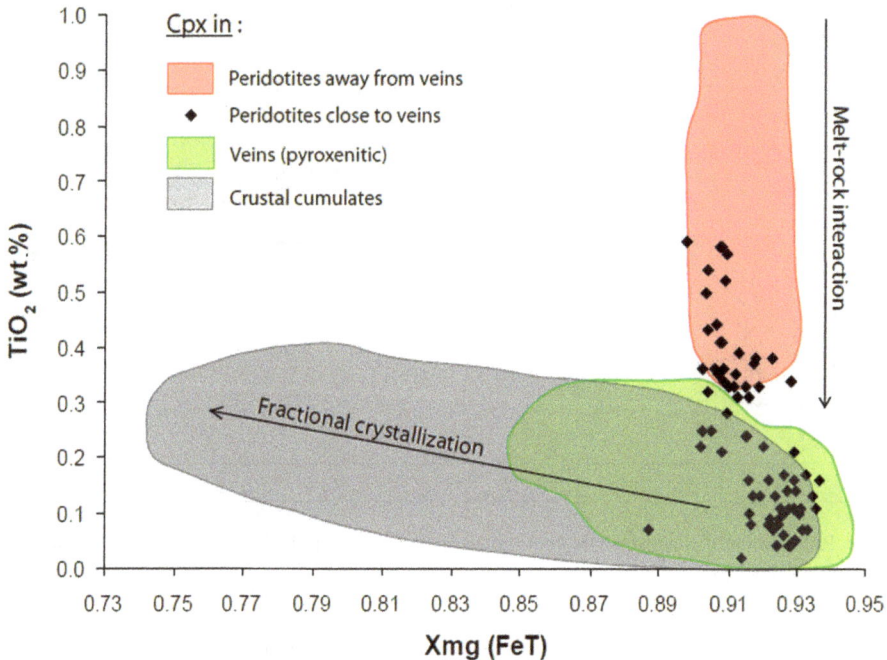

Fig. 8. Ti content *vs.* Mg# of clinopyroxenes from various lithological groups in the Trinity ophiolite. After Ceuleneer and le Sueur (2008).

The same conclusions can be deduced from the comparison of the REE patterns in cpx according to their mode of occurrence (Fig. 9). The cpx from plagioclase-lherzolite away from pyroxenite veins show a typical pattern for cpx in residues of moderate degrees of partial melting. The cpx in gabbroic segregations show the same pattern but with greater concentrations, which is consistent with the view that they represent products of low-degree melting of these lherzolites followed by *in situ* crystallization. The cpx in the crustal ultramafic cumulates show highly depleted patterns characteristic of primitive island arc andesitic-boninitic parent melts. Cpx from the pyroxenite veins show a high scatter in REE concentrations consistent with interaction between the primitive andesite and the host lherzolite.

This evidence indicates that the interaction process highlighted in the sample studied in detail is not a local feature but acted all across the Trinity massif. Peridotite described in the field as 'away from veins' also present a significant scatter in the Ti contents of their clnopyroxenes. It is likely that this also reflects variable degrees of cryptic metasomatism in situations where the percolating melt left no footprint behind it in the form of clinopyr-oxenite vein, or, simply, that the vein does not crop out or was eroded. We can propose that low degrees of decompression melting of the lherzolite left a rather homogeneous, Ti-rich (~1.0% TiO_2) residual cpx and that the scatter in TiO_2 concentration in cpx towards smaller values is mostly attributable to the interaction with the depleted melt,

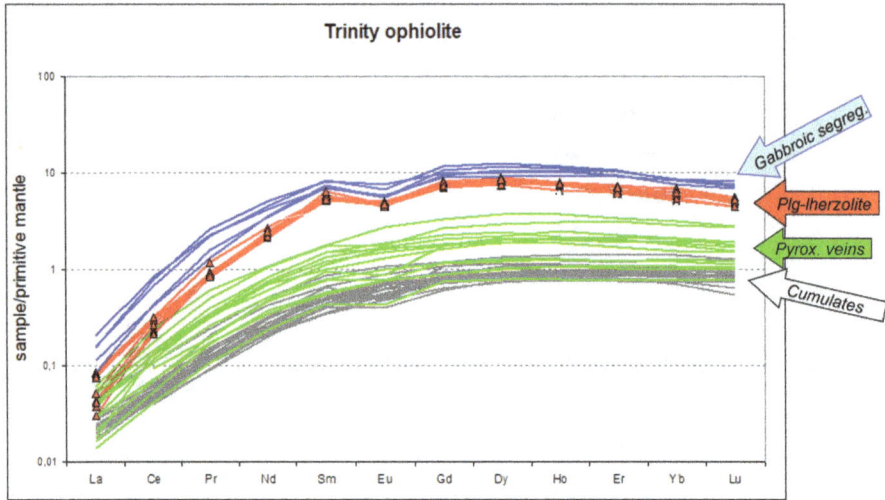

Fig. 9. REE patterns of clinopyroxenes from various lithological groups in the Trinity ophiolite (LA-ICP-MS unpublished data).

regardless of the presence of field evidence for melt percolation in the vicinity of the peridotite.

2.3. From porous flow to dyking: a major transition

Melt extraction from the mantle starts with intergranular porous flow and ends with fracturing ('dyking') according to cooling conditions. However, how, where and when the transition between these two endmembers occurs is still poorly constrained. In other words, the question is to determine if porous flow is rapidly superseded by dyking, as soon as the melt is produced (Fig. 10a) (*e.g.* Waff and Holdren, 1981; Nicolas, 1986; Sleep, 1988; Maaloe, 2003), or if dyking is restricted to shallow depths in which case porous flow would be the dominant mode of melt migration almost up to the surface (Fig. 10b) (*e.g.* Scott and Stevenson, 1986; Turcotte and Phipps Morgan, 1992; McKenzie and Bickle, 1988; Khodakovskii *et al.*, 1995; Rabinowicz *et al.* 2001; Rabinowicz and Ceuleneer, 2005). It primarily governs the composition of a magma, *i.e.* mostly inherited from the deep source with minor modification during migration, or being a blend of all the sources with mixing during its long journey to the surface. To locate where and how the transition from porous flow to dyking occurs also has an impact in the field of geophysics as it governs the mantle rheology, the electrical conductivity and the seismic wave propagation in the partial-melting region.

Dykes are not only the clearest but also the most widespread and abundant melt migration structures in the mantle section of ophiolites: an outcrop devoid of dyke is an uncommon situation, not the reverse. The preponderance of mantle dykes in the ophiolitic record, compared to the more tenuous and, apparently, less abundant porous flow

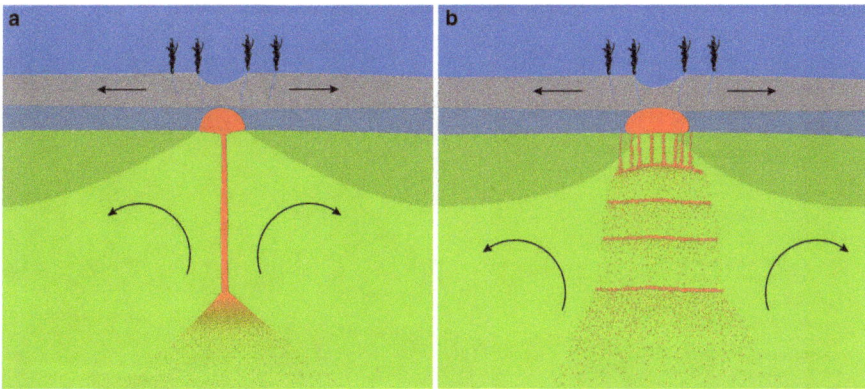

Fig. 10. Two conceptual models of melt migration in the mantle beneath oceanic spreading centres. (*a*) Dykes root directly in the partial melting region at depth and are the main avenues of melt transfer from the asthenosphere to crustal magma chamber (*e.g.* Nicolas, 1986). (*b*) Most of the melt transport occurs by porous flow. The process may imply the development of compaction waves (*e.g.* Scott and Stevenson, 1986; Khodakovskii *et al.*, 1995; Rabinowicz *et al.* 2001; Rabinowicz and Ceuleneer, 2005). Migration occurs in porous flow channels up to the axial magma chamber. Dykes are slightly off-axis features restricted to the lithospheric mantle (Ceuleneer *et al.*, 1996).

structures, has been considered as evidence of a predominantly brittle mode of melt transfer from the partial melting zone to the crust, hydraulic fracturing taking over porous flow at great depths in the asthenospheric mantle (Fig. 10a) (Nicolas, 1986). We will see below how this evidence may be misleading.

The interpretation of geological features can be quite tricky because an outcrop integrates potentially a long history of structural, magmatic or other processes. To quote Carol Cleland (2013), "The observational data (…) are collected in the messy, uncontrollable world of nature (…)". In the particular case of the porous flow *vs.* dyking transition, the challenge is to determine the precise structural setting and mantle temperature at the time a melt migration feature observed in the field was active. It is generally impossible to do. As a matter of fact, the mantle section of ophiolite is, in most cases, a remnant of the oceanic lithosphere accreted progressively when cooling away from an ocean ridge and thus integrates magmatic events during a relatively long, although unknown, time lapse. Luckily, detailed mapping of the plastic flow structures in the Maqsad area of the Oman ophiolite has revealed the existence of an exceptional situation. There, an asthenospheric upwelling was preserved thanks to initiation of detachment of the future ophiolite in a near-ridge context (Ceuleneer, 1986; Rabinowicz *et al.*, 1987; Ceuleneer *et al.*, 1988; Ceuleneer, 1991). This so-called "Maqsad diapir" offers the rare opportunity to observe as near as we get to a snapshot of the melt plumbing system below oceanic spreading centres. It includes structures recorded before the corner flow of the asthenosphere and their eventual transposition in an off-axis position.

Our survey (Ceuleneer *et al.*, 1996) has revealed that the type, orientation, texture and composition of the melt migration features are systematically related and display a concentric zoning centred on the Maqsad diapir (Fig. 11). The mantle in the very centre of the

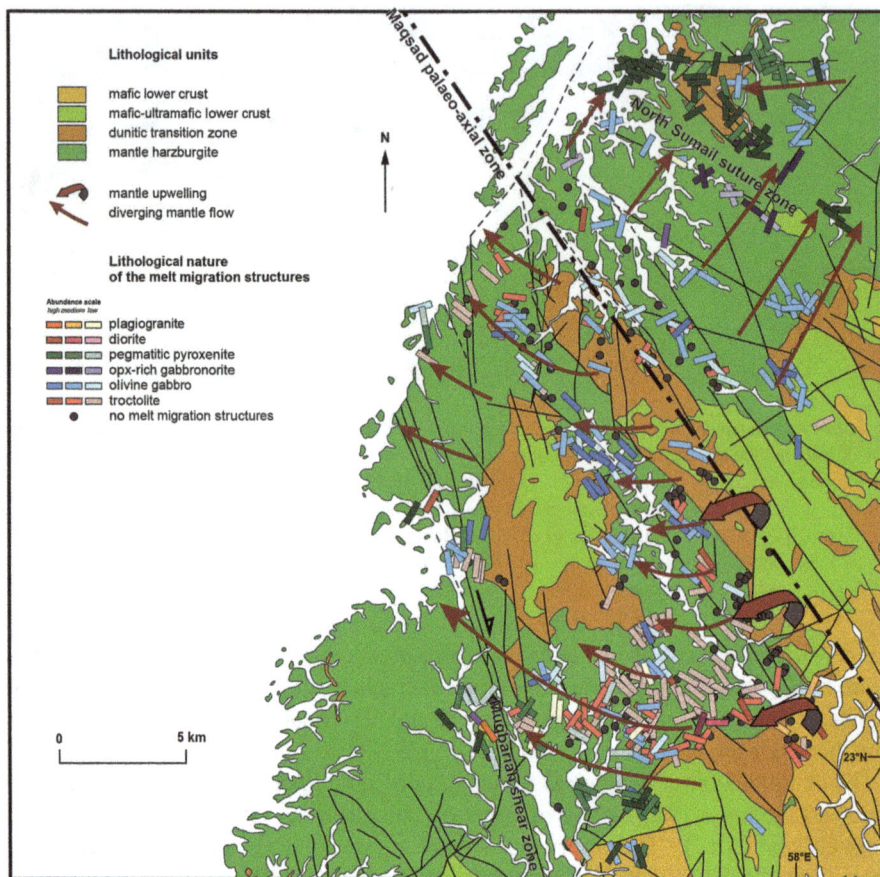

Fig. 11. Distribution of various lithologies of mantle dykes and porous flow channels in and around the Maqsad diapir. Source of the data: Ceuleneer (1991) and Ceuleneer *et al.* (1996).

upwelling contains an unusually small abundance of crystallization products from migrating melts. Only dunite channels are common as bands a few centimetres to a few decimetres thick either transposed into parallelism with the plastic foliation of the peridotite or discordant to this foliation, showing that they can predate or post-date the late stages of asthenospheric flow before the accretion of the asthenospheric mantle to the base of the lithosphere. The rare crystallization products are troctolitic in composition and are scattered in a shallow-dipping dunitic horizon discordant with respect to the foliation (Fig. 12). The contact relationships of these troctolites with their host dunite are gradational over distances of a few centimetres to a few decimetres. Away from the contact, the troctolitic segregations are typically made of 60–70% plagioclase and 30–40% olivine, corresponding to the proportions predicted on experimental grounds for the crystallization of primitive MORB along the olivine-plagioclase low pressure cotectic (*e.g.* Grove *et al.*, 1992). The texture of these troctolites is, in most cases, recrystallized, *i. e.* a mosaic of equant grains with frequent 120° triple junctions. The geochemical data indicate that they were pure adcumulates with virtually no trapped melt (Benoit *et al.*, 1996). Contacts between olivine and plagioclase are underlined by a reactional corona of orthopyroxene, clinopyroxene or hornblende (Fig. 12). These features indicate that troctolites experienced extensive solid-state grain boundary adjustment made possible by intra-crystalline diffusion at high (close to solidus) temperatures, in a slightly hydrous system. The absence of shear deformation texture in the troctolites indicates static conditions of recrystallization.

Moving away from the centre of the diapir, troctolitic segregations become increasingly abundant and adopt progressively a vertical dip but show no preferred azimuth (Fig. 13). On the other hand, their contacts with the host harzburgite become sharper, evolving to dykes (they can be described as 'proto-dykes'). This evolution goes with an increasing abundance of interstitial to poikilitic clinopyroxene. Recrystallization textures are less developed in these cpx-rich troctolites than in the troctolites devoid of clinopyroxene.

Further away from the upwelling, in the zone of diverging flow, olivine-gabbro dykes become abundant. Gabbro-filling the dykes are generally finer grained than troctolites and display adcumulate textures indicative of the simultaneous crystallization of olivine, plagioclase and clinopyroxene. Orthopyroxene, hornblende and Fe-Ti oxides, diagnostic of more evolved melts, become more and more abundant when moving away from the Maqsad diapir. In contrast with the troctolites, which are generally isolated features, olivine-gabbro dykes occur in swarms of many individuals (Fig. 12). Their azimuth is restricted to a NW-SE sector, roughly parallel to the sheeted dyke complex, and their dip is sub-vertical (Fig. 13)

To sum up, there is progressive compositional evolution of the melt migration structures from dunite to troctolites, gabbros and gabbronorites from the centre to the border of the diapir, which is correlated to an evolution in their texture, abundance and orientation. This zoning reflects the thermal structure of the diapir and of the associated melt plumbing system at the time the spreading stopped. Within the diapir, at "Moho" pressures (~0.2 GPa), the troctolites crystallized in porous flow channels at temperatures close to 1200°C assuming a MORB-like primitive olivine tholeiite parental melt. Although not quantifiable, the host mantle temperature was certainly very high also as attested by the static recrystallization textures described above. At the periphery of the diapir, clinopyroxene joined olivine and plagioclase as a cumulus phase pointing to melt temperatures of $<\sim$1150–1190°C depending on the melt composition (*e.g.* Grove *et al.*, 1992; Yang *et al.*, 1996). At this stage, the melt was channelled into dykes and crystallized as olivine-gabbros. Cross-cutting relationships between troctolitic porous flow channels and gabbroic veinlets are not common but can be observed in the transition between the centre and the periphery of the diapir; they confirm that the injection of gabbro dykes post-date the crystallization of troctolite in the dunitic porous flow channels, as expected (Fig. 12).

This case study shows that the transition between the porous flow and the brittle dyking mode of melt migration occurs, at least in the oceanic spreading centre context,

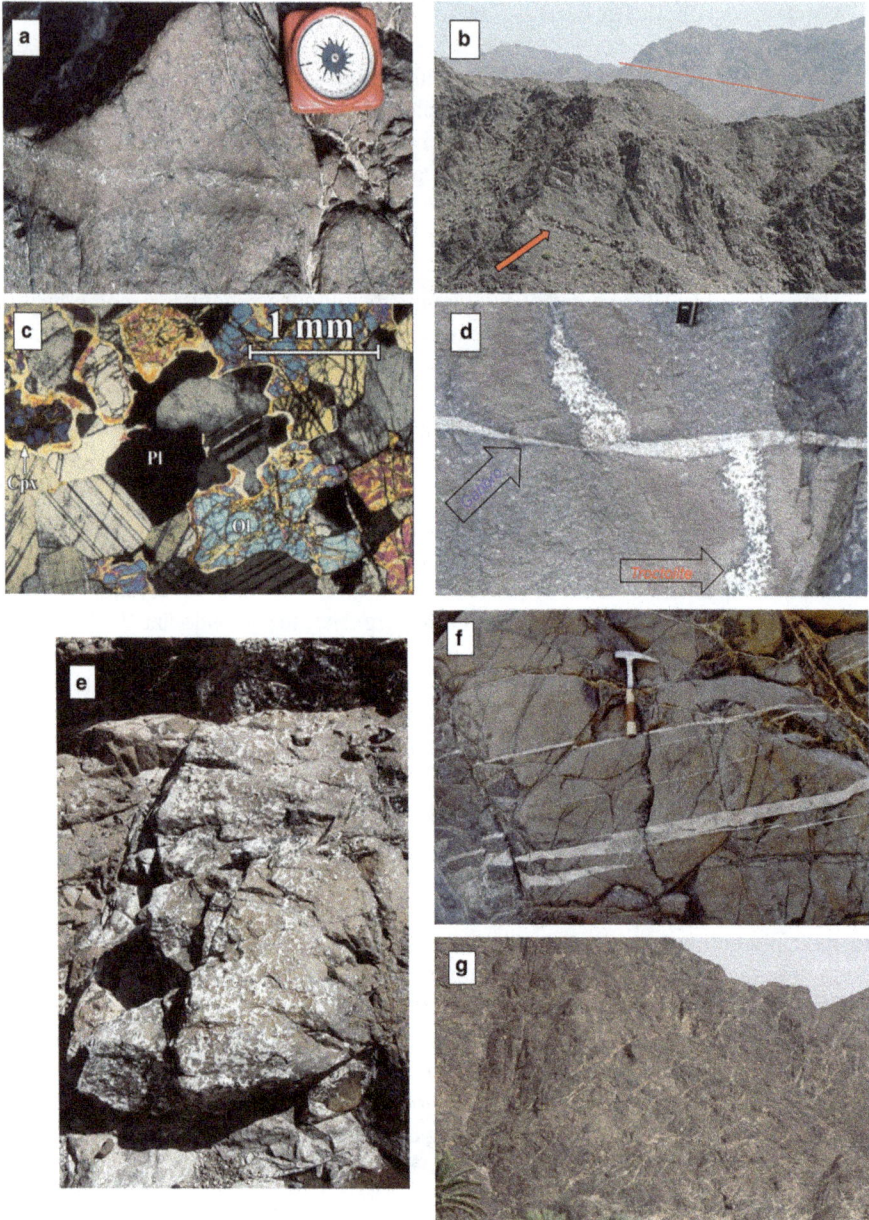

Fig. 12. Porous flow/dyking transition in the Maqsad mantle diapir, Oman ophiolite. (*a*) Typical porous flow channel from the inner part of the diapir: the troctolitic assemblage in its centre marks the location of the melt at the time of crystallization; screens of dunites separate the troctolite from the host harzburgite and attest to infiltration of a melt that induced the melting out of orthopyroxene. (*b*) Troctolite horizon ~1 m thick (red arrow), sub-parallel to the palaeo-Moho (red line) in the harzburgites from the inner part of the diapir, possible footprint of compaction waves. (*c*) Thin section of the troctolite of part b showing evidence of static

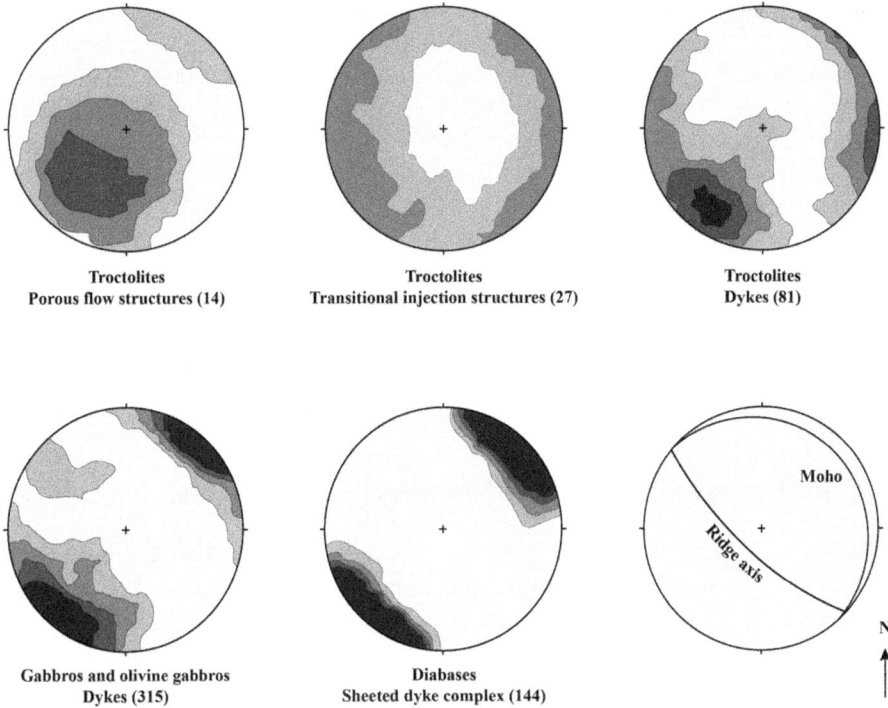

Fig. 13. Orientation of the melt migration structures within and around the Maqsad diapir. The orientation of the palaeo-Moho was measured in the field and the orientation of the palaeo-ridge axis from the mean orientation of the diabase dykes in the sheeted dyke complex. The stereonets are lower-hemisphere equal-area projection with contours at 1, 2, 3 and >4%. The number of orientation measurements is given in parentheses. For explanations, see the text. Redrawn from Ceuleneer *et al.* (1996) originally published in the journal *Nature*.

in a narrow temperature interval bracketed by the crystallization of troctolites, around 1200°C. This temperature is classically attributed by geophysicists to the base of the thermal lithosphere (Parker and Oldenburg, 1973). Therefore, we conclude that porous flow is the mode of melt migration in the asthenosphere and that the transition to dyking occurs at the base of the lithosphere. Consistently, the orientation of the melt

Fig. 12. *Continued.* recrystallization and pyroxene coronas issued from sub-solidus re-equilibration between plagioclase and olivine. (*d*) Rare intersection relationships at the periphery of the diapir between a troctolitic channel and an olivine-gabbro dyke; note the presence of dunitic reactional margins around troctolite and their absence around the olivine-gabbro dyke. The troctolite crystallized in the temperature range of 1190–1210°C and the olivine-gabbro slightly below 1190°C, which allows constraint of the temperature of the transition between the porous flow and brittle modes of melt migration at shallow depth beneath oceanic spreading centres. (*e*) Transitional structure between a porous flow channel and a dyke, filled with cpx-troctolite. The structure is ~40 cm wide in the foreground. (*f*) Olivine-gabbro dykes fracturing the harzburgite at the periphery of the diapir. (*g*) Swarm of close-spaced olivine-gabbro dykes in the same area.

migration structures presents a clear evolution according to their composition and contact relationships with the host mantle: the porous flow structures are shallow dipping, the troctolitic proto-dykes become vertical but have no preferential azimuth, the troctolite and olivine-gabbro dykes are vertical and their azimuth parallels, on average, that of the sheeted dykes complex, *i.e.* the azimuth of the former spreading ridge. This evolution reflects the increasing influence of the lithospheric stress field on the geometry of the melt plumbing system as temperature decreases.

In conclusion, the fact that dykes are the most abundant macroscopic melt migration features observed in the mantle section of ophiolites cannot be used as an evidence for dyking in the asthenosphere. It simply results from the fact that a large proportion of the volume of cumulates crystallizing from a MORB at low pressure occurs along the three phases (olivine, plagioclase, clinopyroxene) cotectic, when the melt is already focused into dykes, the volume of melt crystallizing in porous flow structures being much less.

3. Influence of lithological heterogeneities on the melt trajectory

Melt migration in the asthenosphere is governed by the laws of two-phase flow involving a ductile solid (peridotite matrix) and an interstitial silicate melt. The non-linear dependence of the bulk viscosity of this medium on the melt/rock ratio induces a positive feedback loop that will enhance any local porosity variation and hence focus and maintain the melt in porous flow channels (Scott and Stevenson, 1986; Khodakovskii *et al.*, 1995).

Field observations and numerical simulations show that the formation of these channels is triggered by infiltration melting (Aharonov *et al.*, 1995; Morgan and Liang, 2005; Rabinowicz and Ceuleneer, 2005; Pec *et al.*, 2017) resulting from the disequilibrium between the liquid and the solid phases (mostly orthopyroxene), a process that will significantly increase the porosity and the permeability. The possible origins of the disequilibrium leading to the formation of dunitic channels (the most common among the melt channels) will be discussed in the next section.

Numerical models and laboratory experiments show that the trajectory of the interstitial melt is determined by the deviatoric stress field associated with the convective flow and compaction of the solid matrix (*e.g.* Scott and Stevenson, 1986; Spiegelman and McKenzie, 1987; Rabinowicz *et al.*, 1987; Liang *et al.*, 2010; Baltzel *et al.*, 2015; Rees Jones *et al.*, 2021). Therefore, the idea that the geometry of the melt plumbing system can be predicted, governed solely by the stress field "and not by any pattern inherent in the host rock" (Holtzman *et al.*, 2003), became a common view. In contrast, the observation of melt migration structures has contributed to the view that the melt trajectory in the mantle is strongly influenced by such a pattern.

In most situations, the melt migration features we observe in the field are only small segments of a much larger melt migration network. It is exceptional to have the opportunity to follow the former melt path on distances exceeding a few meters and to observe branching relationships allowing us to better figure out the general architecture of the melt plumbing system. In this context, the discovery of a "Rosetta stone" such as

Fig. 14. Outcrop in the Trinity peridotite showing the impact of lithological heterogeneities (here pyroxenite layering) on the melt trajectories. For explanations, see the text. Redrawn from a paper by Ceuleneer *et al.* (2022), published in the journal *Geology*.

the outcrop shown in Fig. 14, where nicely preserved structures are exposed and speak by themselves, is quite exceptional.

This outcrop is located in the mantle section of the Trinity ophiolite, where the interpretation of the structures is made easier thanks to the lack of regional metamorphism and deformation and to the global preservation of the bottom-to-top polarity (Quick, 1981a; le Sueur *et al.*, 1984). Here is the message delivered by this outcrop (Ceuleneer *et al.*, 2022).

A migrating melt in disequilibrium with the lherzolite (rock with a rugged surface in Fig. 14) has locally dissolved the pyroxenes leaving a dunitic residue (rock with a smooth surface). Dunite distribution reveals the melt trajectory assuming that the general tendency of the buoyant melt was to find its way to the surface. In the lower part of the outcrop, we see that the liquid pooled in rounded structures, which may result from the self-organization of the interstitial melt in a compacting matrix (*e.g.* Rabinowicz and Ceuleneer, 2005). A thin dunitic channel highly oblique to the horizontal ($\sim$45°) rooted in one of these pockets and drained the melt upward. The channel orientation is consistent with the stress field induced by weak sinistral shearing experienced by the lherzolite there (le Sueur *et al.*, 1984; see also Fig. 2b). The orientation of the dunitic vein changes suddenly where it reaches a sub-horizontal pyroxenite layer, becoming parallel to this layer. This behaviour is consistent with the abrupt permeability increase caused by the dissolution of the pyroxenite layer, consisting entirely of crystals soluble in the migrating melt.

Generally, models consider a homogeneous distribution of pyroxenes in the host peridotite. Variations in the pyroxene content of the host rock have been taken into account in some simulations (Schimenz *et al.*, 2011) but they do not concern such abrupt contrasts as in the case of pyroxenite layering. As a consequence, the orientation of the high-permeability channels predicted by the models is still governed by the stress field.

The outcrop shown in Fig. 14 reveals that, in a natural situation, the melt has favoured a trajectory following the high-permeability channel it has itself generated, *i.e.* the former pyroxenite layer, as long as the disequilibrium between the melt and the pyroxenes has not been neutralized by the reaction. The outcrop reveals another interesting feature: the melt has progressed towards a single direction (to the right) which is consistent with the pressure gradient induced by the sinistral shearing. Accordingly, the stress field related to the plastic deformation did not lose control completely on the melt trajectory.

This unique preservation of a snapshot of a transient situation allowed us to understand the origin of more diffuse, partly resorbed, pyroxenite layers that are abundant in the thick tabular dunites of Trinity. They are relics of the process described above, although frozen at a later, more advanced stage of development (Ceuleneer *et al.*, 2022).

Thanks to this outcrop, we have learned that melt trajectory in the mantle is governed by the trade-off between the imposed stress field and the petrological processes. The result is the development of a network of interweaved channels. The thick horizon of tabular dunite commonly observed in ophiolites (*e.g.* Quick, 1981b) may result from the dissolution of closely spaced pyroxenite layers that are common features in mantle peridotite (*e.g.* Le Roux *et al.*, 2016; Tilhac *et al.*, 2016). One of the consequences of this double control on the melt trajectory is that the topology of the plumbing system is hard to predict, as the nature, shape, orientation and distribution of the lithological heterogeneities are inherited from a long petrological and structural history.

4. Initiation of dunitic porous flow channels: the concept of 'pioneer melts'

Returning to the petrological origin of the dunitic channels, the simplest hypothesis is that the melt responsible for the incongruent melting of pyroxenes is the melt produced in the melting region and feeding the crustal section, *i.e.* tholeiites, andesite, boninites or other magma types, depending on the tectono-magmatic setting of formation. The minerals making up the dunite (olivine, chrome spinel $\pm$ minor interstitial pyroxene and plagioclase) are broadly in chemical equilibrium with such common melts (*e.g.* Kelemen *et al.*, 1995; Suhr *et al.*, 2003; Ceuleneer and le Sueur, 2008; Abily and Ceuleneer, 2013).

Initially, a partial melt is in equilibrium with its mantle source. When the porosity exceeds a permeability threshold, which is achieved for a melt amount that does not exceed a few percent (*e.g.* Zhu *et al.*, 2011), the melt will migrate upwards at a faster rate than the upwelling rate of its host solid mantle. Due to the pressure dependence of its multiple saturation composition, the melt becomes progressively out of equilibrium with the peridotite assemblage from which it was issued. In the case of primitive MORB (olivine tholeiites) the pressure dependence concerns essentially silica and magnesium: the lower the pressure, the greater are the SiO_2 and the lesser are the MgO concentrations in the equilibrium, multiply saturated melt (Stolper, 1980). To reach a new equilibrium, the melt will permanently readjust its composition through the incongruent melting of orthopyroxene associated with the precipitation of olivine, a process that will increase the SiO_2 concentration and reduce the MgO content of the melt and leave a solid residue made of dunite.

This hypothesis of 'self-digestion' of the mantle by its own melt is currently the favoured paradigm for the formation of dunite channels. It is a simple and elegant mechanism but recent data on mineral inclusions in chromite grains scattered in the dunites (Rospabé *et al.*, 2017, 2021; Ceuleneer *et al.*, 2022) and bulk-rock ultra-trace elements data on dunite (Rospabé *et al.*, 2018) show that the natural rocks record a more complex history, at least in the case of dunites that formed in the shallowest part of the mantle as those exposed in ophiolites.

Pyroxene has significant contents in various minor and trace elements that will be released during the incongruent melting reaction. Among these elements is chromium, the solubility of which in silicate melts is quite small (Roeder and Reynolds, 1991) and it is not well accommodated by the crystallographic network of olivine. Consequently, as soon as liberated, it will crystallize as chromite on, or close to, the site of the reaction. These neo-formed chromite grains can encapsulate tiny minerals as microliths present in the melt, and/or the melt itself, during their growth and preserve these inclusions from further modification at the contact with later percolating melts. These inclusions are thus privileged witnesses of the nature of the corrosive melt that acted at the earliest stage of formation of the dunite channels.

An exhaustive survey of these mineral inclusions in chromite from the dunites of the Oman ophiolite, in a MORB magmatic context has been performed (Rospabé *et al.*, 2017, 2018, 2019a,b, 2020, 2021), and from the Trinity ophiolite, in an andesitic-boninitic context (Ceuleneer *et al.*, 2022). Surprisingly, in both cases, the minerals included contrast significantly in nature and relative abundance with interstitial ones present in the dunitic matrix. In Oman, the minerals in inclusions are, excluding low-temperature alteration products such as serpentine and chlorite, by decreasing order of abundance, pargasitic amphibole, Mg-Fe clinopyroxene, pure magnesian diopside, micas (both sodic and potassic), olivine, orthopyroxene and much less frequently garnet, plagioclase, jadeite, nepheline, albite, chlorapatite and pectolite (Fig. 15). In Trinity, minerals included in chromite are, by decreasing abundance, pargasitic amphibole, potassic and sodic micas, clinopyroxene and pure diopside, orthopyroxene, olivine and plagioclase.

These minerals assemblages, however they were entrapped as solid inclusions, crystallized from entrapped melt, or a mixture of melt and microcrystals (*cf.* discussion by Rospabé *et al.*, 2021), cannot be considered as common crystallization products of the dominant melt that percolated through the mantle. Those authors pointed out the existence of a fluid rich in water, silica and alkali elements. The common presence of micas is

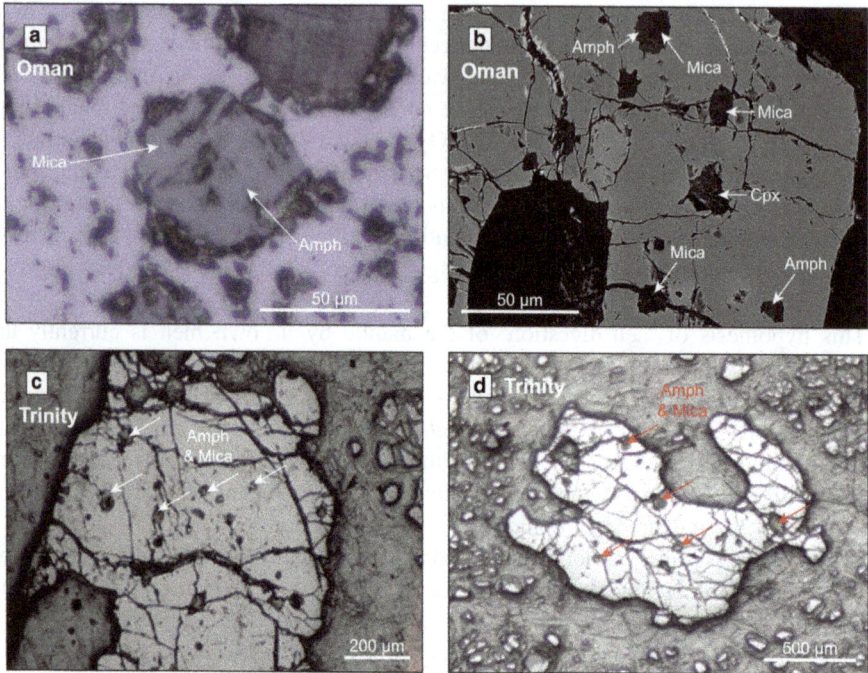

Fig. 15. Mineral inclusions in scattered chrome-spinel grains in the dunite of Oman (*a,b*) and Trinity (*c,d*). Part a is modified from Rospabé *et al.* (2020) and part b from Rospabé *et al.* (2021). Parts c and d, unpublished, are associated with the publication by Ceuleneer *et al.* (2022).

probably the most puzzling observation as these minerals are totally absent from the interstitial minerals in the host dunite and peridotite, as well as in crustal cumulates in both ophiolites. Moreover, the crystallization of potassic micas implies a K-rich source that cannot be the MORB melt or the high-Mg andesite melt, which are both strongly depleted in this element. To solve this paradox, we have proposed that the dunites are generated through the action of an "exotic agent" issued from the hybridization between a common melt and a hydrous fluid rich in alkali elements (Rospabé *et al.*, 2017).

The existence of mineral inclusions in chrome spinel pointing to the involvement of a hydrous fluid during their crystallization was well documented by a wealth of previous studies (see review by Gonzalez-Jimenez *et al.*, 2014). However, that study was considered to be specific to the chromite ore bodies while our survey has revealed that the chrome spinel scattered in reactive dunites contained the same sort, diversity and abundance of included minerals, with a significantly more variable chemical composition, and thus probably shared a common igneous origin with chromitites (Rospabé *et al.*, 2021).

For the same melt/rock ratio, a dunite has a greater permeability than a peridotite due to differences in wetting properties of olivine and pyroxenes (Toramaru and Fujii, 1986). Once formed, a dunite becomes a preferential channel for melt migration in the mantle. Accordingly, the melt forming the dunite can be considered a "pioneer melt" (Ceuleneer *et al.*, 2022) opening the gates to the eventual migration of the volumetrically dominant, basaltic or andesitic magmas formed in the deeper partial melting regions. Consequently, the mineral composition of dunite will be permanently readjusted during this long history of melt percolation (*i.e.* interstitial minerals crystallize or re-equilibrate with the last migrating melt fraction), although inclusions protected in chrome spinel will maintain the memory of the initial stage of dunite genesis, hence the difficulty in determining their origin.

The pioneer melt may be considered a 'universal solvent' that can be produced in various settings, based on the similar nature and composition of inclusions encapsulated in chrome spinel in contrasting Oman and Trinity ophiolites, as discussed above. The condition needed for its formation is an environment favourable to the hybridization between a mafic melt, whatever its precise composition, and a hydrous fluid or a water-rich silicic melt rich in alkali elements, whatever its origin. The concept of pioneer melts allows us to explain the ubiquitous occurrence of dunite, the initial formation of which would not be related to the composition and thermodynamic properties of a specific magma type and thus to a specific tectonic setting. As a matter of fact, dunites are common in many tectonic settings, including mid-ocean ridges, subduction zones and intra-plate ocean island setting.

The origin of alkali-rich hydrous fluid/melt could be either deep or shallow. For example, it can be liberated at the very early stages of decompression melting in a mantle upwelling, and eventually mix with the melt fractions issued from different degrees of melting, forming a network of anastomosing dunite channels preceding the progress of the upwelling (Fig. 16a). At the other extremity of the spectrum, the origin of the fluids may be hydrothermal and the hybridization occurs at Moho level (Fig. 16b).

5. The Moho as a major reactive filter between the depth and the surface

We have seen that dunites are major actors of the genesis of mafic magmas. They play a double role as their formation impacts both the composition and the migration of the partial melts. In ophiolites, dunites are particularly abundant at Moho level where they form a spectacular geological body, the so-called "Dunitic Transition Zone" or "DTZ" (Abily and Ceuleneer, 2013), at the interface between the mantle peridotites and the layered cumulates from the lower crustal section (Fig. 17a).

The DTZ thickness is extremely variable from one ophiolite to the others and along a single ophiolite, ranging from a few meters to > 1 km (*e.g.* Moores and Vine, 1971; Greenbaum, 1972; Ceuleneer, 1991; Boudier and Nicolas, 1995; Akizawa and Arai, 2009; Marchesi *et al.*, 2009; Negishi *et al.*, 2013; Rospabé, 2018). In the Oman ophiolite, these variations have been related to mantle diapirism. A thick DTZ develops at the top of diapirs soaked with melt and impinging the axial lithospheric lid (Rabinowicz *et al.*, 1987; Boudier and Nicolas, 1995). A thinner DTZ results either from tectonic thinning during its transposition off-axis or from lesser magma supply away from the centre of the diapirs (Ceuleneer, 1991; Jousselin *et al.*, 2000).

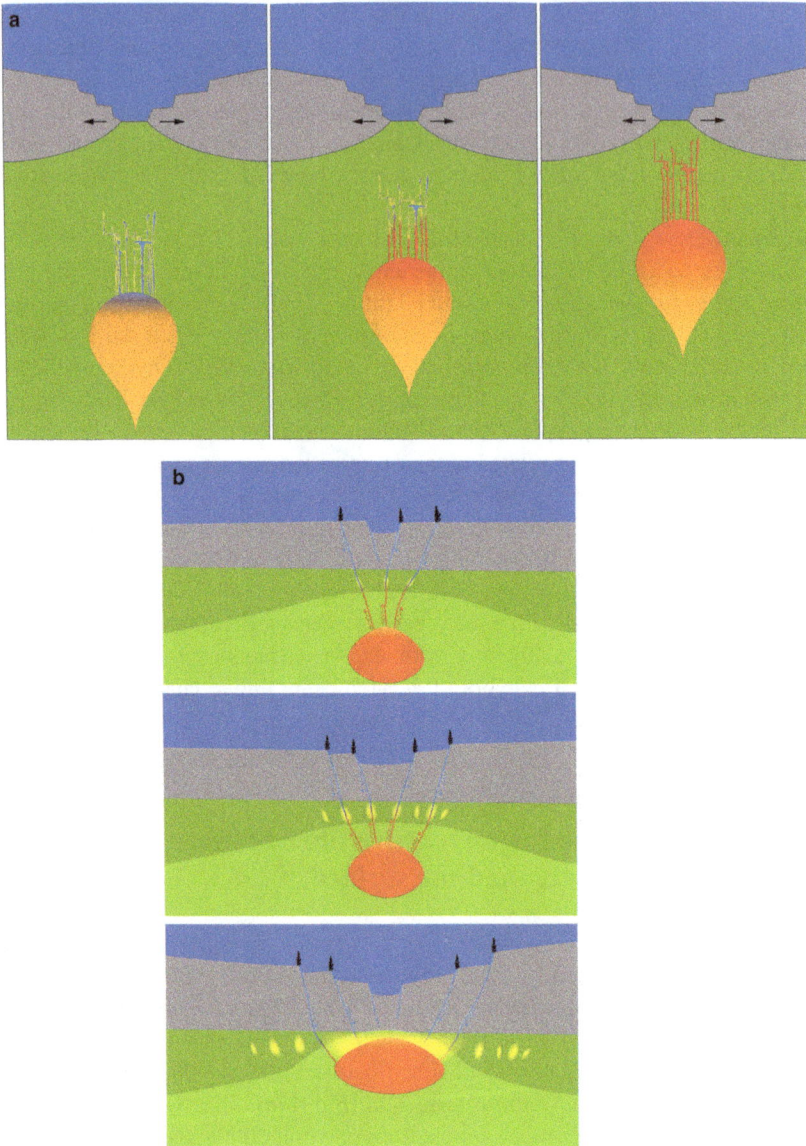

Fig. 16. Sketches illustrating the two possible origins of the hybrid 'pioneer melt': (*a*) mafic melts mix with hydrous fluids or melts saturated in water issued from very low-degree melting at the top of a rising diapir; (*b*) mafic melts mix with hydrothermal fluids reaching the Moho level along syn-magmatic faults.

5.1. Dry origin of the Dunitic Transition Zone

Initially, the genesis of the DTZ was envisioned in a dry igneous system and the debate concerning their origin was focused on two competing hypotheses (Nicolas and

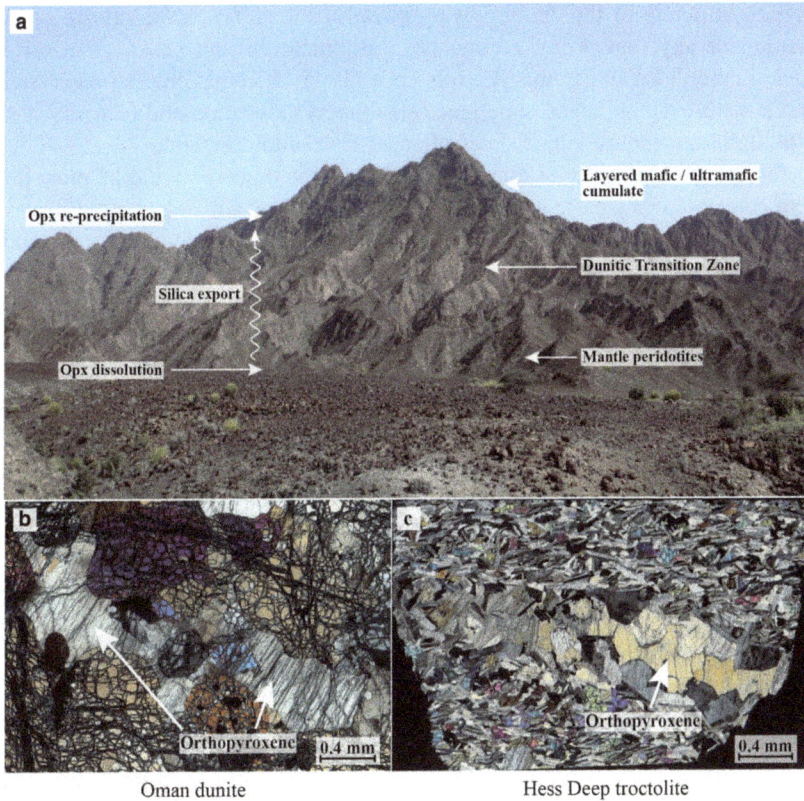

Fig. 17. (*a*) The dunitic transition zone (DTZ) in the Oman ophiolite. The altitude difference between the foothill and the summits is ~400 m. (*b*) Interstitial orthopyroxene in dunite at the top of the DTZ (reproduced from a paper by Rospabé *et al.*, 2017, published in the journal *Geology*). (*c*) Orthopyroxene with an identical texture in the lower troctolites form the East-Pacific Rise (Hess Deep Area). Picture taken onboard the JOIDES Resolution during IODP-Expedition 345 (Gillis *et al.* 2014).

Prinzhofer, 1983): (1) the DTZ is a thick pile of ultramafic cumulates composed essentially of olivine (±minor chromite, pyroxene and/or plagioclase) and lining the bottom of magma chambers; (2) the DTZ is former mantle harzburgites that lost their orthopyroxene due to reaction with a mafic magma undersaturated in this mineral.

According to the first hypothesis, the mantle must be able to produce, by decompression melting, a highly magnesian (MgO > 15%), picritic melt that calls for high potential mantle temperatures, equal or exceeding 1450°C (O'Hara, 1965; Elthon, 1979); the massive crystallization of olivine before eruption, including the formation of the DTZ, would explain why such melts never reach the surface. According to the second hypothesis, there is no need to invoke a picritic parent melt, primary MORB could be close in composition to the most magnesian glasses sampled along present-day mid-ocean ridges (~10% MgO) (Presnall *et al.*, 1979) and the mantle potential temperature away from hot spots would not exceed ~1300°C (McKenzie and Bickle, 1988).

The microstructure of the dunite is very useful for identifying the imprint of plastic deformation, if any, but is not a conclusive argument for the petrological origin of dunite, as a cumulate can eventually experience plastic deformation after its crystallization and a mantle tectonite can experience static recrystallization and recovery if maintained at high temperature after the cessation of plastic deformation.

In theory, the composition of minerals making up the dunites should be more helpful but these compositions are usually quite variable and present a marked scatter in classical diagrams designed to decipher between partial melting and fractional crystallization so that, here again, it is difficult to conclude (Fig. 18a).

A petrological profile through a 330 m-thick DTZ in the Oman ophiolite has revealed that the mineral composition data deliver a much clearer message when obtained on samples well localized vertically along a cross section than on samples collected randomly. To illustrate this by a simple example inspired by the study of Abily and Ceuleneer (2013) (Fig. 18b), the forsterite content in olivine (Fo) increases from ~88 at the base of the DTZ to ~91 about 50 m above. Along the same interval, the NiO content of olivine decreases from ~0.35 to ~0.20. This 'spatial' negative correlation allows us to exclude a cumulative origin for this horizon but is consistent with the reaction leading to the incongruent melting of opx. As a matter of fact, opx has a slightly higher #Mg than an olivine with which it is in equilibrium but the Ni content of opx is much lower than that of olivine.

Olivine in a dunite issued from infiltration melting of a peridotite has three possible origins: (1) pristine olivine from the residual mantle; (2) cumulate olivine product of the direct crystallization from the infiltrating melt; and (3) olivine neo-formed during the incongruent melting of opx. If the infiltrating melt is a MORB, residual mantle olivine (case 1) and early cumulate olivine from the primitive interstitial melt (case 2) will share similar ranges of Fo and NiO contents. However, the neo-formed olivine after opx dissolution (case 3) will have larger Fo and smaller NiO contents.

Due to efficient diffusion of bivalent ions in olivine at high temperature and in the interstitial melt, the olivine composition will be rapidly homogenized at a scale largely exceeding the grain size. As a matter of fact, the olivine composition determined in a thin section of dunite is homogeneous and theoretically corresponds to the averaged contribution of the olivine of different origins. In this frame, the evolution of the dunite mineral composition in the 50-thick basal section of the DTZ records an increasing contribution of neo-formed olivine from bottom to top.

At the top of the DTZ, the olivine composition in a 50 m-thick upper layer shows a reverse trend: the classical decrease in Fo positively correlated to a decrease in the NiO content predicted for fractional crystallization from a primitive MORB. In between, the dunite shows globally constant Fo and NiO on a thickness of ~200 m. This allows us to propose that both processes, fractional crystallization and melt-rock reaction concur to form the DTZ, but not at the same level. The upper layer can be considered as a ~50 m pile of ultramafic cumulate crystallized from a primitive melt before the melt is fractionated sufficiently to reach the crystallization field of gabbros. The basal layer fossilized a snapshot of the transformation of mantle peridotite into dunite. The 200 m-thick dunite sandwiched by the basal and top layers was likely a zone of intense melt

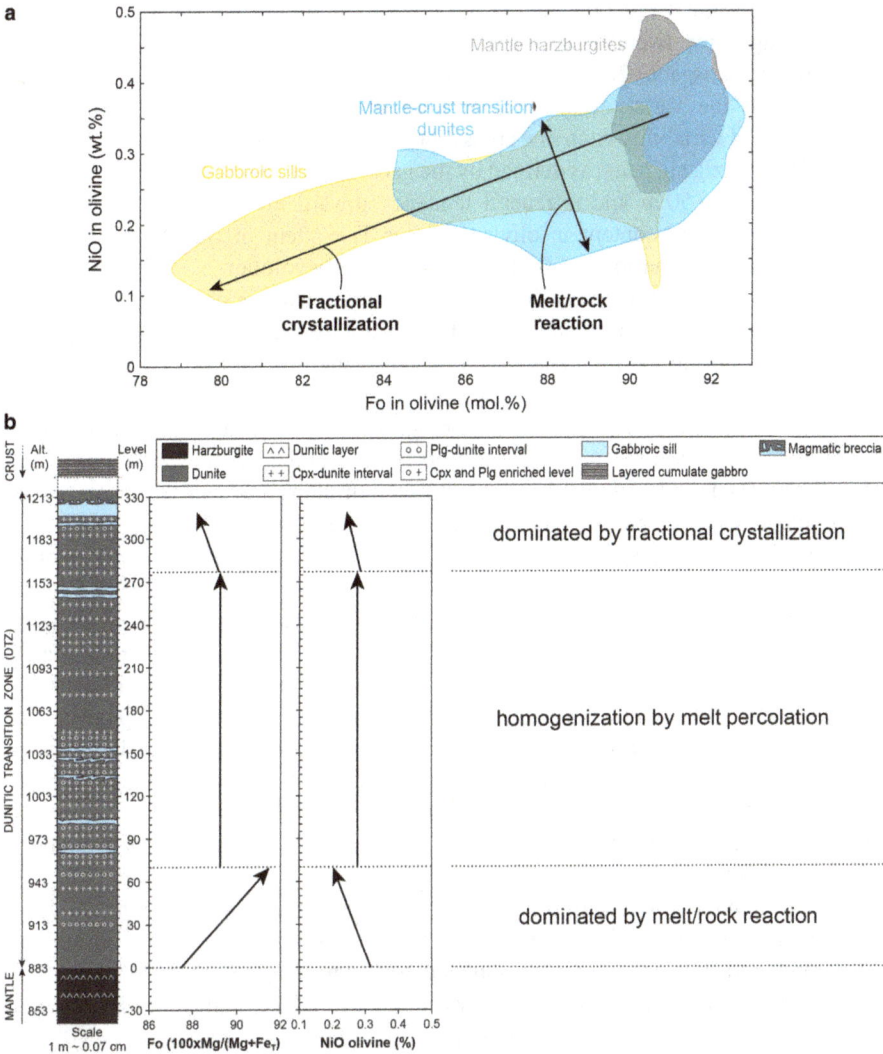

Fig. 18. Olivine composition in the Maqsad DTZ showing how the relationships between the Fo and the Ni variations along the vertical bring simple arguments to decipher between melt–rock reaction and fractional crystallization. For explanations, see the text and the work by Abily and Ceuleneer (2013).

percolation where the olivine composition was buffered and in equilibrium with moderately evolved MORB (see Abily and Ceuleneer, 2013 for further arguments).

5.2. Wet origin of the Dunitic Transition Zone

The "dry" scenario proposed by Abily and Ceuleneer (2013) was inspired by the petrological evolution along a single cross section. With the aim of testing its general validity, at least for the MORB igneous context of the Maqsad area of Oman, we extended the exercise along some twenty additional cross sections at different distances from the Maqsad diapir (Rospabé, 2018; Rospabé *et al.*, 2017, 2018, 2019a,b, 2021). The conclusion is that the variation patterns of both mineral and bulk-rock compositions present a marked variability from one cross section to the others. There are some invariant characteristics as zigzag patterns defined by the geochemical indexes with a characteristic vertical interval of 50 m and a general tendency toward more evolved compositions upward (*i.e.* lower Ni content in olivine, higher Ti content in spinel and pyroxene), but in the detail, a scenario specific to the genesis and evolution of each cross section could be proposed. The ultimate cause of this geochemical variability is to be looked for in the complexity, in space and time, of the coupling between physical and petrological processes during the final stages of transfer of the magma from the mantle to the crust.

Although it has some exceptions, another general conclusion of our survey is that the presence of interstitial minerals between olivine grains (plagioclase, clinopyroxene, orthopyroxene, amphibole) increases up-section along most of the cross-sections we sampled, with a greater abundance of phases that crystallized from melts richer in silica (orthopyroxene) and water (amphibole) approaching the top of the DTZ.

Moreover, among the characteristics shared by all cross sections is the widespread occurrence of mineral inclusions in chromite grains that led us to conclude that the initial stages of dunite formation should be attributed to an exotic melt different from the MORB that migrated in the diapir and fed the crustal section. This 'pioneer melt' evoked above has the characteristics of a hybrid melt between mafic silicate melts and hydrous, silica and alkali-rich fluids. There is a body of corroborating evidence, mineralogical, geochemical and structural, for an origin of these fluids '*per descensum*'. Consequently, a scenario for the formation of the DTZ must welcome a new component: hydrothermal fluids.

In Oman, geological and petrological observations provided evidence of the early, synmagmatic development of faults. Figure 19 shows an example of a ductile normal fault rooting and refracting in a partly crystallized troctolite intrusion in the Maqsad DTZ. Plastic deformation is intense in the steeply dipping part of the fault and evolves to magmatic deformation in the shallow-dipping part of the fault.

Such faults both disturbed the surrounding crystallizing molten rocks and permitted the deep introduction of seawater-derived fluids (Abily *et al.*, 2011; Rospabé, 2018; Rosapbé *et al.*, 2019a; Sauter *et al.*, 2021; Abily *et al.*, 2022). A compelling example displays adjacent gabbroic blocks with variably dipping and folded layers, showing sutured contacts (*i.e.* reassembled under magmatic conditions), associated with recrystallized pegmatitic gabbros and anorthite content in plagioclase buffered to high values indicating a (re)crystallization under hydrous conditions (Abily *et al.*, 2011). Syn-magmatic faults deformed the crystallizing gabbros, influenced the crystallization paths and chemical compositions by introducing hydrothermal fluids along fault planes, and are responsible for the local copper mineralization at lower-greenschist-facies conditions following the magmatic activity.

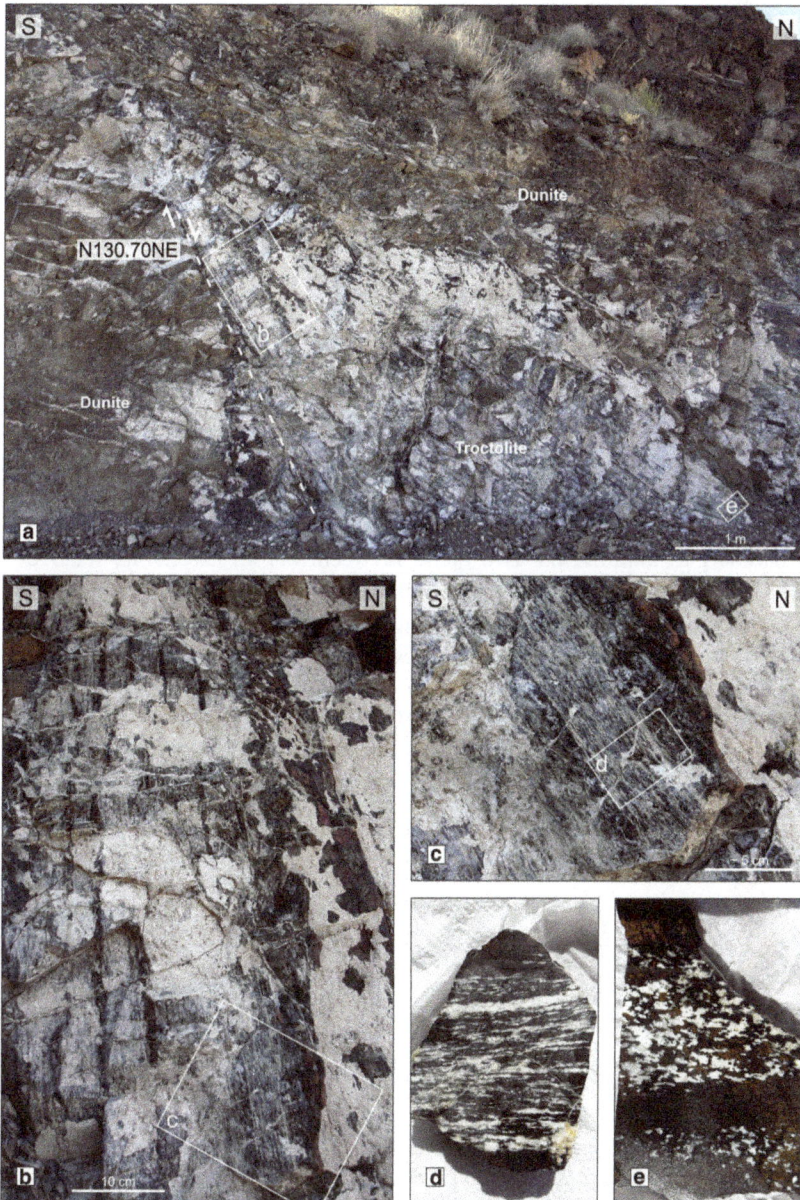

Fig. 19. Ductile to syn-magmatic fault in the Maqsad DTZ at the contact between the dunites and a troctolitic intrusion (modified after Rospabé, 2018). (*a*) General view of the outcrop: the fault strikes N130°E, parallel to the former ridge direction; the shear sense is clearly normal. (*b,c*) Details of the gabbros deformed plastically. (*d,e*) Hand specimens showing the contrast between the flaser texture of the troctolite from the steeply dipping part of the fault and the fluidal texture (with euhedral plagioclase laths) of the basal, weakly dipping part of the troctolite intrusion, away from the ductile fault.

In the case of the DTZ, the ~50 m characteristic distance in the zigzag pattern of the geochemical indexes appears to be controlled by faults (Fig. 20). For example, well-defined and continuous evolutions (increase or decrease) of REE are observed over tens of metres approaching the faults, including in 'pure' dunites (*i.e.* only olivine and minor chromite, thus the whole-rock REE signature is assumed to be hosted by olivine) (Fig. 20). This concerns also elements such as Ti supposedly immobile or not mobilized by low temperature alteration processes (*e.g.* Pearce and Norry, 1979). In the context of cryptic metasomatism (see above section 2.2.1), this simple observation allowed these faults to be qualified as synmagmatic that influenced heavily melt migration at the mantle-crust transition zone, which ultimately influenced the geochemical signatures of dunites. A strong increase in the hydrated imprint approaching the base of the crust is evidenced by the greater abundance, up-section, of both: (1) interstitial minerals indicative of a variably hydrous parent melt (orthopyroxene, amphibole) (Rospabé *et al.*, 2018, 2019a); and (2) exotic silicate inclusions in chromite (*e.g.* sodic nepheline, pure albite, chlorapatite, orthopyroxene, in addition to the widespread occurrence of amphibole and clinopyroxene) (Rospabé *et al.*, 2020, 2021). This mineralogy is absent deeper in the mantle section, allowing interpretation of the pioneer melt involved in the harzburgite dunitization as a blend between: (1) an ascending, relatively dry, MORB issued from the deep decompression melting within the Maqsad diapir; and (2) descending seawater-derived fluids introduced by early faulting (or silica-rich melt/ fluids issued from hydrated partial melting of country rocks thanks to the presence of such fluids).

The capacity of high temperature (> 600°C, *i.e.* above the stability field of serpentine) hydrous fluids to create dunite by dissolving the pyroxene contained in a peridotite was first demonstrated experimentally by Bowen and Tuttle (1949). This major discovery was

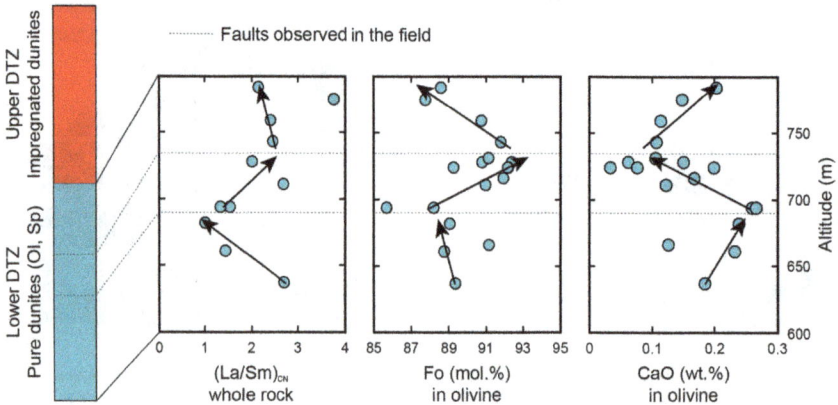

Fig. 20. Vertical chemical evolution along the lower part of the DTZ in the Maqsad area, Oman. Compositions evolve over a scale of a few tens of metres, increasing or decreasing, then shifting to the reverse trend. Shifts correlate with the location of faults observed in the field, interpreted as faults developed syn-magmatically (modified after Rospabé *et al.*, 2019a).

actually fortuitous: "We pointed out that silica was abstracted from some of our charges by water vapour and that when it was imperative that no change of composition of the charge should occur special precautions were necessary to prevent this transport of SiO_2. Thus after heating synthetic enstatite at 725°C and 22,500 lbs/in^2 pressure of water vapour for two days it was found that some of the enstatite was transformed to forsterite, when no precautions were taken against removal of silica by water vapour." (Bowen and Tuttle, 1949).

This process of silica removal assisted by water vapour perfectly reproduces what we observe in nature at a larger scale (hundreds of metres): the opx experiences incongruent melting due to percolation of hydrous fluids in the mantle peridotite as attested by inclusions in neo-formed chromite. Silica is transported upward *via* a supercritical phase (blend of water and silicate melt) and combined with olivine to form new magmatic opx in the uppermost part of the DTZ where and when the temperature is low enough (Fig. 17a). This opx contrasts in texture and composition with the residual, porphyroclastic opx from mantle peridotite: it has an interstitial habit (Fig. 17b) and is richer in TiO_2 (Rospabé *et al.* 2017). Large opx crystals sharing the same interstitial habit occur in the lower troctolitic and gabbroic primitive cumulates from the East Pacific Rise drilled at Hess Deep (Gillis *et al.*, 2014). As in the Oman DTZ they cannot crystallize from primitive MORB and are interpreted as products of an interstitial melt the origin of which is discussed, but that cannot be attributed to a subduction zone environment as currently supposed in the case of ophiolite (see Benoit *et al.*, 1999; Nonnotte *et al.*, 2005). Bowen and Tuttle's experiments were performed in a simplified, synthetic system (in this case, $MgO-SiO_2-H_2O$). Our data on natural systems show that the fluids were not only rich in water and silica but also in many incompatible elements of double origin, *e.g.* hydrothermal for potassium and magmatic for titanium.

6. Some consequences for chemical geodynamics

This paper is a contribution to an EMU volume devoted to the exploration of "New paradigms in chemical geodynamics". Geodynamics was defined as a branch of Earth sciences more than one century ago (Love, 1911). The concept of chemical geodynamics was introduced more recently with the stated goal of "(…) taking another look on the Earth's evolution, the one of the chemist" (translated from Allègre, 1980, p. 87). This cross-fertilization between geophysics and geochemistry resulted in major advances in our representation of the inner parts of the Earth.

The pioneers of chemical geodynamics were primarily isotope geochemists. As a main working hypothesis, they considered that the isotopic composition of a lava mirrors that of its deep mantle sources. This assumption was justified by the absence of measurable (at that time) radiogenic isotope fractionation of the heavy elements during partial melting and by the slowness of diffusion in the solid mantle (*e.g.* Hofmann and Hart, 1978; Zindler and Hart, 1986). The melt migration and petrological processes were treated as black boxes, which was reasonable as a first approach.

We have seen that the real world is far more complex: the composition of a melt can evolve during its long journey to the surface for various reasons including mixing of incremental melt fractions, interaction with the wall rocks and crystallization and/or resorption of mineral phases. Some of these processes have, potentially, an impact on the isotopic composition of the melt. It is thus likely that the geochemical signature of lavas or plutonic rocks is not entirely inherited from a single deep source but is acquired all along the ascent and evolution of their parent melt.

Mixing between various mantle reservoirs was envisioned in early chemical geodynamics models but not (or rarely, *e.g.* Wood, 1979; Hirschmann and Stolper, 1996) connected to the lithological nature of these reservoir (mostly peridotite *vs.* pyroxenite and eclogite) nor to the actual petrogenetic reactions themselves whose impact was underestimated, when not ignored, as stressed by O'Hara and Herzberg (2002). In fact, the message brought by the rocks exposed in ophiolites was considered inextricable and the geological observations too qualitative to be integrated in simple but global and quantitative models, which is partly true, and of local and anecdotic relevance, which is definitely untrue.

There is a general consensus that infiltration melting is a major way for a partial melt to migrate in the mantle. We have seen that the reaction triggering infiltration occurs essentially at the expense of pyroxenes, which are among the main hosts of the trace elements whose isotopic compositions are used in chemical geodynamics. Pyroxenes have two contrasting modes of occurrence in the mantle: scattered in the peridotite and concentrated in pyroxenite layers. It is well established that pyroxenite and peridotite coexisting in outcrops of mantle rocks have in most cases different isotopic signatures (*e.g.* Rampone and Hofmann, 2012; Borghini *et al.*, 2021). Various degrees of mixing between melts issued from the peridotite and the pyroxenite will contribute to the variability of the isotopic signature of lavas and of their crystallization products. A general difficulty is to decipher which part of the isotopic variability of a lava is inherited from the intrinsic heterogeneity of the peridotite and which part must be attributed to the contribution of pyroxenites, the isotopic composition of which is highly variable too (cf. Lambart *et al.*, 2019; Borghini *et al.*, 2021).

When envisioned in chemical geodynamic models, the contribution of a pyroxenite to the composition and isotopic signature of a melt is supposed to occur by mixing in the partial melting region as initially proposed by Wood (1979). A pyroxenite, being less refractory than a peridotite, can potentially be a major source of melt even present in minor amounts in the source; Lambart *et al.* (2009) estimated, on the basis of partial melting experiments, that just 5% pyroxenite layers in a mantle peridotite may account for 40% of the melt production. Accordingly, pyroxenite and peridotite might be equivalent contributors to the isotopic signature of lavas.

Field evidence presented above show that direct, decompression melting, is not the only way in which a pyroxenite may contribute to melt composition. We have seen that, due to the coupling between the petrological processes (infiltration melting) and the physical parameters (porosity and permeability), pyroxenite layers present at shallow level in the mantle may lead to deviation of the melt trajectory and become preferential 'avenues' for melt migration. As a consequence, the contribution of the pyroxenite source to the melt composition will be even greater than that predicted in the case of melting and mixing in the deep-seated regions of partial melting. In addition, it is most likely that the shallow pyroxenite layers at the base of the lithosphere and invaded by melts from below will have an isotopic signature different from those of the peridotite and pyroxenite melted at depth. The isotopic variability should, in that case, be interpreted *a minima* in terms of a three-component source, *i.e.* result from mixing of two deep heterogeneous reservoirs and a shallower one (heterogeneous also).

The network of interweaved channels induced by the competition between the stress field and infiltration melting of the pyroxenite layers or other lithological heterogeneities will considerably increase the reaction surface between the migrating melt and its solid host, enhancing the role of reactive filter of the shallow mantle.

As far as we know from field observations, the development of these reactions culminates at Moho level. There, the formation of the dunitic transition zone implies destabilization of huge amounts of pyroxenes. It will mobilize a wealth of minor and trace elements preferentially hosted by these minerals. The dissolution of pyroxenes is initiated by what we have referred to as the pioneer melt which is, at these shallow levels, a hybrid between mafic magmas and hydrous fluids of hydrothermal origin. Accordingly, the seawater reservoir must be integrated into the scenarios interpreting the isotopic signature of the igneous rocks formed at oceanic spreading centres. This was initially suggested by Benoit *et al.* (1999) and Nonnotte *et al.* (2005) in the case of the Sr isotopic signature of cumulates from the Oman ophiolite and from the Mid-Atlantic Ridge.

Acknowledgments

The authors are grateful to the organizers of the first MEREMA school in Pavia, held in February 2017, for their invitation to give a lecture and to publish this chapter. Many thanks to AnneMarie Cousin for her precious help in designing the figures. We thank the two anonymous reviewers and the editor, Kevin Murphy, for their comments and advice which helped to improve and clarify some parts of the manuscript, as well as Constanza Bonadiman and Elisabetta Rampone for editorial handling.

References

Abily, B., Ceuleneer, G. and Launeau, P. (2011) Syn-magmatic normal faulting in the lower oceanic crust: evidence from the Oman ophiolite. *Geology*, **39**, 391–394.

Abily, B. and Ceuleneer, G. (2013) The dunitic mantle-crust transition zone in the Oman ophiolite: Residue of melt-rock interaction, cumulates from high-MgO melts, or both? *Geology*, **41**, 67–70.

Abily, B., Ceuleneer, G., Rospabé, M., Kaczmarek, M.-A., Python, M., Grégoire, M., Benoit, M. and Rioux, M. (2022) Ocean crust accretion along a high-temperature detachment fault in the Oman ophiolite: a structural and petrological study of the Bahla massif. *Tectonophysics*, **822**, 229160, doi.org/10.1016/j.tecto.2021.229160.

Aharonov, E., Whitehead, J.A., Kelemen, P.B. and Spiegelman, M. (1995) Channeling instability of upwelling melt in the mantle. *Journal of Geophysical Research*, **100**, 20433–20450.

Akizawa, N. and Arai, S. (2009) Petrologic profile of peridotite layers under a possible Moho in the northern Oman ophiolite: an example from wadi Fizh. *Journal of Mineralogical and Petrological Sciences*, **104**, 389–394.

Allègre, C.J. (1980) *La géodynamique chimique*. Livre jubilaire de la Société Géologique de France, Mém. h.-s. n°10, 87–104.

Baltzell, C., Parmentier, E.M., Liang, Y. and Tirupathi, S. (2015) A high-order numerical study of reactive dissolution in an upwelling heterogeneous mantle: 2. Effect of shear deformation. *Geochemistry, Geophysics, Geosystems*, **16**, 3855–3869.

Barrat, J.-A., Zanda, B., Moynier, F., Bollinger, C., Liorzou, C. and Bayon, G. (2012) Geochemistry of CI chondrites: major and trace elements, and Cu and Zn Isotopes. *Geochimica et Cosmochimica Acta*, **83**, 79–92.

Belgrano, T.M., Diamond, L.W., Vogt, Y., Biedermann, A.R., Gilgen, S.A. and Al-Tobi, K. (2019) A revised map of volcanic units in the Oman ophiolite: insights into the architecture of an oceanic proto-arc volcanic sequence. *Solid Earth*, **10**, 1181–1217.

Benoit, M., Polvé, M. and Ceuleneer, G. (1996) Trace element and isotopic characterization of mafic cumulates in a fossil mantle diapir (Oman ophiolite). *Chemical Geology*, **134**, 199–214.

Benoit, M., Ceuleneer, G. and Polvé, M. (1999) The remelting of hydrothermally altered peridotite at mid-ocean ridges by intruding mantle diapir. *Nature*, **402**, 514–518.

Bodinier, J.-L., Vasseur, G., Vernieres, J., Dupuy, C. and Fabries, J. (1990) Mechanisms of mantle metasomatism: Geochemical evidence from the Lherz orogenic peridotite. *Journal of Petrology*, **31**, 597–628.

Borghini, G., Rampone, E., Class, C., Goldstein, S., Cai, Y., Cipriani, A., Hofmann, A.W. and Bolge, L. (2021) Enriched Hf-Nd isotopic signature of veined pyroxenite-infiltrated peridotite as a possible source for E-MORB. *Chemical Geology*, **586**, 120591.

Boudier, F. and Coleman, R.G. (1981) Cross-section through the peridotite in the Samail ophiolite, Southeastern Oman mountains. *Journal of Geophysical Research*, **86**, 2573–2592.

Boudier, F. and Nicolas, A. (1995) Nature of the Moho transition zone in the Oman ophiolite. *Journal of Petrology*, **36**, 777–796.

Bouilhol, P., Burg, J.P., Bodinier, J.L., Schmidt, M.W., Dawood, H. and Hussain, S. (2009) Magma and fluid percolation in arc to forearc mantle: evidence from Sapat (Kohistan, Northern Pakistan). *Lithos*, **107**, 17–37.

Bowen, N.L. (1928) *The Evolution of the Igneous Rocks*. Princeton University Press, Princeton, New Jersey, USA, 333 pp.

Bowen, N.L. and Tuttle, O.F. (1949) The system $MgO-SiO_2-H_2O$. *Geological Society of America Bulletin*, **60**, 439–460.

Braun, M.G. and Kelemen, P.B. (2002) Dunite distribution in the Oman ophiolite: implications for melt flux through porous dunite conduits. *Geochemistry, Geophysics, Geosystems*, **3**, 1–21.

Ceuleneer, G. (1986) *Structure des ophiolites d'Oman: flux mantellaire sous un centre d'expansion océanique et charriage à la dorsale*. PhD thesis, Nantes University, France 349 pp.

Ceuleneer, G. (1991) Evidence for a paleo-spreading center in the Oman ophiolite: mantle structures in the Maqsad area. Pp. 149–175 in: *Ophiolite Genesis and Evolution of Oceanic Lithosphere* (T.J. Peters, editor). Kluwer Academic Press, Dordrecht, The Netherlands.

Ceuleneer, G. and le Sueur, E. (2008) The Trinity ophiolite (California): the strange association of fertile mantle peridotite with ultra-depleted crustal cumulates. *Bulletin de la Société Géologique de France*, **179**, 503–516.

Ceuleneer, G., Monnereau, M. and Amri, I. (1996) Thermal structure of a fossil mantle diapir inferred from the distribution of mafic cumulates. *Nature*, **379**, 149–153.

Ceuleneer, G., Nicolas, A. and Boudier, F. (1998) Mantle flow patterns at an oceanic spreading centre: the Oman peridotites record. *Tectonophysics*, **151**, 1–26.

Ceuleneer, G., Rospabé, M., Chatelin, T., Henry, H., Tilhac, R., Kaczmarek, M.-A. and le Sueur, E.A (2022) Rosetta stone linking melt trajectories in the mantle to the stress field and lithological heterogeneities (Trinity ophiolite, California). *Geology*, **50**, 1192–1196.

Cleland, C.E. (2013) Common cause explanation and the search for a smoking gun. In: *Rethinking the Fabric of Geology* (V.R. Baker, editor). *Geological Society of America Special Papers*, **502**, https://doi.org/10.1130/2013.2502(01).

Clénet, H., Ceuleneer, G., Pinet, P., Abily, B., Daydou, Y., Harris, E., Amri, I. and Dantas, C. (2010) Thick sections of layered ultramafic cumulates in the Oman ophiolite revealed by an airborne hyperspectral survey: petrogenesis and relationship to mantle diapirism. *Lithos*, **114**, 265–281.

Collier, M.L. and Kelemen, P.B. (2010) The case for reactive crystallization at mid-ocean ridges. *Journal of Petrology*, **51**, 1913–1940.

Daly, R.A. (1914) *Igneous Rocks and their Origin*. McGraw Hill, New York, 563 pp.

Dawson, J.B. (1984) Contrasting types of upper-mantle metasomatism? *Developments in Petrology*, **11**, 289–294.

Dick, H.J.B. (1977) Evidence of partial melting in the Josephine Peridotite. Pp. 59–62 in: *Magma Genesis 1977: Proceedings of the American Geophysical Union Chapman Conference on Partial Melting in the Earth's Upper Mantle* (H.J.B. Dick, editor). State of Oregon Department of Geology and Mineral Industries, USA.

Dilek, Y. and Furnes, H. (2011) Ophiolite genesis and global tectonics: geochemical and tectonic fingerprinting of ancient oceanic lithosphere. *Geological Society of America Bulletin*, 123, 387-411.

Dygert, N., Liang, Y. and Kelemen, P.B. (2016) Formation of plagioclase lherzolite and associated dunite-harzburgite-lherzolite sequences by multiple episodes of melt percolation and melt-rock reaction: an example from the Trinity ophiolite, California, USA. *Journal of Petrology*, **57**, 815–838.

Elthon, D. (1979) High magnesia liquids as the parental magma for ocean floor basalts. *Nature*, **278**, 514–518.

Gerbert-Gaillard, L. (2002) *Caractérisation Géochimique des Péridotites de l'ophiolite d'Oman: Processus Magmatiques aux Limites Lithosphère/Asthénosphère*. PhD Thesis, Université Montpellier II - Sciences et Techniques du Languedoc, France, 238 pp.

Gillis, K.M., Snow, J.E., Klaus, A., Abe, N., de Brito Adriao, A., Akizawa, N., Ceuleneer, G., Cheadle, M., Faak, K., Falloon, T., Friedman, S., Godard, M., Guerin, G., Harigane, Y., Horst, A., Hoshide, T., Ildefonse, B., Jean, M.M., John, B.E., Koepke, J.H., Machi, S., Maeda, J., Marks, N.E., McCaig, A.M., Meyer, R., Morris, A., Nozaka, T., Python, M., Saha, A. and Wintsch, R.P. (2014) Primitive layered gabbros from fast-spreading lower oceanic crust. *Nature*, **505**, 204–207.

Girardeau, J., Monnier, C., Le Mée, L. and Quatrevaux, F. (2002) The Wuqbah peridotite, central Oman Ophiolite: petrological characteristics of the mantle in a fossil overlapping ridge setting. *Marine Geophysical Research*, **23**, 43–56.

Godard, M., Bodinier, J.L. and Vasseur, G. (1995) Effects of mineralogical reactions on trace element redistributions in mantle rocks during percolation processes: a chromatographic approach. *Earth and Planetary Science Letters*, **133**, 449–461.

Godard, M., Jousselin, D. and Bodinier, J.L. (2000) Relationships between geochemistry and structure beneath a palaeo-spreading centre: a study of the mantle section in the Oman ophiolite. *Earth and Planetary Science Letters*, **180**, 133–148.

Godard, M., Lagabrielle, Y., Alard, O. and Harvey, J. (2008) Geochemistry of the highly depleted peridotites drilled at ODP Sites 1272 and 1274 (Fifteen-Twenty Fracture Zone, Mid-Atlantic Ridge): Implications for mantle dynamics beneath a slow spreading ridge. *Earth and Planetary Science Letters*, **267**, 410–425.

Gonzalez-Jimenez, J.M., Griffin, W.L., Proenza, J.A., Gervilla, F., O'Reilly, S.Y., Akbulut, M., Pearson, N.J. and Arai, S. (2014) Chromitites in ophiolites: how, where, when, why? Part II. The crystallization of chromitites. *Lithos*, **189**, 140–158.

Greenbaum, D. (1972) Magmatic processes at ocean ridges: evidence from Troodos massif, Cyprus, *Nature*, **238**, 18–21.

Grove, T.L., Kinzler, R.J. and Bryan, W.B. (1992) Fractionation of Mid-Ocean Ridge Basalt (MORB). Pp. 281–310 in: *Mantle Flow and Melt Generation at Mid-ocean Ridges* (J.P. Morgan, D.K. Blackman and J. M. Sinton, editors). AGU Geophysical Monograph, **71**.

Gruau, G., Bernard-Griffiths, J. and Lécuyer, C. (1998) The origin of U-shaped rare earth patterns in ophiolite peridotites: assessing the role of secondary alteration and melt/rock reaction. *Geochimica et Cosmochimica Acta*, **62**, 3545–3560.

Hanghøj, K., Kelemen, P.B., Hassler, D. and Godard, M. (2010) Composition and genesis of depleted mantle peridotites from the Wadi Tayin massif, Oman ophiolite; major and trace element geochemistry, and Os isotope and PGE systematics. *Journal of Petrology*, **51**, 201–227.

Hirschmann, M.M and Stolper, E.M. (1996) A possible role for garnet pyroxenite in the origin of the "garnet signature" in MORB. *Contributions to Mineralogy and Petrology*, **124**, 185–208.

Hofmann, A.W. and Hart, S.R. (1978) Assessment of local and regional isotopic equilibrium in mantle. *Earth and Planetary Science Letters*, **38**, 44–62.

Holtzman, B.K., Groebner, N.J., Zimmerman, M.E., Ginsberg, S.B. and Kohlstedt, D.L. (2003) Stress-driven melt segregation in partially molten rocks. *Geochemistry, Geophysics, Geosystems*, **4**, 8607.

Hutton, J. (1788) Theory of the Earth. *Transactions of the Royal Society of Edinburgh*, **1**, 209–305.

Jousselin, D. and Nicolas, A. (2000) The Moho transition zone in the Oman ophiolite-relation with wehrlites in the crust and dunites in the mantle. *Marine Geophysical Research*, **21**, 229–241.

Kelemen, P.B., Dick, H.J.B. and Quick, J.E. (1992) Formation of harzburgite by pervasive melt/rock reaction in the upper mantle. *Nature*, **358**, 635–641.

Kelemen, P.B., Shimizu, N. and Salters, V.J.M. (1995) Extraction of mid-ocean-ridge basalt from the upwelling mantle by focused flow of melt in dunite channels. *Nature*, **375**, 747–753.

Khodakovskii, G., Rabinowicz, M., Ceuleneer, G. and Trubitsyn, V.P. (1995) Melt percolation in a partially molten mantle mush: effect of a variable viscosity. *Earth and Planetary Science Letters*, **134**, 267–281.

Kourim, F., Rospabé, M., Dygert, N., Chatterjee, S., Takazawa, E., Wang, K.L., Godard, M., Benoit, M., Giampouras, M., Ishii, K., Teagle, D.A.-H. and Kelemen, P. (2022) Geochemical Characterization of the Oman Crust-Mantle Transition Zone, OmanDP Holes CM1A and CM2B. *Journal of Geophysical Research Solid Earth*, **127**, e2021JB022694.

Lambart, S., Laporte, D. and Schiano, P. (2009) An experimental study of pyroxenite partiel melts at 1 and 1.5 GPa: implications for the major-element composition of Mid-Ocean Ridge Basalts. *Earth and Planetary Science Letters*, **288**, 335–347.

Lambart, S., Koornneef, J.M., Millet, M.-A., Davies, G.R., Cook, M., and Lissenberg, C.J. (2019) Highly heterogeneous depleted mantle recorded in the lower oceanic crust. *Nature Geoscience*, **12**, 482–486.s

Le Roux, V., Nielsen, S.G., Sun, C. and Yao, L. (2016) Dating layered websterite formation in the lithospheric mantle: *Earth and Planetary Science Letters*, **454**, 103–112.

Le Sueur, E., Boudier, F., Cannat, M., Ceuleneer, G. and Nicolas, A. (1984) The Trinity mafic-ultramafic complex: first results of the structural study of an untypical ophiolite. *Ofioliti*, **9**, 487–498.

Liang, Y., Schiemenz, A., Hesse, M.A., Parmentier, E.M. and Hesthaven, J.S. (2010) High-porosity channels for melt migration in the mantle: Top is the dunite and bottom is the harzburgite and lherzolite. *Geophysical Research Letters*, **37**, L15306.

Lippard, S.J., Shelton, A.W. and Gass, I.G. (1986) *The ophiolite of Northern Oman*. Geological Society of London Memoir, 11, 178 pp.

Love, A.E.H. (1911) *Some Problems of Geodynamics*. Cambridge University Press, Cambridge, UK.

Lyell, C. (1830) *Principles of Geology*, Vol. I. John Murray Ed. Albemarle Street, London, 511 pp..

Maaloe, S. (2003) Melt dynamics of a partially molten mantle with randomly oriented veins. *Journal of Petrology*, **44**, 1193–1210.

Maaloe, S. (2005) The dunite bodies, websterite and orthopyroxenite dikes of the Leka ophiolite complex, Norway. *Mineralogy and Petrology*, **85**, 163–204.

Malpas, J. (1978) Magma generation in the upper mantle, field evidence from ophiolite suites and application to the generation of oceanic lithosphere. *Philosophical Transactions of the Royal Society of London A*, **288**, 527–546.

Marchesi, C., Garrido, C.J., Godard, M., Belley, F. and Ferré, E. (2009) Migration and accumulation of ultra-depleted subduction-related melts in the Massif du Sud ophiolite (New Caledonia). *Chemical Geology*, **266**, 171–186.

McKenzie, D.P. and Bickle, M.J. (1988) The volume and composition of melt generated by extension of the lithosphere. *Journal of Petrology*, **25**, 713–765.

Monnier, C., Girardeau, J., Le Mée, L. and Polvé, M. (2006) Along-ridge petrological segmentation of the mantle in the Oman ophiolite. *Geochemistry, Geophysics, Geosystems*, **7**. https://doi.org/10.1029/2006GC001320

Moores, E.M. and Vine, J.F. (1971) The Troodos massif, Cyprus and other ophiolites as oceanic crust – evaluation and implications. *Philosophical Transactions of the Royal Society of London*, **268**, 443–467.

Morgan, Z. and Liang, Y. (2005) An experimental study of the kinetics of lherzolite reactive dissolution with application to melt channel formation. *Contributions to Mineralogy and Petrology*, **150**, 369–385.

Navon, O. and Stolper, E. (1987) Geochemical consequences of melt percolation: the upper mantle as a chromatographic column. *Journal of Geology*, **95**, 285–307.

Negishi, H., Arai, S., Yurimoto, H., Ito, S., Ishimaru, S., Tamura, A. and Akizawa, N. (2013) Sulfide-rich dunite within a thick Moho transition zone of the northern Oman ophiolite: Implications for the origin of Cyprus-type sulfide deposits. *Lithos*, **164–167**, 22–35.

Nicolas, A. (1986) A melt extraction model based on structural studies in mantle peridotites. *Journal of Petrology*, **27**, 999–1022.

Nicolas, A. and Jackson, M. (1982) High-temperature dikes in peridotites: origin by hydraulic fracturing. *Journal of Petrology*, **23**, 568–582.

Nicolas, A. and Prinzhofer, A. (1983) Cumulative or residual origin for the transition zone in ophiolites – structural evidence. *Journal of Petrology*, **24**, 188–206.

Nonnotte, P., Ceuleneer, G. and Benoit, M. (2005) Genesis of andesitic-boninitic magmas at mid-ocean ridges by melting of hydrated peridotites: geochemical evidence from DSDP Site 334 gabbronorites. *Earth and Planetary Science Letters*, **236**, 632–653.

O'Hara, M.J. (1965) Primary magmas and the origin of basalts. *Scottish Journal of Geology*, **1**, 19–40.

O'Hara, M.J. and Herzberg, C. (2002) Interpretation of trace element and isotope features of basalts: Relevance of field relations, petrology, major element data, phase equilibria, and magma chamber modeling in basalt petrogenesis. *Geochimica et Cosmochimica Acta*, **66**, 2167–2191.

Parker, R.L. and Oldenburg, D.W. (1973) Thermal model of ocean ridges. *Nature*, **242**, 137–139.

Pearce, J.A. and Norry, M.J. (1979) Petrogenetic implications of Ti, Zr, Y and Nb variations in volcanic rocks. *Contributions to Mineralogy and Petrology*, **69**, 33–47.

Pec, M., Holtzman, B.K., Zimmerman, M.E. and Kohlstedt, D.L. (2017) Reaction infiltration instabilities in mantle rocks: an experimental investigation: *Journal of Petrology*, **58**, 979–1004.

Presnall, D.C., Dixon, J.R., O'Donnell, T.H. and Dixons, S.A. (1979) The generation of mid-ocean ridge tholeiites. *Journal of Petrology*, **20**, 3–35.

Prinzhofer, A. and Allègre, C.J. (1985) Residual peridotites and the mechanisms of partial melting. *Earth and Planetary Science Letters*, **74**, 251–265.

Python, M. and Ceuleneer, G. (2003) Nature and distribution of dykes and related melt migration structures in the mantle section of the Oman ophiolite. *Geochemistry, Geophysics, Geosystems*, https://doi.org/10.1029/2002GC000354

Quick, J.E. (1981a) Petrology and petrogenesis of the Trinity peridotite, an upper mantle diapir in the eastern Klamath Mountains, northern California. *Journal of Geophysical Research*, **86**, 11837–11863.

Quick, J.E. (1981b) The origin and significance of large, tabular dunite bodies in the Trinity peridotite, northern California, Contributions to Mineralogy and Petrology, **78**, 413–422.

Rabinowicz, M. and Ceuleneer, G. (2005) The effect of sloped isotherms on melt migration in the shallow mantle: a physical and numerical model based on observations in the Oman ophiolite. *Earth and Planetary Science Letters*, **229**, 231–246.

Rabinowicz, M., Ceuleneer, G. and Nicolas, A. (1987) Melt segregation and asthenospheric flow in diapirs below spreading centers: evidence from the Oman ophiolite. *Journal of Geophysical Research*, **92**, 3475–3486.

Rabinowicz, M., Genthon, P., Ceuleneer, G. and Hillairet, M. (2001) Compaction in a mantle mush with high melt concentrations and the generation of magma chambers. *Earth and Planetary Science Letters*, **188**, 313–328.

Rampone, E. and Hofmann, A.W. (2012) A global overview of isotopic heterogeneities in the oceanic mantle. *Lithos*, **148**, 247–261.

Rees Jones, D.W., Zhang, H. and Katz, R.F. (2021) Magmatic channelization by reactive and shear-driven instabilities at mid-ocean ridges: a combined analysis. *Geophysical Journal International*, **226**, 582–609.

Roeder, P.L. and Reynolds, I. (1991) Crystallization of chromite and chromium solubility in basaltic melts. *Journal of Petrology*, **32**, 909–934.

Rospabé, M. (2018) *Etude pétrologique, géochimique et structurale de la zone de transition dunitique dans l'ophiolite d'Oman: identification des processus pétrogénétiques à l'interface manteau/croûte.* PhD thesis, Université Paul Sabatier, Toulouse III, France, 628 pp.

Rospabé, M., Ceuleneer, G., Benoit, M., Abily, B. and Pinet, P. (2017) Origin of the dunitic mantle-crust transition zone in the Oman ophiolite: The interplay between percolating magmas and high-temperature hydrous fluids. *Geology*, **45**, 471–474.

Rospabé, M., Benoit, M., Ceuleneer, G., Hodel, F. and Kaczmarek, M.-A. (2018) Extreme geochemical variability through the dunitic transition zone of the Oman ophiolite: implications for melt/fluid-rock reactions at Moho level beneath oceanic spreading centers. *Geochimica et Cosmochimica Acta*, **234**, 1–23.

Rospabé, M., Benoit, M., Ceuleneer, G., Kaczmarek, M.-A. and Hodel, F. (2019a) Melt hybridization and metasomatism triggered by syn-magmatic faults within the Oman ophiolite: a clue to understand the genesis of the dunitic mantle-crust transition zone. *Earth and Planetary Science Letters*, **516**, 108–121.

Rospabé, M., Ceuleneer, G., Granier, N., Arai, S. and Borisova, A.Y. (2019b) Multi-scale development of a stratiform chromite ore body at the base of the dunitic mantle-crust transition zone (Maqsad diapir, Oman ophiolite): the role of repeated melt and fluid influxes. *Lithos*, **350–351**, 105235.

Rospabé, M., Ceuleneer, G., Benoit, M. and Kaczmarek, M.A. (2020) Composition gradients in silicate inclusions in chromites from the dunitic mantle-crust transition (Oman ophiolite) reveal high temperature fluid-melt-rock interaction controlled by faulting. *Ofioliti*, **45**, 103–114.

Rospabé, M., Ceuleneer, G., Le Guluche, V., Benoit, M. and Kaczmarek, M.-A. (2021) The chicken and egg dilemma linking dunites and chromitites in the Mantle–Crust Transition Zone beneath oceanic spreading centres: a case study of chromite-hosted silicate inclusions in dunites formed at the top of a mantle diapir (Oman ophiolite). *Journal of Petrology*, **62**, egab026.

Rubin, A. (1995) Propagation of magma-filled cracks. *Annual Review of Earth and Planetary Sciences*, **23**, 287–336.

Sauter, D., Werner, P., Ceuleneer, G., Manatschal, G., Rospabé, M., Tugend, J., Gillard, M., Autin, J. and Ulrich, M. (2021) Sub-axial deformation in oceanic lower crust: insights from seismic reflection profiles in the Enderby Basin and comparison with the Oman ophiolite. *Earth and Planetary Science Letters*, **554**, 116698.

Schiemenz, A., Liang, Y. and Parmentier, E.M. (2011) A high-order numerical study of reactive dissolution in an upwelling heterogeneous mantle—I. Channelization, channel lithology and channel geometry: *Geophysical Journal International*, **186**, 641–664.

Scott, D.R. and Stevenson, D.J. (1986) Magma ascent by porous flow. *Journal of Geophysical Research*, **97**, 9283–9296.

Sleep, N.H. (1988) Tapping melts by veins and dykes. *Journal of Geophysical Research*, **93**, 10255–10272.

Spiegelman, M. and McKenzie, D.P. (1987) Simple 2-D models for melt extraction at mid-ocean ridges and island arcs. *Earth and Planetary Science Letters*, **83**, 137–152.

Stolper, E. (1980) A phase diagram for Mid-Ocean Ridge basalts: preliminary results and implications for petrogenesis. *Contributions to Mineralogy and Petrology*, **74**, 13–27.

Suhr, G., Hellebrand, E., Snow, J.E., Seck, H.A. and Hofmann, A.W. (2003) Significance of large, refractory dunite bodies in the upper mantle of the Bay of Islands ophiolite. *Geochemistry, Geophysics, Geosystems*, https://doi.org/10.1029/2001GC000277.

Takahashi, E. (1986) Melting of a dry peridotite KLB-1 up to 14 GPa: implications on the origin of peridotitic upper mantle. *Journal of Geophysical Research*, **91**, 9367–9382.

Takazawa, E., Okayasu, T. and Satoh, K. (2003) Geochemistry and origin of the basal lherzolites from the northern Oman ophiolite (northern Fizh block). *Geochemistry, Geophysics, Geosystems*, https://doi.org/10.1029/2001GC000232.

Tilhac, R., Ceuleneer, G., Griffin, W.L., O'Reilly, S.Y., Pearson, N.J., Benoit, M., Henry, H., Girardeau, J. and Grégoire, M. (2016) Primitive arc magmatism and delamination: petrology and geochemistry of pyroxenites from the Cabo Ortegal Complex, Spain. *Journal of Petrology*, **57**, 1921–1954.

Toramaru, A. and Fujii, N. (1986) Connectivity of melt phase in a partially molten peridotite. *Journal of Geophysical Research*, **91**, 9239–9252.

Turcotte, D.L. and Phipps Morgan, J. (1992) The physics of magma migration and mantle flow beneath a mid-ocean ridge. Pp. 155-182 in: *Mantle Flow and Melt Generation at Mid-ocean Ridges* (J.P. Morgan, D.K. Blackman and J. M. Sinton, editors). AGU Geophysical Monograph, **71**.

Varfalvy, V., Hébert, R. and Bédard, J.H. (1996) Interactions between melt and upper-mantle peridotites in the North Arm mountain massif, Bay of Islands ophiolite, Newfoundland, Canada: implications for the genesis of boninitic and related magmas. *Chemical Geology*, **129**, 71–90.

Vernières, J., Godard, M. and Bodinier, J.L. (1997) A plate model for the simulation of trace element fractionation during partial melting and magma transport in the Earth's upper mantle. *Journal of Geophysical Research Solid Earth*, **102**, 24,771–24,784.

Waff, H.S. and Holdren, G.R. (1981) The nature of grain boundaries in dunite and lherzolite xenoliths: implication for magma transport in refractory upper mantle material. *Journal of Geophysical Research*, **86**, 3677–3683.

Wager, L.R., Brown, G.M. and Wadsworth, W.J. (1960) Types of igneous cumulates. *Journal of Petrology*, **1**, 73–85.

Wood, D.A. (1979) A variably veined suboceanic upper mantle – genetic significance for mid-ocean ridge basalts from geochemical evidence. *Geology*, **7**, 499–503.

Yang, H.-J., Kinzler, R. and Grove, T.L. (1996) Experiments and models of anhydrous, basaltic olivine-plagioclase-augite saturated melts from 0.001 to 10 kbar. *Contributions to Mineralogy and Petrology*, **124**, 1–18.

Zhu, W., Gaetani, G.A., Fusseis, F., Montési, L.G.J. and De Carlo, F. (2011) Microtomography of partially molten rocks: three-dimensional melt distribution in mantle peridotite. *Science*, **332**, 88–91.

Zindler, A. and Hart, S. (1986) Chemical geodynamics. *Annual Reviews in Earth and Planetary Science*, **14**, 493–571.

The role of H_2O in the deformation and microstructural evolution of the upper mantle

KÁROLY HIDAS[1] and JOSÉ ALBERTO PADRÓN-NAVARTA[2,3]

[1]*Instituto Geológico y Minero de España (IGME), CSIC, E-18006 Granada (Granada), Spain, e-mail k.hidas@igme.es*
[2]*Instituto Andaluz de Ciencias de la Tierra (IACT), CSIC, E-18100 Armilla (Granada), Spain, e-mail alberto.padron@csic.es*
[3]*Géosciences Montpellier, Université de Montpellier & CNRS, F-34095 Montpellier cedex 5, France*

In this chapter, we evaluate how the incorporation of H_2O as a thermodynamic component influences phase relations in a peridotite composition. This component – present either in the form of hydrous minerals, aqueous fluids and hydrous melts, or as a structurally-bonded trace element at defect sites of nominally anhydrous minerals (NAMs) – may influence upper-mantle rheology in diverse ways. By presenting various natural cases, we identify key incorporation mechanisms and assess their role in the microstructural evolution of ultramafic rocks at different depths in the Earth's interior. These data suggest that the influence of either aqueous fluids or hydrous melts on rheology outmatches that of NAMs or stable hydrous phases across much of the lithospheric mantle. Consequently, future research is expected to shift towards a better understanding of the transient conditions in the lithosphere that control the availability and transport of aqueous fluids and hydrous melts. These transient conditions are likely to play a more dominant role than the sole ability of hydrous defects in NAMs – a role that is currently less well-constrained experimentally – in controlling the ductile deformation of the upper mantle.

1. Introduction: the addition of H_2O as a component to the upper mantle

Mantle deformation governs the most important large-scale geological processes, such as plate tectonics (Fig. 1), on Earth and in other terrestrial planets because the rheological properties of the mantle have a first-order influence on how this deformation is accommodated at great depth (*e.g.* Karato, 2010). It is widely accepted that the rheology of subsolidus peridotite is essentially controlled by the crystal plasticity of olivine, which is the most abundant and the weakest mineral phase in all prevalent lithologies in the upper mantle (Karato and Wu, 1993; Bürgmann and Dresen, 2008). However, despite decades of intense research, processes allowing for ductile deformation of the upper mantle remain poorly understood, yet the role of volatiles has been of interest since the earliest experimental works. The reason is that even the smallest quantities of fluids potentially weaken the rheology (see Kohlstedt and Hansen, 2015 for an overview), and – given that a completely dry and refractory upper mantle can be considered unlikely – the

DOI: 10.1180/EMU-notes.21.6

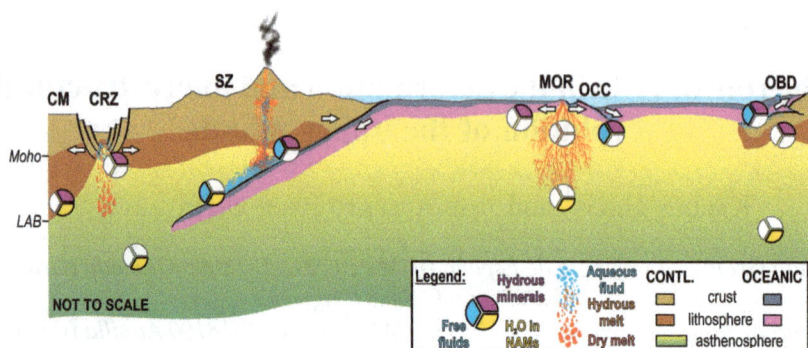

Fig. 1. Schematic cross section of some of the most important plate tectonic settings on Earth. Pie charts indicate the expected dominant forms of the H_2O component to occur in the underlying upper mantle (see text for further details); white slice = irrelevant occurrence. The distribution and provenance of aqueous fluids, hydrous or nominally dry melts along the cross section is theoretical in a given setting. In the present chapter, the actual contribution of various hydrous species to mantle rheology is discussed based on a representative geological record. The figure is not to scale and it may not illustrate the exact tectonic setting of the natural examples cited in the text. Labels of geodynamic environments – CM: cratonic mantle (deep hydrous minerals are mostly controlled by the availability of potassium); CRZ: mantle in continental rift zones; MOR: mantle beneath mid-ocean ridges; OBD: mantle rocks in obduction settings; OCC: mantle in oceanic core-complexes ('megamullions'); SZ: mantle wedge of subduction zones (hydrous minerals at greater depths than indicated are subject to the availability of potassium). Natural records of the continental mantle are typically exposed in CM, CRZ, OBD and SZ settings, while mantle rocks of the oceanic lithosphere would occur in MOR, OBD, OCC and SZ settings as members of either subduction-unrelated or subduction-accretion (*i.e.* down-going plate) and suprasubduction (*i.e.* upper plate) ophiolites. Other abbreviations – CONTL.: continental; LAB: lithosphere–asthenosphere boundary; Moho: Mohorovičić discontinuity (crust-mantle boundary); NAMs: nominally anhydrous minerals. White arrows show the direction of the main tectonic movements; black lines are undifferentiated brittle, brittle-ductile or ductile tectonic interfaces.

response of the shallow mantle to deformation is probably influenced by the distribution of aqueous fluids and hydrous melts[1] in a fundamental way.

The dominant volatiles of the Earth's upper mantle are species in the C-O-H system, mostly in the form of H_2O and CO_2 molecular species. However, minor CO, CH_4, H_2, N_2, H_2S and SO_2, electrolytic species, as well as halogens and noble gasses can be present locally (Wyllie and Ryabchikov, 2000; Zhang *et al.*, 2009; Klemme and Stalder, 2018; Mysen, 2022). In this chapter, the attention will be focused on phases containing the H_2O component, and we therefore refer to H_2O as a thermodynamic component of the system. The main pathways to supply this hydrous component to the upper mantle are:

---

[1]Aqueous fluids are H_2O-rich solutions, and hydrous melts are melts with dissolved H_2O content. The two behave as immiscible "phases" at low to intermediate pressure. In general, with increasing pressure, aqueous fluids tend to dissolve more silicate components and hydrous melts more H_2O, until reaching the second critical endpoint of the system, beyond which no difference exists between hydrous melts and aqueous fluids (*e.g.* Bureau and Keppler, 1999; Manning, 2004a,b; Hermann *et al.*, 2006). For the particular case of the H_2O-mantle system, the pressure and temperature conditions of the second critical endpoint are poorly constrained and range from ~3 to 13 GPa and from 950 to 1150°C (*e.g.* Melekhova *et al.*, 2007; Mibe *et al.*, 2007; Wang *et al.*, 2020 and references therein).

(1) the direct crystallization of hydrous minerals, (2) the incorporation of hydrogen (more specifically compensated protons) through point defects in the crystal structure of nominally anhydrous minerals (NAMs), and (3) the percolation of aqueous fluids or hydrous melts at mantle depths (Fig. 1). Ultimately, the source and distribution of this H_2O component is largely controlled by subduction zones and dehydration processes in the slab.

The exact role of each of the above processes in controlling the rheology of the upper mantle is, however, controversial. Some of these controversies stem from an unclear identification of the H_2O-bearing phases that might actually control the deformation processes, both in nature and in experiments. From an experimental perspective (*e.g.* Hirth and Kohlstedt, 2003) the effect of H_2O on the strain rate ($\dot{\epsilon}$) for a given stress (σ) in a 'wet' rheology is usually accounted for by the water fugacity f_{H_2O} term. For example (disregarding the effect of partial melts or oxygen fugacity, Kohlstedt and Hansen, 2015),

$$\dot{\epsilon}(T, P) = A_w \sigma^n d^{-p} f_{H_2O}^r (T, P) \exp\left(-\frac{E_w^* + PV_w^*}{RT}\right) \qquad (1)$$

where A_w is a constant for the 'wet' rheology, n is the exponent of the stress term, d is the grain size with a p exponent, r is the exponent of the H_2O fugacity term, E_w^* and V_w^* are the activation energy and volume terms under 'wet' conditions, respectively (Mei and Kohlstedt 2000a,b) and R is the universal gas constant. While the f_{H_2O} term offers a quantitative means to constrain the influence of H_2O on deformation (*cf.* Eq. 1) – which is highly appreciated, *e.g.* in geodynamic modelling – it may obscure the underlying physical processes that govern weakening. Moreover, the widespread use of ambiguous terms in the literature – such as water-saturated and water-undersaturated conditions, water activity, 'hydrous' deformation, 'water' or hydrolytic weakening, and notably, the direct substitution of the f_{H_2O} term in Eq. 1 by 'water' content in olivine (C_{OH}, as discussed by Karato and Jung 2003) – contributes to the overall lack of clarity regarding the actual weakening mechanism. This results in a limited significance of its use as a quantitative parameter in flow laws. The recognition that the term f_{H_2O} in a fluid-rock system – arbitrarily defined from an intensive variable, the H_2O chemical potential (μH_2O) – does not depend solely on pressure and temperature (as implicitly stated in Eq. 1), but fundamentally relies on the relative proportion of H_2O content in the system under consideration, serves as a cautionary note when applying Eq. 1. In an attempt to clarify these ambiguities, a set of definitions illustrated in simple binary and ternary systems including H_2O as a component is presented below.

1.1. H₂O as a thermodynamic component

The thermodynamic description of a system where chemical reactions – in the present case, hydration, dehydration and hydrous melting – are feasible, requires the definition of a 'compositional space'. This compositional space constitutes the minimum set of components necessary to adequately describe (or, more accurately, 'model') the chemical variability of the system and its constituent phases (Spear, 1995). The definition of the system components is arbitrary, with the only constraint that they need to be linearly independent (*i.e.* linear combinations among components are not allowed). For example, in pure molecular fluids (*i.e.* those consisting of species without charge), the choice of including the two components H_2 and O_2 or, alternatively, the single component H_2O as part of the compositional space depends on whether one intends to model or not, respectively, redox reactions among fluid/melt species and solids. Assuming electrical neutrality and ignoring redox reactions, H_2O serves as a convenient thermodynamic component to describe phase relations among H_2O-bearing melts, aqueous fluids, hydrous minerals, and even NAMs as discussed below. The common use of the term 'water' to refer to the H_2O component, although colloquial, is not appropriate because (1) 'thermodynamic components' are just entities of the compositional space and lack of any physical meaning, and (2) 'water' is a phase with a well-defined and limited pressure and temperature stability, which falls well below the conditions commonly encountered in the upper mantle. Therefore, in this book chapter, we prefer to apply the H_2O tag, which should be understood as the colloquial 'water' term, widely – and perhaps unfortunately – used in the mantle literature.

There is no consensus in the literature regarding the terminology for expressing the amount of the H_2O component in hydrated minerals like NAMs, where protons typically balance vacancies and associate with oxygen to form hydroxyl groups (OH^-). However, adopting a consistent approach – as employed for nominally hydrous phases such as amphibole and serpentine that also possess structural hydroxyl groups – justifies the use of H_2O as a thermodynamic component for NAMs too. This becomes more evident when considering the representation of hypothetical hydrous endmembers in NAMs, such as $Mg_2H_4\square O_4$ (defined in terms of MgO and H_2O components) in the case of forsterite, where the empty square indicates a vacancy in the tetrahedral site that satisfied charge balance (Jollands *et al.*, 2023).

1.2. The H_2O chemical potential

The computation of equilibrium phase diagrams enables the definition of the chemical potential of all components in the system (*e.g.* Powell *et al.*, 2019), including H_2O. From a computational perspective, a phase diagram can be constructed by changing either the 'quantity' of the H_2O component (as an extensive variable[2]), or the H_2O chemical potential (as an intensive variable[3]). When the amount of the H_2O component is selected as an independent variable, the H_2O chemical potential becomes a dependent variable[4] with a numerical value that is a function of pressure, temperature and, importantly, phase assemblage. Therefore, from that very moment, the H_2O chemical potential

---

[2]an extensive variable (*e.g.* mass, entropy, or volume) scales with the size or extent of the system. For most purposes we are interested in the state of the system and the relative proportion of its phases; thus the mass of the components is normalized to the total mass of the system. For example, the specific variable $X(H_2O)$ – that is the molar fraction of H_2O in the system (see Fig. 2) – is sufficient to compute the stable phases and their relative proportions.

[3]an intensive variable (*e.g.* temperature, pressure, density, concentration, or chemical potential) is not influenced by the size or extent of the system.

[4]a dependent variable changes in the function of independent variables of the system.

is intrinsically linked to the bulk chemical composition and, in our particular case, the amount of H_2O present in the system in question. From an experimental perspective, the H_2O chemical potential is also a dependent variable unless the experimenter deliberately imposes H_2O fluid saturation, or induces saturation with a fluid containing another non-reacting component in the solution (*e.g.* Ar; for a comprehensive discussion see the seminal work of Greenwood, 1961). In this latter case, the non-reacting component results in a fixed chemical potential of the H_2O component at the run conditions, which would be lower than the chemical potential of a pure fluid containing only H_2O at the same conditions (*i.e.* imposing an H_2O activity lower than one). Thus, the H_2O chemical potential is equal to that of pure H_2O – or, in the buffered experiments, the chosen fixed chemical potential of the H_2O component – exclusively in this situation. It is also experimentally feasible to establish a numerical value for the H_2O chemical potential by introducing hydrous solid phases as H_2O-buffers, as discussed by Otsuka and Karato (2011). However, this approach is constrained to discontinuous values, dictated by the buffering effect observed in specific H_2O-undersaturated mineral assemblages. Notably, Otsuka and Karato (2011) investigated such buffering effects in mineral combinations like clinohumite–periclase–forsterite and brucite–periclase (see the following section).

1.2.1. The H₂O component in a simple MgO-H₂O system

Figure 2 shows a simple MgO-H_2O system computed at constant pressure and temperature conditions to illustrate: (1) the geometric definition of the μH_2O (*i.e.* the H_2O chemical potential); (2) derive corresponding values for a_{H_2O} (*i.e.* the H_2O activity) and f_{H_2O} (*i.e.* the H_2O fugacity); and (3) show their dependence on $X(H_2O)$ (*i.e.* the relative amount of H_2O). In this system, there are no solid solutions, and the determination of stable phases involves the computation of their Gibbs energies per mol of components in the system (number of mols of H_2O and MgO in each phase) under the specified pressure and temperature (P-T) conditions. Subsequently, the identification of stable phases is achieved by locating the minimum surface defined by two phases: periclase-brucite and brucite-H_2O in Fig. 2a, and periclase-H_2O in Fig. 2b. Notably, in the latter case, the appearance of brucite above the line connecting periclase-H_2O indicates that the univariant reaction brucite $=$ periclase $+ H_2O$ has been crossed at the corresponding P-T conditions, and brucite is no longer stable. Once the stable phases are found, it is possible to compute the two chemical potentials by extending the line that joins the two phases until its intersection with the vertical axis of the two system components (*i.e.* MgO-H_2O) at $X(H_2O) = 0$ for the MgO component, and at $X(H_2O) = 1$ for the H_2O component. By definition, both chemical potentials (μMgO and μH_2O) are equal for each of the individual two-phase assemblages that are stable (*e.g.* per-br and br-H_2O in Fig. 2a define a total of two different pairs of μH_2O and μMgO). Each two-phase assemblage defines, therefore, a unique μH_2O. In the illustrated example in Fig. 2a, considering the periclase-brucite assemblage, the line joining these two stable phases (solid black line) is extended until it intersects with the vertical axis at $X(H_2O) = 1$ (dashed violet line). This intersection point occurs at $-268,844$ J/ mol (depicted as a violet dot in Fig. 2a), and it defines the H_2O chemical potential of periclase-brucite ($\mu_{H_2O}^{per-br}$) at 2 GPa and 500°C. Note that $\mu_{H_2O}^{per-br}$ is lower than the chemical

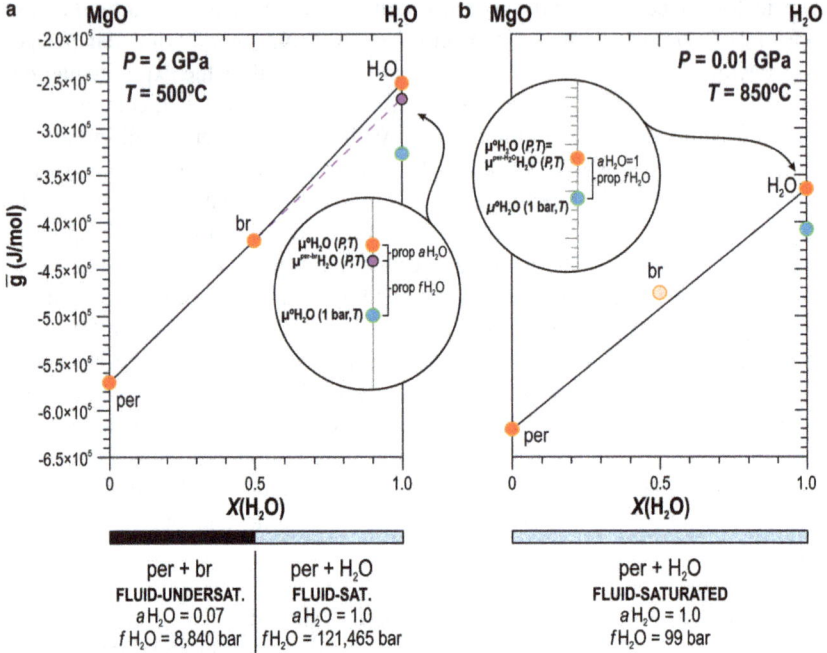

Fig. 2. Definition of the thermodynamic H_2O chemical potential (μH_2O) for two distinct two-phase assemblages within the MgO-H_2O system: periclase-brucite, $\mu_{H_2O}^{per-br}$ at 2 GPa and 500°C (773 K) in (*a*) and $\mu_{H_2O}^{per-H_2O}$ at 0.01 GPa and 850°C (1123 K) in (*b*). Also depicted are their two arbitrarily derived quantities: H_2O activity (a_{H_2O}) proportional to the difference between the μH_2O of the phase-assemblage and the chemical potential of the pure H_2O at the same P,T conditions, $\mu_{H_2O}^{o}$ (P,T) (see Eq. 2); and H_2O fugacity (f_{H_2O}), relative to 1 bar and the same temperature, $\mu_{H_2O}^{o}$(1 bar, T) (see Eq. 3). Note that the μH_2O (and all derived quantities) is composition-dependent – expressed as X(H_2O) or H_2O/(H_2O + MgO) in molar fractions – in (**a**), but not in (**b**), where fluid saturation occurs for all compositions, and periclase and H_2O are more stable than brucite (partially transparent circle). The vertical axis ($\bar{g}$ in J/mol) represents the Gibbs energy of the phase per mol of components in the system (*i.e.* the sum of mole of H_2O and MgO in the stoichiometric formula of the phase). The figure was computed using *Perple_X* (Connolly, 2009) and the Holland and Powell (2011) thermodynamic dataset. The equation of state for H_2O is from Pitzer and Steiner (1994). prop: proportional to.

potential of pure H_2O ($\mu_{H_2O}^{o}$) at the same conditions, which is situated at –252,003 J/mol and is depicted as a red dot at X(H_2O) = 1 in Fig. 2a. Consequently, all compositions ranging from X(H_2O) = 0 to 0.5 (corresponding to the compositional periclase-brucite joint) are fluid undersaturated and exhibit identical μH_2O. The difference in energy between the pure H_2O at the given P-T conditions – denoted as $\mu_{H_2O}^{o}$(P,T) and $\mu_{H_2O}^{per-br}$ (P,T) – defines the H_2O activity (a_{H_2O}), as given by Eq. 2.

$$a_{H_2O} = \exp\left(\frac{\mu_{H_2O}^{per-br}(P,T) - \mu_{H_2O}^{o}(P,T)}{RT}\right) \qquad (2)$$

In the specific case depicted in Fig. 2a,

$$a_{H_2O} = \exp\left(\frac{-268{,}844 \text{ J}\cdot\text{mol}^{-1} - (-252{,}003 \text{ J}\cdot\text{mol}^{-1})}{8.3145 \text{ J}\cdot\text{mol}^{-1}\cdot\text{K}^{-1}\cdot 773 \text{ K}}\right) = 0.073$$

Figure 2b shows the same MgO-H$_2$O system at lower P (0.01 GPa) and higher T (850°C) conditions, where H$_2$O saturation is achieved. As evident in the figure, at H$_2$O saturation, both H$_2$O chemical potentials – *i.e.* $\mu^{\circ}_{H_2O}(P, T)$ and $\mu^{per-H_2O}_{H_2O}(P,T)$, depicted as a red dot at $X(H_2O) = 1$ – are identical, hence the H$_2$O activity equals to one ($a_{H_2O} = 1$), according to Eq. 2.

In contrast to the reference used in the definition of H$_2$O activity (a_{H_2O}) at the P-T conditions of interest, the H$_2$O fugacity (f_{H_2O}) is defined relative to the H$_2$O chemical potential at the temperature of interest, yet at a reference pressure of 1 bar, denoted as $\mu^{\circ}_{H_2O}(1 \text{ bar}, T)$. In the studied MgO-H$_2$O system at 2 GPa and 500°C (Fig. 2a), the f_{H_2O} can be calculated as shown in Eq. 3.

$$f_{H_2O} = f^{\circ}_{H_2O} \exp\left(\frac{\mu^{per-br}_{H_2O}(P,T) - \mu^{\circ}_{H_2O}(1 \text{ bar}, T)}{RT}\right) \tag{3}$$

where $f^{\circ}_{H_2O}$ is the standard state fugacity at the reference pressure and equals 1 bar. Thus, for the H$_2$O-undersaturated conditions shown in Fig. 2a (at $X(H_2O) = 0.0$–0.5), we have:

$$f_{H_2O} = 1 \text{ bar}\cdot\exp\left(\frac{-268{,}844 \text{ J}\cdot\text{mol}^{-1} - (-327{,}247 \text{ J}\cdot\text{mol}^{-1})}{8.3145 \text{ J}\cdot\text{mol}^{-1}\cdot\text{K}^{-1}\cdot 773 \text{ K}}\right)$$

$$= 8840 \text{ bar } (0.88 \text{ GPa})$$

At the H$_2$O saturation condition shown in Fig. 2b, the f_{H_2O} can be computed directly by determining the difference between $\mu^{\circ}_{H_2O}(P,T)$ and $\mu^{\circ}_{H_2O}(1 \text{ bar}, T)$. Importantly, this calculation is solely dependent on the pressure and temperature of the system, and it remains unaffected by $X(H_2O)$ (*i.e.* for $X(H_2O) > 0.5$ in Fig. 2a, and for all $X(H_2O)$ values in Fig. 2b). It is crucial to realize that without knowledge of the stability conditions of brucite – specifically without Gibbs energy minimization of the full system – establishing *a priori* whether fluid saturation conditions are met is not possible. Consequently, μ_{H_2O}, f_{H_2O} and a_{H_2O} cannot be computed solely based on a chosen P-T condition if this information is lacking. In experimental works, employing Eq. 1 directly, using f_{H_2O} from the equation of state (EoS) for pure H$_2$O at the running pressure and temperature conditions assumes that hydrous phases, hydrous melt or H$_2$O in NAMs are equilibrated with excess H$_2$O, and such an approach implicitly considers H$_2$O saturation.

1.2.2. The H$_2$O component in solid solutions in the MgO-SiO$_2$-H$_2$O system

The aforementioned approach gains greater relevance for a mantle composition when extended to the MgO-SiO$_2$-H$_2$O system. This extension enables the exploration of the implications of incorporating solid solutions, such as hydrous melts and NAMs with variable compositions. In this context and for illustrative purposes, the continuous Gibbs energy is piecewise linearized using pseudocomponents (*i.e.* discrete compositions along the solid solution that are considered as discrete phases), as outlined by Connolly and Kerrick (1987). Solid solution models for NAMs are still under development (Padrón-Navarta, 2019).

For didactic purposes, a simplified example for forsterite is presented in a sandbox environment, considering only protons associated with oxygen in silicon vacancies ($Mg_2H_4\square O_4$, where the empty square indicates a vacancy; see Jollands *et al.*, 2023 for further details), and referred here to as hydrous forsterite (hfo), a hypothetical endmember. To facilitate a more straightforward analysis in 2D plots, the system can be projected conveniently from enstatite[5], as illustrated in Fig. 3 (purple arrows in the ternary diagrams showing the direction of projection from enstatite). Despite its simplicity, this system allows the definition of three crucial types of phase relations. At extremely low H_2O concentrations, the only stable phase is hydrous forsterite (and enstatite) at all temperatures, as depicted by the blue region in the compatibility diagrams in Fig. 3. It is noteworthy that in this single-phase field (hfo), there is a continuous increase in H_2O content of hydrous forsterite with increasing $X(H_2O)$ (see inset in Fig. 3a for an enlarged view; the same process applies to Fig. 3b–c), with tangents to its energy ω-surface approximated by two pseudocomponents. In the example, the tangent line shown in Fig. 3a is defined by hfo_1 and hfo_2 (pale blue circles labeled accordingly in the inset of the figure). In any other two-phase assemblage involving hydrous forsterite – *i.e.* hfo + atg in Fig. 3a, hfo + H_2O in Fig. 3b and hfo + melt in Fig. 3c –, the H_2O content in forsterite does not increase with increasing $X(H_2O)$, and the composition of hfo is fixed. At low temperatures, the hydrous phase antigorite is stable. Antigorite has a lower Gibbs energy than pure H_2O and one of the pseudocomponents from hydrous forsterite (*i.e.* hfo_3 in Fig. 3a). Thus, it plots in a lower position in the diagram than a metastable line joining them (Fig. 3a). As temperatures increase, antigorite is no longer stable, resulting in a simplified compatibility diagram (Fig. 3b) with only two fields (+enstatite): hydrous forsterite, and hydrous forsterite+H_2O, depicting H_2O-undersaturated and H_2O-saturated conditions, respectively. These regions are divided by one pseudocomponent (equivalent to hfo_3[6]). This scenario is important, as only a relatively small amount of H_2O – greater than the solubility of H_2O in the NAMs – would be enough to reach H_2O saturation, which has significant implications for the interpretation of hydrous deformation experiments.

At the highest temperatures, the hydrous melt surface intersects the H_2O-hfo joint, resulting in four regions. As a consequence, hydrous forsterite is no longer stable with pure H_2O (Fig. 3c). In the hfo + melt region, the H_2O content in olivine is buffered by the hydrous melt and does not increase with the H_2O content of the system, because

---

[5]although this has the drawback that certain fields at lower temperatures are metastable relative to talc, and compositions of melts with silica contents exceeding those in enstatite cannot be considered. The avid reader would find that the latter issue renders the melt + H_2O (+en) field represented in Fig. 3c metastable in the full MgO-SiO_2-H_2O system.
[6]it is customary to give a special term to the composition of this pseudocomponent as it represents the 'solubility' or maximum H_2O content in the NAMs at H_2O-saturation at these pressure and temperature conditions.

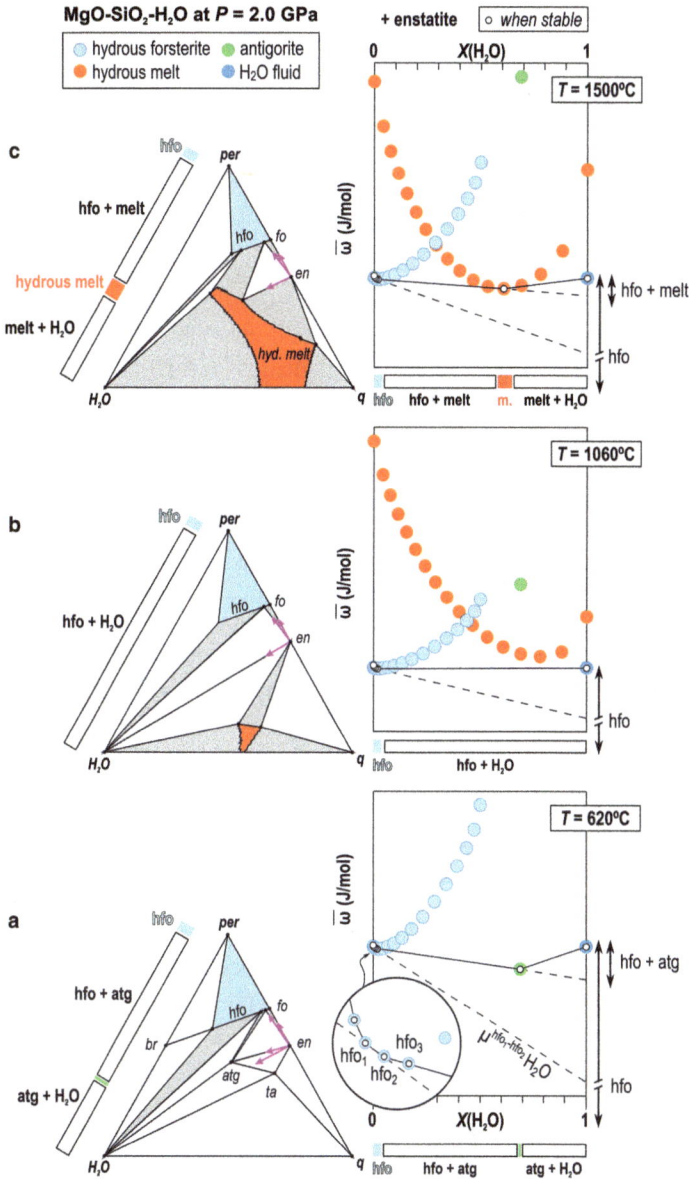

Fig. 3. General phase relations for an enstatite-saturated MgO(forsterite)-H₂O system at constant 2 GPa pressure and various temperatures, involving a H₂O phase, a hydrous phase (antigorite), a hydrous melt, and a hydrous forsterite (hfo) solid solution (Mg_2SiO_4-$Mg_2H_4 \square O_4$) representing a NAM, created here for didactic purposes only. (*a*) Low temperature (620°C) where antigorite is stable, (*b*) medium temperature (1060°C) above the dehydration of antigorite, and (*c*) high temperature (1500°C) above the wet hydrous melting. Gibbs energies (designated as ω) are plotted after projection from enstatite as a function of $X(H_2O)$, now defined as H₂O /(MgO' + H₂O) in mol units, where MgO' is MgO after projection from enstatite

the system is divariant. This situation may also be relevant for deformation experiments involving a small degree of hydrous partial melting. In such cases, the f_{H_2O} cannot be directly computed from the P-T conditions, but the inclusion of a solubility model for H_2O in the melt – dictating the H_2O chemical potential – is needed (*cf.* Hirth and Kohlstedt, 1996; Asimow *et al.*, 2004).

Figure 4 illustrates a T-$X(H_2O)$ diagram for the entire MgO-SiO_2-H_2O system, maintaining a constant MgO/SiO_2 ratio at a fixed pressure of 2 GPa, within a small range of bulk H_2O contents (0.0–0.4 mol.%; equivalent to 0–1470 ppm wt. H_2O), representing conditions typical of the upper mantle. These simple diagrams bring a novel perspective by examining the impact of H_2O in NAMs on the chemical potential of H_2O (panel 2) and other derived variables under H_2O-undersaturated conditions in a thermodynamically consistent manner, as opposed to iterative calculations (*cf.* Hirth and Kohlstedt, 1996; Asimow *et al.*, 2004). In the H_2O-undersaturated region, where hydrous phases or hydrous melts are not stable (shown as a semi-transparent field in Fig. 4), the chemical potential depends on the incorporation of H_2O in into the hydrous forsterite, which thus controls all dependent quantities (*i.e.* H_2O activity and H_2O fugacity). Suffice to say that experiments aimed at investigating the rheological effect of H_2O incorporated in the structure of olivine should be conducted specifically within this $T(P)$ and X (H_2O) window. Unfortunately, this window is poorly constrained in natural systems, primarily due to the dependence of H_2O incorporation into NAMs on various factors, such as trace element concentration and oxygen fugacity (*e.g.* Tommasi *et al.*, 2017). It has been suggested that the correlation of the strain rate with H_2O fugacity can be used to support the equivalent effect of H_2O concentration in NAMs on strain rate (*e.g.* Hirth and Kohlstedt, 2003; Karato and Jung, 2003; Kohlstedt and Hansen, 2015). The f_{H_2O} is very sensitive to changes in P and T under H_2O-saturated conditions; therefore, the fact that this dependency is identified when considering data at different P-T conditions does not, by itself, demonstrate that the dependency of f_{H_2O} (or H_2O content in olivine) can be quantified (but see Faul *et al.*, 2016).

Fig. 3. *Continued.* ($Mg_2Si_2O_6$). The projection is shown as purple arrows in the ternary diagrams. In this projection, antigorite ($Mg_{48}Si_{34}O_{85}(OH)_{62}$) has the following composition: en = 17, MgO' = 14, and H_2O = 31 in the new component system, and thus plots at $X(H_2O) = 31/45$ (green circles). Stable phases are defined by the minimum ω surface (indicated by white circles), i.e. the Legendre transform of the molar Gibbs energy. For example, g^{atg} is transformed to ω^{atg} through the expression $\omega^{atg} = (g^{atg} - X_{en}\mu^{en})/(1 - X_{en})$, where $X_{en} = en/(en + MgO' + H_2O)$ in mol ratio and g^{atg} is the Gibbs energy of atg in the new component system (en-MgO'-H_2O, 17-14-31). Dashed lines represent extrapolation to $X(H_2O) = 1$ of the indicated two-phase assemblages (or two-pseudocomponents in the case of the single-phase hydrous forsterite, hfo). The vertical arrows show their differences relative to the chemical potential of pure H_2O ($\mu^{\circ}_{H_2O}$), proportional to the activity of H_2O through Eq. 2. Although the diagram is quantitative, the values of ω have been corrected by a factor proportional to $X(H_2O)$ and the difference between the energies of forsterite and H_2O to better illustrate stable phases. At the bottom of each diagram, a schematic compatibility diagram is shown, illustrating the stable phases. In the ternary diagrams, abbreviations are: atg – antigorite, br – brucite, en – enstatite, fo – forsterite, hfo – hydrous forsterite, per – periclase, q – quartz, ta – talc, solid solutions (melt and hydrous forsterite) are represented as pseudocomponents (red and pale blue, respectively).

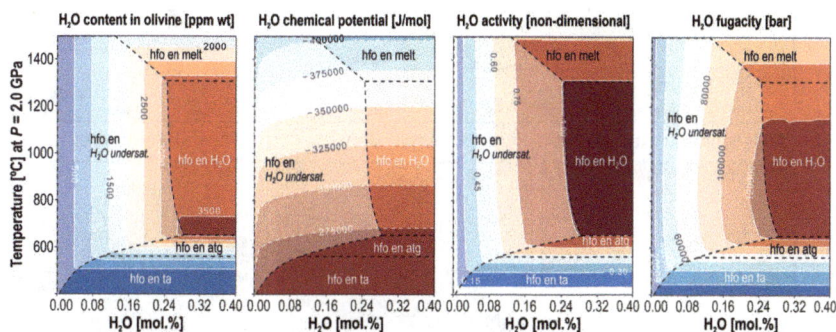

Fig. 4. Pseudosections of temperature *vs.* molar percentage of H_2O at a constant pressure of 2 GPa, maintaining a constant MgO/SiO_2 ratio of 0.55/0.45 mol, and increasing H_2O along the horizontal axis at low bulk H_2O contents (up to 0.40 mol%, equivalent to 1470 ppm wt. H_2O). Superimposed on all diagrams are the boundaries of the equilibrium field assemblages, in which the stable phases are indicated as follows: atg – antigorite; en – enstatite; hfo – hydrous forsterite; ta – talc ($\pm H_2O$ or melt). All fields are divariant (three-phases in a three-component system), except for the two-phase assemblage consisting of hydrous forsterite and enstatite, which is highlighted with semi-transparency (see also Fig. 3). This latter field corresponds to H_2O undersaturated conditions without the presence of hydrous phases or melt. It is important to note that, by definition, only in this two-phase field does the H_2O chemical potential (and derived H_2O activity and H_2O fugacity values) change with bulk H_2O content at constant T (and P). Thus, this is the only relevant field for deformation experiments addressing the rheological effect of the structurally bonded H_2O component in NAMs. The molar percentage of H_2O is defined here as $H_2O/(MgO+SiO_2+H_2O)$ in mol ratio, and expressed in percent.

1.2.3. The H_2O component in the upper mantle

In the following example, we disregard the effect of solid solution for H_2O in NAMs (as these models are not yet fully available; Padrón-Navarta, 2019), and we focus on the Na_2O-CaO-FeO-MgO-Al_2O_3-SiO_2-H_2O-Cr_2O_3 (NCFMASHCr) system to explore processes that take place in a more realistic mantle composition. Among these components, Na_2O is relevant for the stability of pargasitic amphiboles that – together with NCFMASH melts – exert a primary influence on the H_2O chemical potential. Note that the K_2O component – although not considered in the system discussed in this section – is also important in other settings (see Fig. 1) for the stability of phlogopite and K-richerite amphibole, which are stable at depths >180 km (*e.g.* Konzett and Ulmer, 1999). Figure 5 illustrates that regions saturated in a free H_2O fluid phase (fields with water pattern, where the H_2O activity is $= 1$ as shown in Fig. 5b,e) become wider at lower pressures. Only in these regions do the H_2O fugacity, the modal content of hydrous phases, and the melt fraction remain independent of the amount of H_2O in the system, as evidenced by the solid horizontal lines in Fig. 5. For the low bulk H_2O contents (0.0–1.0 wt.% H_2O) considered in Fig. 5, which are feasible in various settings in the shallow upper mantle (Fig. 1), most phase assemblages are H_2O-undersaturated with either hydrous melts at high temperature or hydrous phases – such as amphibole or chlorite – at low temperatures. In complex multicomponent systems, these H_2O-undersaturated fields containing hydrous phases or hydrous melts also constrain the values of the H_2O activity and H_2O fugacity, as is done in simpler systems (Fig. 3a, c).

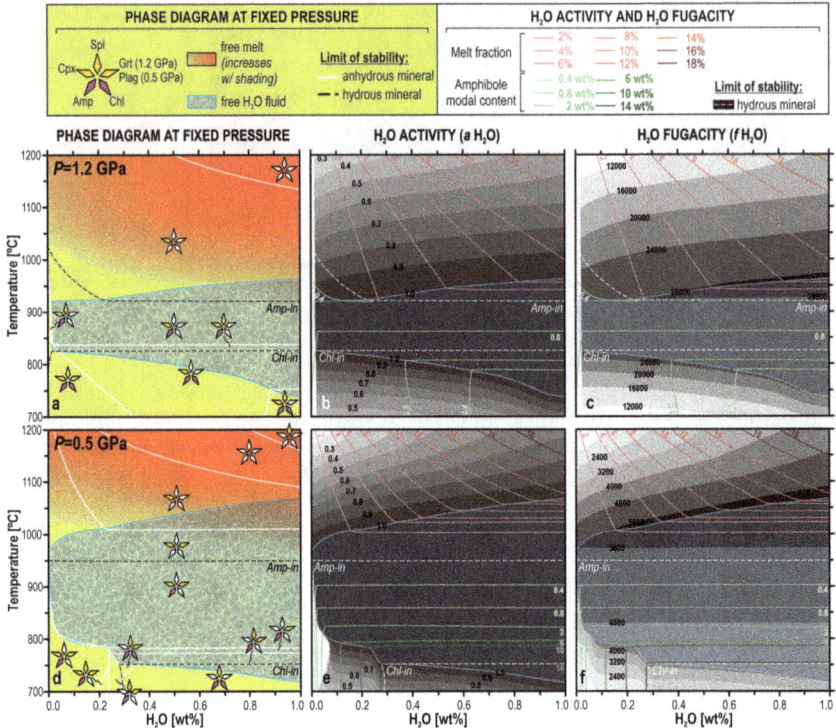

Fig. 5. Calculated pseudosections (colour-shaded diagrams in *a,d*), as well as H₂O activity (grey-shaded diagrams in *b,e*) and H₂O fugacity (grey-shaded diagrams in *c,f*) in the Na₂O-CaO-FeO-MgO-Al₂O₃-SiO₂-Cr₂O₃ system for a harzburgite whole-rock composition (sample RK025 – CaO: 1.74 wt.%; MgO: 41.9 wt. %; Al₂O₃: 1.67 wt.%; SiO₂: 43.8 wt.%; FeO: 7.58 wt.%; Na₂O: 0.02 wt.%; Cr₂O₃: 0.37 wt.%; data from Hidas *et al.,* 2016 from the Ronda massif, Spain) at equilibrium pressures of 1.2 GPa (**a–c**) and 0.5 GPa (**d–f**) as a function of temperature (°C) and H₂O content of the system (wt.%). The computation was carried out in *Perple_X* (Connolly, 2009) using the solid solution models of Holland *et al.* (2018). In (*a,d*), the stable mineral assemblage is indicated by diamond symbols (yellow: nominally anhydrous minerals, purple: hydrous minerals); olivine and orthopyroxene are stable throughout the entire modelled space. Note that at 1.2 GPa, a minor (<30°C) compositional gap between tremolitic and pargasitic amphiboles is predicted by the modelling at low temperature (<770°C) and high bulk H₂O contents (>0.7 wt.%), which is omitted from the figure for simplicity. Increasing melt fraction as a function of temperature and H₂O content is visualized qualitatively in parts **a** and **d** as shading of the melt-bearing field, but quantitative results of the calculation are shown in **b–c** and **e–f**. Black numbers in **b,e** and **c,f** indicate H₂O activity (dimensionless ranging from 0.0 in white shaded fields to 1.0 in dark grey shaded field) and H₂O fugacity (in bars) values, respectively. H₂O-saturation conditions are met at H₂O activity of 1.0, thus the fields with a free H₂O fluid in **a** and **d** are the same as those with a H₂O activity of 1.0 in **b** and **e**. Note that due to their formal definition, H₂O activity (*b,e*) and H₂O fugacity (*c,f*) are not equivalent terms and hence their use is not interchangeable. For more details, see section 1.2.1. As a comparison, the limits of stability of hydrous minerals in **a** and **d** (as black dashed lines) are also represented in **b,c** at 1.2 GPa and in **e,f** at 0.5 GPa (as white dashed lines). In **b**, **c**, **e** and **f**, pink to red coloured solid lines (in the upper half of the figure) and green-shaded solid lines (in the lower half of the figure) represent the volume fraction of melt (vol.%) and the modal content of amphibole (wt.%), respectively. Observe that at relatively low temperatures (~800–900°C), both at pressures of 0.5 GPa and 1.2 GPa, the increase of H₂O component in the bulk-rock composition does not influence the amphibole content and the system remains fluid-saturated, recording relatively small modal amounts of amphibole (<2.0 wt.%).

However, due to the existence of many possible continuous multivariant reactions between all phases (*e.g.* tremolite content in amphibole, anorthite content in plagioclase, chromium in spinel, *etc.*), the values of H_2O activity/fugacity do not vary linearly with the relative amount of H_2O in the system, as opposed to simple systems (*cf.* Fig. 4) because they depend on the amount of hydrous phases or hydrous melts at H_2O-undersaturated conditions (*i.e.* lower half or upper half of the diagrams, respectively). We emphasize that under these H_2O undersaturated conditions, introducing the H_2O fugacity term in rheological constitutive laws (Eq. 1) is unlikely to have any physical meaning without considering all the variables that may influence it, particularly the total H_2O content in the system at the relevant *P-T* conditions of deformation. The H_2O fugacity can only be computed independently from the chemical complexities of the rock in situations of H_2O saturation, which can only be achieved for realistic low H_2O contents when neither hydrous phases nor hydrous melts are stable, and the H_2O content in the system exceeds that required for H_2O saturation of NAMs (Fig. 4). In any case, it is highly probable that under those circumstances, the rheological behaviour would be more influenced by the presence of a stable interstitial free fluid phase than by the numerical value of the H_2O fugacity, or even the actual H_2O content in the NAMs.

These observations suggest the need for further research focused on a better understanding of the conditions in the upper mantle that control the stability of hydrous minerals, as well as the distribution and composition of aqueous fluids and hydrous melts. Additionally, the solubility of H_2O in NAMs (particularly for pyroxenes with mantle composition) and its dependency on other factors, such as trace elements and eventually fO_2, need to be further constrained. This is important because under natural conditions, their impact on rheology may be major and, perhaps, different than previously thought. Moreover, in future constitutive rheological laws accounting for the effect of hydrous components, it is essential to differentiate clearly between direct measurements of extensive variables (*i.e.* $X(H_2O)$ in the experimental system, as illustrated in Figs 2–5), and dependent intensive parameters (such as H_2O fugacity), the latter being a complex function of the reactive deforming system.

2. Effect of H₂O components on the rheological behaviour and microstructural evolution of the upper mantle: experiments *vs.* nature

In recent decades, many efforts have been made to obtain quantitative experimental data on the rheological behaviour of the upper mantle. Below we review the major achievements that – at least marginally – addressed the potential role of H_2O components presented in Fig. 1, and we summarize what the numerous recent natural case studies demonstrate when these components are inferred to be at play during ductile deformation of the upper mantle.

2.1. Hydrous melt and hydrous minerals

In contrast to anhydrous melts (*e.g.* Hirth and Kohlstedt, 1995a,b; Kohlstedt and Zimmerman, 1996; Bai *et al.*, 1997; Zimmerman *et al.*, 1999; Holtzman *et al.*, 2003a,b;

Zimmerman and Kohlstedt, 2004; Scott and Kohlstedt, 2006; Takei and Holtzman, 2009a,b,c; King *et al.,* 2010; Kohlstedt *et al.,* 2010; Soustelle *et al.,* 2014), there are very few experiments on the mechanical behaviour of hydrous melts during deformation, but available data indicate that the presence of H_2O enhances strain rates by approximately two orders of magnitude compared to equivalent anhydrous systems (Mei *et al.,* 2002). Of particular importance in hydrous melts is that they are common metasomatic agents in the upper mantle, which contribute to the generation of chemical and mineralogical heterogeneities (O'Reilly and Griffin, 2013 and references therein). Amphibole and mica are well-known products of such melt percolation in a wide depth range and in various tectonic settings (*e.g.* Dawson and Smith, 1982; Vannucci *et al.,* 1995; Zanetti *et al.,* 1996; Coltorti *et al.,* 2007; Shaw, 2009; Fumagalli and Klemme, 2015; Mandler and Grove, 2016; Bonadiman *et al.,* 2021; Wang *et al.,* 2021). Moreover, amphibole is the most important H_2O-storing mineral in the upper mantle down to ~3 GPa, ~90–100 km depth in an oceanic environment (Green *et al.,* 2010; Green, 2015), while phlogopite – subject to sufficient potassium – plays a similar role to somewhat deeper levels (Mengel and Green, 1989; Wyllie and Ryabchikov, 2000). The breakdown of amphibole causes a remarkable drop in the solidus temperature of fertile lherzolites and, at the same time, reduces the H_2O-storage capacity of the mantle, which is then controlled only by the NAMs under subsolidus conditions at greater depths (Green *et al.,* 2010; Kovács *et al.,* 2021). Although these results demonstrate that the presence of hydrous minerals controls the H_2O-storage capacity of the mantle at lithospheric mantle depths, to date, natural observations are scarce and no experimental data are available to constrain their implication for the rheology while they are present as stable phases in an otherwise nominally anhydrous, four-phase peridotite. Tommasi *et al.* (2017) provided one of the first microstructural observations from coarse-grained pargasite- and phlogopite-bearing peridotites, which record pervasive ductile deformation in the dislocation creep regime coeval to percolation of hydrous Si-rich melts at mantle depths (980–1080°C and <2 GPa) in the Finero massif (southern Alps, Italy). Their study concurs with the available experimental data (Mei *et al.,* 2002) and shows that the synkinematic presence of interstitial hydrous melts leads to significant rheological weakening of mantle rocks by accommodating large amounts of deformation due to a switch of the deformation mechanisms to stress-controlled dissolution-precipitation creep. However, the lack of correlation of olivine fabric strength and the symmetry of crystallographic-preferred orientation (CPO) – or lattice-preferred orientation (LPO) – of olivine to the occurrence of hydrous minerals (pargasite and/or phlogopite) in the peridotites implies little or no direct effect of these phases on the microstructural evolution of mantle rocks (Fig. 6). Even when hydrous minerals are present in peridotites at modal abundances of up to 25 vol.%, the fabric strength remains relatively weak in the peridotites (Fig. 6). This observation indicates that the mere presence of stable hydrous minerals in peridotites during deformation does not exert a significant influence on the rheological behaviour of the upper mantle (Tommasi *et al.,* 2017). It is conceivable that, as long as hydrous minerals are stable, rather than simply present, it is the saturation of rocks in H_2O and/or hydrous melts that controls the rheology of the mantle. Moreover, it is important to note that hydrous metasomatism does not consistently influence the H_2O content of the

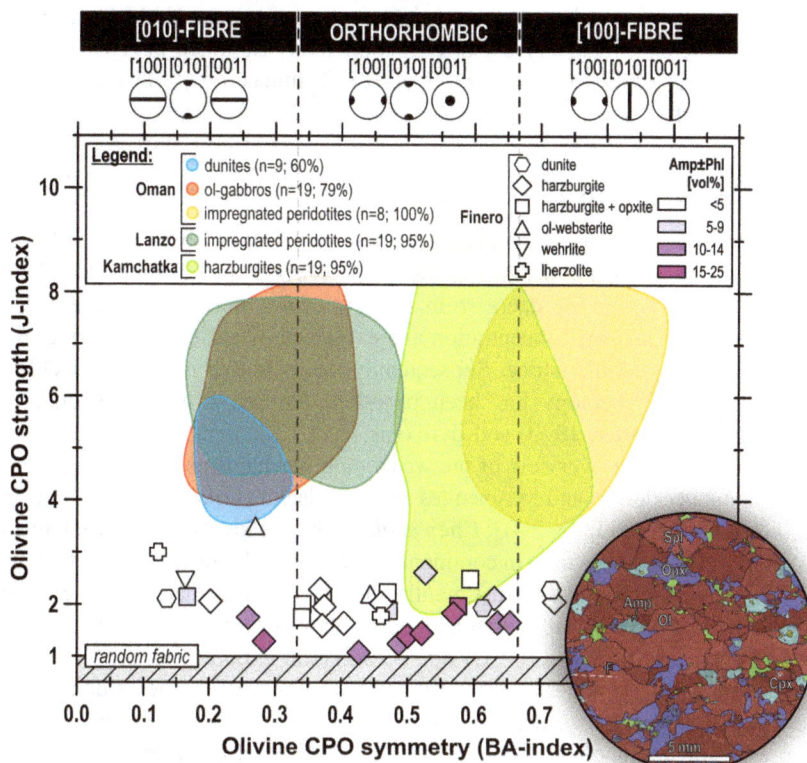

Fig. 6. Olivine crystallographic preferred orientation (CPO) symmetry (BA-index) as a function of the olivine fabric strength (J-index) in pargasite and phlogopite-bearing coarse-grained peridotites from the Finero massif (Tommasi *et al.*, 2017) compared to peridotites deformed in the presence of H₂O-poor basaltic melts (Oman and Lanzo; Higgie and Tommasi, 2012, 2014) or aqueous fluids (Kamchatka; Soustelle *et al.*, 2010), represented as colour-coded fields that cover >60% of the total dataset; *n*: number of samples shown in the field, and the percentage they correspond to compared to the total available dataset. Outliers are not shown for simplicity. Note that the Finero peridotites encompass the entire range of typical olivine CPO symmetries and, despite the microstructure, suggest coeval deformation of anhydrous and hydrous minerals (see inset) and the hydrous minerals display patent CPO (not shown here) consistent with shear deformation in the plane of the foliation (marked by 'F' and dashed line in the inset); the overall weak olivine fabric is not correlated to the modal abundance of hydrous minerals. The J-index (Bunge, 1982) is a dimensionless measure of the fabric strength, which takes a value of 1 in a randomly oriented material and it is infinite for a single-crystal orientation; olivine J-indices in mantle peridotites typically range from 2–20 with a peak at ~5 (Tommasi and Vauchez, 2015). The BA-index is a tool to characterize point and girdle-like distribution of olivine [010] and [100] crystallographic axes, which allow for differentiating the three most common olivine CPO symmetries in mantle peridotites (see Mainprice *et al.*, 2014 for more details). Labels – Amp: amphibole, Cpx: clinopyroxene; Ol: olivine; Opx: orthopyroxene; Spl: spinel (after Whitney and Evans, 2010). The original version of this figure was published in "Hydrous melts weaken the mantle, crystallization of pargasite and phlogopite does not: Insights from a petrostructural study of the Finero peridotites, southern Alps", by Tommasi, A., Langone, A., Padrón-Navarta, J.A., Zanetti, A. & Vauchez, A. in *Earth and Planetary Science Letters,* **477,** 59–72 (2017), Copyright Elsevier.

coexisting NAMs. In cases where hydrous minerals are present, the NAMs – particularly olivine – tend to remain relatively dry or only moderately saturated in H_2O (*e.g.* Denis *et al.*, 2015; Hao *et al.*, 2016; Aradi *et al.*, 2017; Tommasi *et al.*, 2017; Demouchy and Tommasi, 2021).

2.2. Structurally bonded H_2O

The weakening effect of the hydroxyl concentration in NAMs has been known since the earliest experiments on quartz crystals in the 1960s (Griggs and Blacic, 1965; Griggs, 1967; Blacic, 1975), where the decrease in strength of the mineral was observed with increasing hydrogen (proton) concentration and explained by easier movement of dislocations in the wet crystalline lattice. Subsequently, the potential impact of H_2O-linked weakening on mantle rheology has been proposed through experiments on olivine (Avé Lallemant and Carter, 1970) and pyroxene (Avé Lallemant, 1978). Later studies have been focused on the overview of the weakening mechanisms in major constituent phases of mantle peridotites and pyroxenites (*e.g.* Kohlstedt *et al.*, 1995; Mei and Kohlstedt, 2000a,b; Karato and Jung, 2003; Chen *et al.*, 2006; Tielke *et al.*, 2017) and resulted in the widespread acceptance that, in continental environments, the rheology of the lithospheric mantle, in general, and the crust-mantle coupling, in particular, may be controlled by the behaviour of olivine, with variable concentrations of hydrous defects. In the last decade, however, low-T (800–1100°C) dry deformation experiments on olivine challenged this paradigm (Demouchy *et al.*, 2013, 2014). These experiments demonstrated that the inferred important weakening effect of hydrous defects is partly due to the large error in the extrapolation of high-T (>1200°C) deformation experiments to natural conditions prevalent in the shallow lithospheric mantle (~900–1000°C), and so the rheology of dry olivine is indeed significantly weaker at these low-T conditions than previously thought. Thus, numerical modelling of lithospheric strength profiles may not need to operate with wet olivine rheology to weaken the lithospheric mantle (Boioli *et al.*, 2015). Currently, there is a general consensus within the scientific community that hydrous defects incorporated into the olivine structure have a discernible weakening effect, at least up to ~18 ppm wt. H_2O concentrations (Girard *et al.*, 2013; Faul *et al.*, 2016). However, it is also acknowledged that the significance of this weakening effect may have been somewhat overestimated in both diffusion (Fei *et al.*, 2013) and dislocation creep regimes (Demouchy *et al.*, 2012; Girard *et al.*, 2013).

As for the microstructural record of deformation at upper mantle conditions, CPO in olivine is often, although not exclusively, induced by deformation flow that activates different dislocation slip systems (*e.g.* Karato, 2008). The resulting CPO patterns are potentially important tracers in seismogenic modelling of the Earth's interior (Tommasi *et al.*, 1999; Mainprice *et al.*, 2000; Karato *et al.*, 2008; Hansen *et al.*, 2021). Natural record shows that the most common CPO symmetries in mantle rocks – developed by simple shear deformation in the dislocation creep regime – are characterized by the [100] olivine crystallographic axes distributed subparallel to the deformation flow direction and the {0kl} planes around the shear planes, with preferential alignment of (010) planes in the plane of the foliation (*e.g.* Tommasi *et al.*, 2000; Tommasi and Vauchez,

2015; Demouchy *et al.,* 2023). These CPO symmetries are explained by the dominant activation of {0kl}[100] and (010)[100] dislocation slip systems, and are often referred to as D-type (also known as axial-[100], [100]-fibre, or olivine 'pencil-glide') and A-type olivine fabric in the literature, respectively. Deformation experiments have suggested many microscopic and macroscopic factors to control the olivine CPO patterns in the dislocation creep regime but, among the intrinsic processes acting at the intracrystalline level, the biggest emphasis was put on the role of temperature, confining pressure, differential stress and structurally bonded hydrous defects (*e.g.* Carter and Avé Lallemant, 1970; Durham and Goetze, 1977; Bai *et al.,* 1991; Jung and Karato, 2001; Couvy *et al.,* 2004; Katayama *et al.,* 2004; Raterron *et al.,* 2009, 2011, 2012; Tasaka *et al.,* 2016). In the experimental works of the early 2000s, uncommon olivine fabric types were revealed, which indicated [001] slip direction and/or switches in the common slip plane to (001) or (100). The resulting CPO symmetries were consistent with the dominance of (001) [100], (010)[001] or (100)[001] slip (E-, B- and C-type olivine fabric, respectively), and were suggested to be controlled by the differential stress and the concentration of structurally bonded hydrous defects during deformation (Fig. 7a) (Jung and Karato, 2001; Katayama *et al.,* 2004; Jung *et al.,* 2006; Karato, 2008; Karato *et al.,* 2008). These observations have been used widely to interpret a seismic anisotropy signal in the mantle wedge (*e.g.* Mizukami *et al.,* 2004; Kneller *et al.,* 2005; Katayama and Karato, 2006; Lassak *et al.,* 2006; Tasaka *et al.,* 2008) as well as to render deformation conditions, particularly H₂O content and differential stress, to peridotites displaying a specific olivine CPO symmetry (*e.g.* Karato, 2004). However, considering the saturation of olivine in H₂O at the *P-T* conditions of experiments (Fig. 7b), large amounts of aqueous fluids were probably present during deformation, which is also corroborated by the published Fourier-Transform Infrared (FTIR) spectra displaying broad bands instead of sharp OH-stretching related peaks (*e.g.* Jung *et al.,* 2006). This suggests that olivine crystals contained the possible maximum of structurally bonded hydrous defects during most of the deformation experiments (Fig. 7b). Hence, apart from a few experiments where olivine may not have reached the saturation (*e.g.* data points in the yellow-shaded field in Fig. 7b), the dependence of olivine CPO types on the concentration of structurally bonded hydrous defects remains ambiguous.

The lack of correlation between structurally bonded H₂O and olivine CPO symmetry is consistent with the results of a recent comprehensive study of olivine CPO types as a function of H₂O content and stresses in natural mantle samples (Bernard *et al.,* 2019). Those authors found that all olivine CPO types were able to form at very low H₂O contents and stresses, and there was no clear relationship between hydrous components in the NAMs and CPO types in natural rocks. The only agreement with experimental results was that A-type olivine CPO symmetry did not occur at H₂O contents above the experimental A-to-E type boundary, and C-type fabric recorded the highest H₂O concentrations (Bernard *et al.,* 2019). However, in the case of the C-type fabric, the effect of pressure in the activation of [001]-directed olivine slip could not be excluded because the corresponding peridotites were garnet-bearing, hence originated at higher-pressure conditions (Bernard *et al.,* 2019). Although structurally bonded hydrous defects in the olivine crystalline lattice possibly have an impact on slip system activity (*e.g.* Tasaka

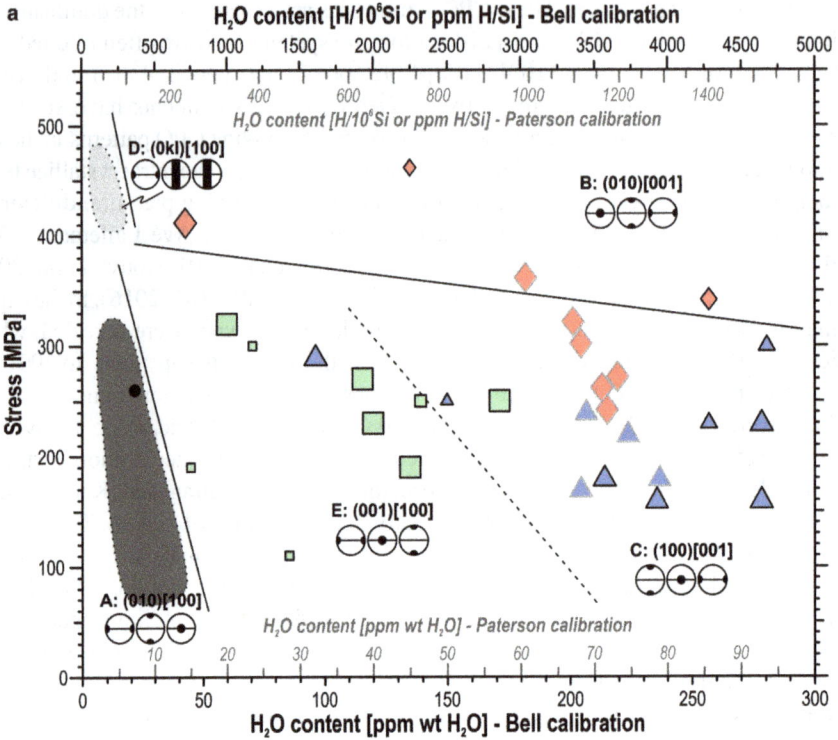

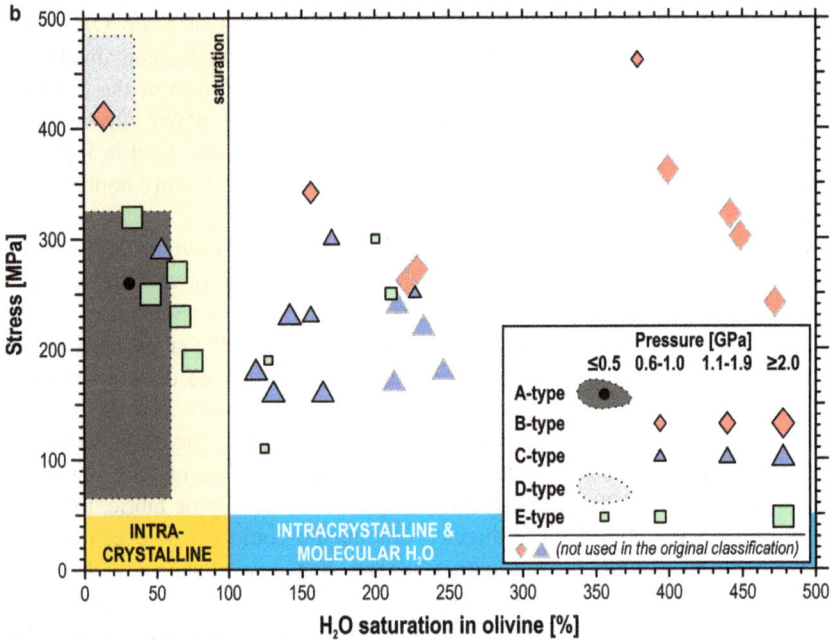

et al., 2016; Faul *et al.*, 2016; Wallis *et al.*, 2019) by altering dislocation self-energy in an anisotropic manner (*e.g.* Karato *et al.*, 2008; Karato, 2013), the natural record indicates that the role of other factors (*e.g.* deformation history, interstitial fluids/melts) in the development of uncommon olivine CPO symmetries in mantle peridotites is significantly larger than that of H_2O concentration and differential stress (*e.g.* Bernard and Behr, 2017; Bernard *et al.*, 2019; Kumamoto *et al.*, 2019). Alternatively, instead of the absolute concentration of hydrous defects bonded in the crystalline structure, it is the incorporation site that may control the microstructural evolution and weakening of H_2O-bearing olivine (Padrón-Navarta *et al.*, 2014, Padrón-Navarta and Hermann, 2017). Recent experimental data are consistent with this hypothesis, suggesting that the type of hydrous defects in olivine and the incorporation site of protons in the crystalline structure coupled to certain trace elements – such as Ti – may influence the hydrolytic weakening of olivine. This influence is pivotal as it ultimately controls the absolute concentration of H_2O under lithospheric conditions when hydrous phases or hydrous melts are not stable (Berry *et al.*, 2005, 2007a,b; Faul *et al.*, 2016). The future direction of research in the field of hydrous weakening of mantle NAMs and the variation of olivine CPO types is thus expected to explore the dependence of rheological behaviour on the incorporation site of structurally bonded protons under truly H_2O-undersaturated conditions.

Fig. 7. (*a*) Variation of olivine crystallographic preferred orientation (CPO) symmetry in the H_2O content [axis at the bottom: ppm wt. H_2O; axis on top: $H/10^6$ Si or ppm H/Si] versus deviatoric stress [MPa] space applied during the high temperature (1100–1300°C) deformation experiments of Bystricky *et al.* (2000), Zhang *et al.* (2000), Jung and Karato (2001), Katayama *et al.* (2004), and Katayama and Karato (2006). The five olivine CPO types (labelled as A-, B-, C-, D- and E-type with a cartoon depicting the corresponding ideal CPO symmetry) are traditionally interpreted as a result of H_2O-induced fabric transitions during dislocation creep deformation (*e.g.* Karato, 2008 and references therein). The schematic cartoons of CPO types are lower hemisphere stereographic projections with the out-of-plane foliation marked by a horizontal solid line. Stretching lineation is at 090°/0° (or 270°/0°) of the pole figures and it develops subparallel to the E–W directed simple shear along the foliation plane; the corresponding dominant dislocation slip system is indicated in the form of (slip plane)[slip direction]. The symbol size of experimental data points is proportional to the confining pressure applied during the deformation experiment; see legend in (**b**). The figure has been modified from Katayama *et al.* (2004); data points with a grey outline were not available at the time the original olivine fabric diagram was published. The H_2O content is adjusted roughly for the Bell *et al.* (2003) calibration, but the method of Paterson (1982) is also provided in grey with italic letters. (**b**) H_2O saturation in olivine [%] versus deviatoric stress [MPa] of the experimental data presented in (**a**). The solubility of H_2O in olivine is calculated following the method of Padrón-Navarta and Hermann (2017) considering TiO_2 concentrations in San Carlos olivine ranging from 12 to 24 ppm wt. TiO_2 (Tollan *et al.*, 2018), and the H_2O saturation in olivine is expressed as the ratio of measured H_2O concentration in olivine (as reported in the original publications, adjusted for the Bell *et al.*, 2003 calibration here) and the theoretical saturation of olivine in H_2O at the pressure-temperature conditions of the corresponding deformation experiments. The 100% saturation (vertical solid line) indicates that all the defects in the olivine crystalline structure are occupied by structurally bonded H_2O, thus further increase in H_2O concentration leads to accommodation of molecular H_2O as visible or invisible fluid inclusions and/or accumulation of excess aqueous fluid at grain boundaries during experiments. Note that most deformation experiments were supersaturated in H_2O (up to ~5×), and the correlation of olivine CPO symmetry type with H_2O content is ambiguous. Only a few high-pressure (>2 GPa) experiments were undersaturated in H_2O, but the concentration of intracrystalline H_2O in olivine in these experiments did not result in the systematic switch of CPO symmetry as suggested by data in (**a**).

2.3. Synkinematic aqueous fluids

Experimental data addressing the effect of synkinematic interstitial aqueous fluids on the rheological behaviour of mantle rocks are scarce. Torsion experiments at 300 MPa and 1200°C on wet olivine polycrystals showed a limited strength decrease by a factor of 2–3 relative to dry samples, which is explained by the presence of both structurally bonded hydrous defects in the olivine structure and importantly by H_2O-derived species in grain boundaries or pores (Demouchy *et al.*, 2012). On the distribution of fluids during ductile deformation, recent high-pressure experiments conducted at 1.2 GPa and 900°C on olivine-pyroxene aggregates have shown that large amounts of aqueous fluids can be concentrated in domains experiencing the highest strain (Précigout *et al.*, 2019). This observation suggests that deformation may enhance fluid pumping into ductile shear zones in the mantle (Précigout *et al.*, 2019), which is primarily driven by dynamically changing pore-fluid pressure at the grain scale during creep cavitation (*e.g.* Fusseis *et al.*, 2009). This potentially has an important feedback on generating shear instabilities, because if melts/fluids can form continuous films at grain boundaries, effective pressure is probably reduced significantly (*e.g.* Karato, 2008). However, unlike interstitial melts, at pressures relevant to the shallow subcontinental upper mantle, H_2O-rich fluids have a large dihedral angle ($\geq 60°$) at triple junctions between mineral grains (*e.g.* Watson and Brenan, 1987; Mibe *et al.*, 1999), which impedes efficient wetting of grain boundaries and promotes the formation of fluid pockets instead of interconnected layers, except for very large pore-fluid volumes. Nevertheless, both deformation (*e.g.* Hier-Majumder and Kohlstedt, 2006) and impurities dissolved in the aqueous fluids (*e.g.* Huang *et al.*, 2019; Mysen, 2022) tend to decrease the size of dihedral angles, leading to the formation of interconnected networks of aqueous fluids so that they may reach the base of the crust. If that is the case, the aqueous fluid-saturated grain boundaries – similar to those wetted by interstitial melts (*e.g.* Bai *et al.*, 1997 and references therein) – are expected to serve as a fast intergranular diffusion path. Therefore, free interstitial aqueous fluids may foster the shift of deformation from grain interiors to grain boundaries, favouring fluid-assisted grain boundary migration (*e.g.* Urai *et al.*, 1986) and/or dissolution-precipitation (pressure-solution) creep, which involves advection of elements by the fluid in response to applied stresses (*e.g.* Rutter, 1976; Spiers *et al.*, 2004). This latter deformation mechanism is common in the crust (*e.g.* Wintsch and Yi, 2002; Bestmann *et al.*, 2004) and has also been described in various rock types exhumed from subduction zones (*e.g.* Wassmann and Stöckhert, 2013), but is usually considered to be of minor importance in the deformation of peridotites due to the limited availability of free aqueous fluids in the low-porosity mantle. Nonetheless, experimental data on the solubility of enstatite and forsterite in aqueous fluids at deep lithospheric conditions (700–900°C and 0.4–1.5 GPa; Newton and Manning, 2002) show that silica solubility is sufficient for dissolution-precipitation creep to be efficient during disequilibrium conditions if time-integrated fluid-rock ratios are high. Thus, if free, impure aqueous fluids are present, it is conceivable that dissolution-precipitation creep could equally be a potential deformation mechanism in the shallow lithospheric mantle.

The microstructural record of aqueous fluid-assisted deformation is rare in fresh (*i.e.* non-serpentinized) mantle peridotites. In a plagioclase-facies subcontinental lithospheric mantle peridotite shear zone from the Ronda massif (Spain), Hidas *et al.* (2016) interpreted the progressive strain localization in the presence of synkinematic aqueous fluids during deformation. The shear zone microstructures (*e.g.* crystallization of secondary orthopyroxene in the pressure shadow of porphyroclasts, Fig. 8a) and major element geochemical compositions (*e.g.* no compositional difference between shear-zone rocks and host peridotite) were consistent with dissolution-precipitation creep in a melt-free system. This process is inferred to be responsible for the formation of olivine-rich mylonitic domains alternating with orthopyroxene-rich mylonitic-ultramylonitic bands (Fig. 8a), leaving the modal composition essentially intact at the bulk rock scale, compared to country host peridotites. Moreover, the olivine CPO symmetry recorded a switch from the country host peridotite (axial-[100] or D-type olivine fabric) to the H$_2$O-impregnated mylonites (so-called E-type olivine fabric) coupled to secondary orthopyroxenes mimicking the olivine CPO within the shear zone (Fig. 8b). This latter observation is inconsistent with dislocation creep deformation of orthopyroxene and indicates oriented crystallization controlled by olivine CPO (Hidas *et al.*, 2016). The origin of the unusual E-type olivine fabric in the plagioclase facies peridotites, however, remained unclear due to the low solubility of hydrous defects in olivine at shallow mantle conditions (<1 GPa; Padrón-Navarta and Hermann, 2017). Nevertheless, from a fossil mantle wedge environment exhumed in the Semail Ophiolite, Oman (Prigent *et al.*, 2018), and from an oceanic transform fault setting (Bickert *et al.*, 2023), similar observations were revealed between porphyroclastic tectonites and intermediate-*T* (700–900°C) mylonitic-ultramylonitic peridotites, where olivine displayed a remarkable switch to (001)[100] CPO symmetry (*i.e.* E-type olivine fabric) associated with synkinematic oriented crystallization of secondary pyroxenes, mimicking the olivine CPO. Thus, the natural record shows that ductile shear zones can channelize fluid flow in the upper mantle (*e.g.* Précigout *et al.*, 2017, 2019), eventually promoting dissolution-precipitation creep assisted by siliceous fluids (Hidas *et al.*, 2016; Bickert *et al.*, 2023).

The conditions at which dissolution-precipitation may be active in the lithospheric mantle, however, are very limited. This process requires the presence of significant amounts of impure aqueous pore fluids in unaltered mantle, which could be achieved locally in a domain constrained by the upper limit of serpentine and chlorite stability (~750–820°C; *e.g.* Vissers *et al.*, 1995; Padrón-Navarta *et al.*, 2010 and dashed line #2 in Fig. 5a,d in this chapter) and the onset of hydrous melting (<1000–1100°C; Green, 1973 and lower limit of free melt domain in Fig. 5a,d in this chapter). So far, the natural record documenting dissolution-precipitation creep in non-altered peridotites is scarce yet these data show that it can be a viable deformation mechanism in the mantle wedge of subduction zones (~700–900°C; Prigent *et al.*, 2018), at oceanic transform faults (~735–945°C; Bickert *et al.*, 2023), and in the shallow subcontinental lithospheric mantle (~750–1000°C; Hidas *et al.*, 2016) (*cf.* Fig. 1), allowing for deformation of peridotites at lower stresses than dislocation creep. If active, dissolution-precipitation creep would imply feedback between deformation and fluid transport and, hence, would promote highly heterogeneous deformation in the shallow upper mantle in diverse

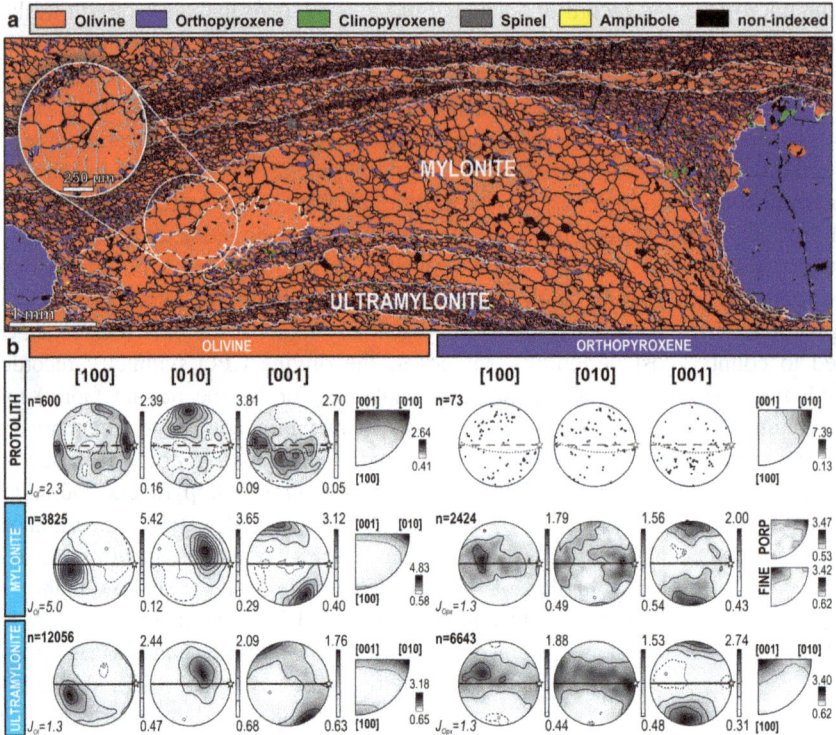

Fig. 8. (*a*) Electron backscatter diffraction (EBSD)-generated phase map of mylonite and ultramylonite microstructures that are interpreted to develop during fluid-assisted ductile strain localization in non-serpentinized peridotites from the Ronda massif (Hidas *et al.*, 2016). The irregular limits between mylonite lenses and ultramylonitic bands are underlined in white. The white dashed line in the mylonite depicts an elongated olivine porphyroclast. Note (i) the corroded grain boundaries of orthopyroxene porphyroclasts and their tails composed of fine-grained, well-mixed olivine-orthopyroxene assemblages that grade into ultramylonitic bands, (ii) the interstitial fine-grained orthopyroxene crystals at the olivine grain boundaries, and (iii) the subgrain boundaries perpendicular to the grain elongation in the inset. The (i) and (ii) are consistent with dissolution-precipitation creep and (iii) indicates activation of dislocation creep in olivine at least in the early stages of deformation. (*b*) Pole figures showing the crystallographic preferred orientation (CPO) of olivine and orthopyroxene in the coarse-grained protolith (microstructure not shown) as well as in the mylonitic and ultramylonitic domains of the shear zone shown in (**a**). The inverse pole figures in the crystal reference frame displaying the rotation axes accommodating low angle (2–15°) misorientations in the two mineral phases are also shown in each microstructural domain. Pole figures are lower hemisphere, equal-area stereographic projections; contours as 0.5 multiples of a uniform distribution; black square: maximum density, white circle: minimum density. Pole figures are plotted using the average orientation of each grain ('one point per grain') to avoid over-representation of large porphyroclasts in the thin sections using the careware software package by David Mainprice. All data are presented in the structural reference frame of the shear zone; the solid horizontal line denotes the average orientation of the foliation and stretching lineation is indicated by a white star. For the protolith, the dashed line indicates the average orientation of the foliation in the shear zone and the dotted line represents the average orientation of the high-T foliation in nearby plagioclase tectonites after Hidas *et al.* (2013). Labels – n: number of measured grains; J_{Ol} and J_{Opx} are the J-indices of olivine and orthopyroxene, respectively, calculated using the MTEX toolbox of MATLAB (https://mtex-toolbox.github.io/, last accessed on 26 October 2023; Bachmann *et al.*,

tectonic settings. Association of highly localized deformation and the very weak olivine CPO produced by dissolution-precipitation creep may also provide an explanation for the very weak seismic anisotropy usually recorded in the fore-arc domains of subduction zones (delay times <0.5 s; *e.g.* Di Leo *et al.*, 2012; Long and Silver, 2008). Altogether, fluid-assisted dissolution-precipitation creep could also be a major rheological process controlling strain localization and faulting at plate boundaries in general (Bickert *et al.*, 2023).

2.4. Fluid-rock interaction at low-*T* deformation

At typical tectonic strain rates (10^{-12} to 10^{-15} s^{-1}), unweathered mantle rocks transition progressively from ductile to semi-brittle behaviour as temperature decreases below ~850–900°C, and they primarily deform by brittle mechanisms below 600°C (*e.g.* Boettcher *et al.*, 2007; Kohli and Warren, 2019; Prigent *et al.*, 2020). In oceanic peridotites, extensive shearing at these low-*T* conditions – yet occurring beyond the serpentinite stability – often produces ultramylonitic shear zones with remarkably fine-grained (<10 μm) minerals, in which strain is predominantly accommodated by grain-size sensitive deformation mechanisms (*e.g.* Jaroslow *et al.*, 1996; Warren and Hirth, 2006). Nevertheless, recent microstructural, petrological and geochemical data indicate that at mid-ocean ridges, the hydrologic cycle might extend past the brittle-ductile transition, and substantial amounts of fluids can reach deep levels of the mantle lithosphere along transform faults (*e.g.* Harigane *et al.*, 2019; Kohli and Warren, 2020; Prigent *et al.*, 2020; Vieira-Duarte *et al.*, 2020; Patterson *et al.*, 2021; Kakihata *et al.*, 2022). Besides ample microstructural evidence for brittle deformation, such as crosscutting intra- and transgranular fractures, as well as pulled apart and/or displaced microboudinage structures, these hydrated low-*T*, non-serpentinized ultramylonites are accompanied by the synkinematic formation of hydrous phases, such as amphibole $\pm$ chlorite (*e.g.* Prigent *et al.*, 2020; Vieira-Duarte *et al.*, 2020; Kakihata *et al.*, 2022), and occasionally they may exhibit weak to moderately developed CPO patterns. In such cases, olivine is observed with weak bimodal (010)[001] and (001)[100] CPO symmetries, potentially indicating that low-temperature plasticity was at play under hydrous conditions at high-stress (Kakihata *et al.*, 2022).

At even lower temperatures, provided that H$_2$O fluid supply is sufficient, and depending on the Si content of the ultramafic rocks (*e.g.* Früh-Green *et al.*, 2004), infiltration along microfractures of the brittle peridotite matrix initiates serpentinization processes, which start with the development of typical mesh textures and culminate in the formation of pure serpentinites at the expense of peridotites (*e.g.* Boudier *et al.*, 2010; Evans *et al.*, 2013; Schwartz *et al.*, 2013; Rouméjon and Cannat, 2014; Rouméjon *et al.*, 2015; Escario *et al.*, 2018). In general, the degree of serpentinization weakens peridotites

---

Fig. 8. Continued. 2010; Mainprice *et al.*, 2014); for the interpretation of J-index values, see the caption of Fig. 3. The original version of this figure was published in 'Fluid-assisted strain localization in the shallow subcontinental lithospheric mantle', by Hidas, K., Tommasi, A., Garrido, C.J., Padrón-Navarta, J.A., Mainprice, D., Vauchez, A., Barou, F. and Marchesi, C. (2016) *Lithos*, **262**, 636–650, Copyright Elsevier.

significantly, and it has been shown that slightly serpentinized peridotites have a rheological behaviour similar to that of pure serpentinites (*e.g.* Escartín *et al.*, 2001). Given the high strength of unweathered mantle minerals at low temperatures, serpentinization is expected to promote strain localization in the alteration products, facilitating the development of major (semi-)brittle faults in both subduction zones and oceanic settings. Furthermore, if interstitial fluids accompany such low-*T* deformation, the deformation is likely to be increasingly accommodated by dissolution and precipitation processes, particularly at the plate interface in subduction zones (*e.g.* Wassmann *et al.*, 2011; Padrón-Navarta *et al.*, 2012; Wassmann and Stöckhert, 2013).

3. Conclusions

In this contribution, we briefly reviewed the effect of different H_2O components on the rheology and microstructural evolution of the upper mantle by contrasting theoretical considerations with available experimental data and the natural record. We recalled that H_2O activity (a_{H_2O}), H_2O fugacity (f_{H_2O}), and H_2O chemical potential (μH_2O) are all related terms but that only μH_2O has a true thermodynamic meaning as it emerges once the stable phases are determined through Gibbs energy minimization. Through thermodynamic calculations in both simple and chemically complex systems, we emphasize that H_2O activity is equal to one only when the system is saturated in H_2O, and this is the only situation in which H_2O fugacity can be computed at any pressure and temperature conditions independently of the bulk amount of H_2O present in the deforming rock. However, under H_2O-saturated conditions, it is more likely that the rheological behaviour of any rock is governed by grain boundary processes rather than the potentially weakening effect of intracrystalline H_2O or the numerical value of f_{H_2O}. Furthermore, available laboratory data and natural observations suggest that:

- The formation of hydrous minerals is very dependent on the whole-rock geochemical composition (*e.g.* availability of Na, K, *etc.*), as well as the prevailing equilibrium pressure and temperature conditions. Consequently, the presence of only small modal amounts (<1–2 vol.%) of these minerals in refractory peridotites at lithospheric mantle conditions is entirely compatible with the percolation of substantial quantities of interstitial aqueous fluids within the rock, as fluids in excess may not lead to an increased crystallization of hydrous phases if other physicochemical conditions in the system are unfavourable. Nevertheless, compared to the dry mantle, the presence of small modal amounts of amphibole indeed implies greater H_2O fugacity and, therefore, larger H_2O contents in NAMs. However, there are situations, such as during mantle exhumation or low-pressure mantle metasomatism, where this expectation is not met. This discrepancy is primarily due to the combination of the low-pressure conditions and, particularly in the case of olivine, the complex behaviour of trace elements in the presence of hydrous phases that limit the potential for the incorporation of extrinsic hydrous defects in the crystalline structure. Overall, small modal amounts of stable hydrous phases and small H_2O

contents in coexisting NAMs in refractory peridotites of the lithospheric mantle cannot be taken as direct evidence of a dry mantle environment.

- Stable, coarse-grained hydrous minerals, such as amphibole and mica, control the H_2O storage capacity of the upper mantle, but their presence alone has little or no impact on the microstructural evolution of their host peridotite during deformation in the dislocation creep regime. Based on natural samples, this observation seems to apply to relatively large modal amounts of such hydrous phases, at least up to 25 vol.%.

- Structurally bonded hydrous defects might weaken rheology, but olivine may not need to be wet to effectively weaken the rheology of the shallow upper mantle.

- Predicted correlation of experimentally deformed olivine CPO-types with H_2O content and stress may not directly apply to nature, where other factors – such as temperature, pressure, deformation mechanism, history and geometry of deformation, strain magnitude and presence of interstitial fluids/melts – play a more significant role in controlling CPO symmetry.

- Synkinematic hydrous or anhydrous melts and aqueous fluids significantly weaken the rheology but their effect is transient, *i.e.* deformation probably remains active as long as a liquid phase is present and it may cease under liquid-absent conditions. This indicates that: (1) in certain tectonic environments (*e.g.* subduction zones, mid-ocean ridges, continental rifts), response of mantle rocks to deformation may be subject to the availability of melts/fluids and, as such, can rapidly change at the geological time-scale; and (2) in these settings the relatively short intermittent periods of active, fluid/melt-assisted deformation may be responsible for the permanent microstructural changes rather than the long periods of tectonic inactivity of the otherwise resistant mantle.

- Interstitial, free aqueous fluids promote potential switches in the dominant deformation slip systems and/or mechanisms, and dissolution-precipitation creep may be a viable deformation mechanism if specific physico-chemical conditions meet in the shallow upper mantle. The presence of synkinematic aqueous fluids, however, is difficult to recognize in most cases because such fluids may not leave an obvious trace in the rock and determining them requires detailed microstructural, petrological and geochemical analyses.

- At low temperatures, but above the stability field of serpentinites, deformation of the upper mantle in the presence of synkinematic H_2O components is progressively dominated by grainsize-sensitive deformation mechanisms, leading to the development of brittle-ductile shear zones that often contain hydrous phases (typically amphibole ± chlorite). Reaching the temperatures of serpentinite stability, H_2O-assisted deformation is essentially controlled by serpentinite minerals, which develop within a resistant matrix composed of original mantle minerals, promoting (semi-)brittle faults and brittle-ductile strain localization at various scales.

We conclude that further experiments and carefully documented natural studies are needed for a better understanding of the effect of the H_2O component on upper mantle rheology. In the future, the role of hydrous volatile species on the microstructural evolution of the mantle must be explored through thermodynamically controlled models, which take into account the availability, speciation, concentration and incorporation into fluids (?) of a wide range of volatile species at different levels of the Earth's deep interior.

Acknowledgements

The authors appreciate the constructive reviews by Marguerite Godard and Alberto Zanetti, as well as the kind assistance and careful editorial work of Costanza Bonadiman and Elisabetta Rampone throughout the writing of this manuscript. José Alberto Padrón-Navarta acknowledges a Ramón y Cajal fellowship (RYC2018-024363-I) funded by the Spanish Ministry of Science and Innovation (MICINN/AEI/10.13039/501100011033) and the FSE program 'FSE invierte en tu futuro', and support by the Junta de Andalucía research project ProyExcel 00757. Károly Hidas acknowledges funding by the Agencia Estatal de Investigación of the MICINN (PID2020-119651RB-I00/AEI/10.13039/501100011033).

References

Aradi, L.E., Hidas, K., Kovács, I.J., Tommasi, A., Klébesz, R., Garrido, C.J. and Szabó, C. (2017) Fluid-enhanced annealing in the subcontinental lithospheric mantle beneath the westernmost margin of the Carpathian–Pannonian extensional basin system. *Tectonics*, **36**, 2987–3011.

Asimow, P.D., Dixon, J.E. and Langmuir, C.H. (2004) A hydrous melting and fractionation model for mid-ocean ridge basalts: Application to the Mid-Atlantic Ridge near the Azores. *Geochemistry, Geophysics, Geosystems*, **5**, Q01E16.

Avé Lallemant, H.G. (1978) Experimental deformation of diopside and websterite. *Tectonophysics*, **48**, 1–27.

Avé Lallemant, H. and Carter, N. (1970) Syntectonic recrystallization of olivine and modes of flow in the upper mantle. *Bulletin of the Geological Society of America*, **81**, 2203–2220.

Bachmann, F., Hielscher, R. and Schaeben, H. (2010) Texture analysis with MTEX - free and open source software toolbox. *Solid State Phenomena*, **160**, 63–68.

Bai, Q., Mackwell, S.J. and Kohlstedt, D.L. (1991) High-Temperature Creep of Olivine Single-Crystals. 1. Mechanical Results for Buffered Samples. *Journal of Geophysical Research–Solid Earth and Planets*, **96**, 2441–2463.

Bai, Q., Jin, Z.-M. and Green, H.W. (1997) Experimental investigation of the rheology of partially molten peridotite at upper mantle pressures and temperatures. Pp. 40–61 in: *Deformation-enhanced Fluid Transport in the Earth's Crust and Mantle* (M.B. Holness, editor). Chapman & Hall, London.

Bell, D.R., Rossman, G.R., Maldener, J., Endisch, D. and Rauch, F. (2003) Hydroxide in olivine: A quantitative determination of the absolute amount and calibration of the IR spectrum. *Journal of Geophysical Research–Solid Earth*, **108**, No. B2, 2105, https://dx.doi.org/10.1029/2001JB000679.

Bernard, R.E. and Behr, W.M. (2017) Fabric heterogeneity in the Mojave lower crust and lithospheric mantle in Southern California. *Journal of Geophysical Research: Solid Earth*, **122**, 5000–5025.

Bernard, R.E., Behr, W.M., Becker, T.W. and Young, D.J. (2019) Relationships between olivine CPO and deformation parameters in naturally deformed rocks and implications for mantle seismic anisotropy. *Geochemistry, Geophysics, Geosystems*, **20**, 3469–3494.

Berry, A.J., Hermann, J., O'Neill, H.S.C. and Foran, G.J. (2005) Fingerprinting the water site in mantle olivine. *Geology*, **33**, 869–872.

Berry, A.J., O'Neill, H.S.C., Hermann, J. and Scott, D.R. (2007a) The infrared signature of water associated with trivalent cations in olivine. *Earth and Planetary Science Letters*, **261**, 134–142.

Berry, A.J., Walker, A.M., Hermann, J., O'Neill, H.S.C., Foran, G.J. and Gale, J.D. (2007b) Titanium substitution mechanisms in forsterite. *Chemical Geology*, **242**, 176–186.

Bestmann, M., Prior, D.J. and Veltkamp, K.T.A. (2004) Development of single-crystal σ-shaped quartz porphyroclasts by dissolution-precipitation creep in a calcite marble shear zone. *Journal of Structural Geology*, **26**, 869–883.

Bickert, M., Kaczmarek, M.-A., Brunelli, D., Maia, M., Campos, T.F.C. and Sichel, S.E. (2023) Fluid-assisted grain size reduction leads to strain localization in oceanic transform faults. *Nature Communications*, **14**, 4087.

Blacic, J.D. (1975) Plastic-deformation mechanisms in quartz: The effect of water. *Tectonophysics*, **27**, 271–294.

Boettcher, M.S., Hirth, G., Evans, B. (2007) Olivine friction at the base of oceanic seismogenic zones. *Journal of Geophysical Research: Solid Earth*, **112**, B01205.

Boioli, F., Tommasi, A., Cordier, P., Demouchy, S. and Mussi, A. (2015) Low steady-state stresses in the cold lithospheric mantle inferred from dislocation dynamics models of dislocation creep in olivine. *Earth and Planetary Science Letters*, **432**, 232–242.

Bonadiman, C., Brombin, V., Andreozzi, G.B., Benna, P., Coltorti, M., Curetti, N., Faccini, B., Merli, M., Pelorosso, B., Stagno, V., Tesauro, M. and Pavese, A. (2021) Phlogopite-pargasite coexistence in an oxygen reduced spinel-peridotite ambient. *Scientific Reports*, **11**, 11829.

Boudier, F., Baronnet, A. and Mainprice, D. (2010) Serpentine mineral replacements of natural olivine and their seismic implications: oceanic lizardite versus subduction-related antigorite. *Journal of Petrology*, **51**, 495–512.

Bunge, H.J. (1982) *Texture Analysis in Materials Science*. Butterworth, London, 593 pp.

Bureau, H. and Keppler, H. (1999) Complete miscibility between silicate melts and hydrous fluids in the upper mantle: experimental evidence and geochemical implications. *Earth and Planetary Science Letters*, **165**, 187–196.

Bürgmann, R. and Dresen, G. (2008) Rheology of the Lower Crust and Upper Mantle: Evidence from Rock Mechanics, Geodesy, and Field Observations. *Annual Review of Earth and Planetary Sciences*, **36**, 531–567.

Bystricky, M., Kunze, K., Burlini, L. and Burg, J.P. (2000) High shear strain of olivine aggregates; rheological and seismic consequences. *Science*, **290**, 1564–1567.

Carter, N. and Avé Lallemant, H. (1970) High temperature flow of dunite and peridotite. *Bulletin of the Geological Society of America*, **81**, 2181–2202.

Chen, S., Hiraga, T. and Kohlstedt, D.L. (2006) Water weakening of clinopyroxene in the dislocation creep regime. *Journal of Geophysical Research*, **111**, B08203, https://dx.doi.org/10.1029/2005JB003885.

Coltorti, M., Bonadiman, C., Faccini, B., Gregoire, M., O'Reilly, S.Y. and Powell, W. (2007) Amphiboles from suprasubduction and intraplate lithospheric mantle. *Lithos*, **99**, 68–84.

Connolly, J.A.D. (2009) The geodynamic equation of state: What and how. *Geochemistry Geophysics Geosystems*, **10**, Q10014.

Connolly, J.A.D. and Kerrick, D.M. (1987) An algorithm and computer-program for calculating composition phase-diagrams. *CALPHAD*, **11**, 1–55.

Couvy, H., Frost, D., Heidelbach, F., Nyilas, K., Ungár, T., Mackwell, S. and Cordier, P. (2004) Shear deformation experiments of forsterite at 11 GPa-1400°C in the multianvil apparatus. *European Journal of Mineralogy*, **16**, 877–889.

Dawson, J.B. and Smith, J.V. (1982) Upper mantle amphiboles: a review. *Mineralogical Magazine*, **45**, 35–46.

Demouchy, S. and Tommasi, A. (2021) From dry to damp and stiff mantle lithosphere by reactive melt percolation atop the Hawaiian plume. *Earth and Planetary Science Letters*, **574**, 117159.

Demouchy, S., Tommasi, A., Barou, F., Mainprice, D. and Cordier, P. (2012) Deformation of olivine in torsion under hydrous conditions. *Physics of the Earth and Planetary Interiors*, **202**, 56–70.

Demouchy, S., Tommasi, A., Ballaran, T.B. and Cordier, P. (2013) Low strength of Earth's uppermost mantle inferred from tri-axial deformation experiments on dry olivine crystals. *Physics of the Earth and Planetary Interiors*, **220**, 37–49.

Demouchy, S., Mussi, A., Barou, F., Tommasi, A. and Cordier, P. (2014) Viscoplasticity of polycrystalline olivine experimentally deformed at high pressure and 900°C. *Tectonophysics*, **623**, 123–135.

Demouchy, S., Wang, Q. and Tommasi, A. (2023) Deforming the upper mantle — Olivine mechanical properties and anisotropy. *Elements*, **19**, 151–157.

Denis, C.M.M., Alard, O. and Demouchy, S. (2015) Water content and hydrogen behaviour during metasomatism in the uppermost mantle beneath Ray Pic volcano (Massif Central, France). *Lithos*, **236–237**, 256–274.

Di Leo, J.F., Wookey, J., Hammond, J.O.S., Kendall, J.M., Kaneshima, S., Inoue, H., Yamashina, T. and Harjadi, P. (2012) Deformation and mantle flow beneath the Sangihe subduction zone from seismic anisotropy. *Physics of the Earth and Planetary Interiors*, **194–195**, 38–54.

Durham, W.B. and Goetze, C. (1977) Plastic flow of oriented single crystals of olivine: 1. Mechanical data. *Journal of Geophysical Research*, **82**, 5737–5753.

Escario, S., Godard, M., Gouze, P. and Leprovost, R. (2018) Experimental study of the effects of solute transport on reaction paths during incipient serpentinization. *Lithos*, **323**, 191–207.

Escartín, J., Hirth, G., Evans, B. (2001) Strength of slightly serpentinized peridotites: Implications for the tectonics of oceanic lithosphere. *Geology*, **29**, 1023-1026.

Evans, B.W., Hattori, K. and Baronnet, A. (2013) Serpentinite: what, why, where? *Elements*, **9**, 99–106.

Faul, U.H., Cline Ii, C.J., David, E.C., Berry, A.J. and Jackson, I. (2016) Titanium-hydroxyl defect-controlled rheology of the Earth's upper mantle. *Earth and Planetary Science Letters*, **452**, 227–237.

Fei, H., Wiedenbeck, M., Yamazaki, D. and Katsura, T. (2013) Small effect of water on upper-mantle rheology based on silicon self-diffusion coefficients. *Nature*, **498**, 213–215.

Früh-Green, G.L., Connolly, J.A.D., Plas, A., Kelley, D.S. and Grobéty, B. (2004) Serpentinization of oceanic peridotites: implications for geochemical cycles and biological activity. Pp. 119–136 in: *The Subseafloor Biosphere at Mid-Ocean Ridges* (W.S.D. Wilcock, E.F. Delong, D.S. Kelley, J.A. Baross and S.C. Cary, editors). Geophysical Monograph Series. American Geophysical Union.

Fumagalli, P. and Klemme, S. (2015) 2.02 – Mineralogy of the Earth: Phase Transitions and Mineralogy of the Upper Mantle. Pp. 7–31 in: *Treatise on Geophysics* (2nd edition). Elsevier, Oxford, UK.

Fusseis, F., Regenauer-Lieb, K., Liu, J., Hough, R.M. and De Carlo, F. (2009) Creep cavitation can establish a dynamic granular fluid pump in ductile shear zones. *Nature*, **459**, 974–977.

Girard, J., Chen, J., Raterron, P. and Holyoke, C.W. (2013) Hydrolytic weakening of olivine at mantle pressure: Evidence of [100](010) slip system softening from single-crystal deformation experiments. *Physics of the Earth and Planetary Interiors*, **216**, 12–20.

Green, D.H. (1973) Experimental melting studies on a model upper mantle composition at high pressure under water-saturated and water-undersaturated conditions. *Earth and Planetary Science Letters*, **19**, 37–53.

Green, D.H. (2015) Experimental petrology of peridotites, including effects of water and carbon on melting in the Earth's upper mantle. *Physics and Chemistry of Minerals*, **42**, 95–122.

Green, D.H., Hibberson, W.O., Kovács, I. and Rosenthal, A. (2010) Water and its influence on the lithosphere-asthenosphere boundary. *Nature*, **467**, 448–451.

Greenwood, H.J. (1961) The system $NaAlSi_2O_6$-H_2O-argon: Total pressure and water pressure in metamorphism. *Journal of Geophysical Research (1896-1977)*, **66**, 3923–3946.

Griggs, D. (1967) Hydrolytic weakening of quartz and other silicates. *Geophysical Journal of the Royal Astronomical Society*, **14**, 19–31.

Griggs, D.T. and Blacic, J.D. (1965) Quartz: anomalous weakness of synthetic crystals. *Science*, **147**, 292–295.

Hansen, L.N., Faccenda, M. and Warren, J.M. (2021) A review of mechanisms generating seismic anisotropy in the upper mantle. *Physics of the Earth and Planetary Interiors*, **313**, 106662.

Hao, Y.-T., Xia, Q.-K., Tian, Z.-Z. and Liu, J. (2016) Mantle metasomatism did not modify the initial H_2O content in peridotite xenoliths from the Tianchang basalts of eastern China. *Lithos*, **260**, 315–327.

Harigane, Y., Okamoto, A., Morishita, T., Snow, J.E., Tamura, A., Yamashita, H., Michibayashi, K., Ohara, Y. and Arai, S. (2019) Melt–fluid infiltration along detachment shear zones in oceanic core complexes: Insights from amphiboles in gabbro mylonites from the Godzilla Megamullion, Parece Vela Basin, the Philippine Sea. *Lithos*, **344–345**, 217–231.

Hermann, J., Spandler, C., Hack, A. and Korsakov, A.V. (2006) Aqueous fluids and hydrous melts in high-pressure and ultra-high pressure rocks: Implications for element transfer in subduction zones. *Lithos*, **92**, 399–417.

Hidas, K., Booth-Rea, G., Garrido, C.J., Martínez-Martínez, J.M., Padrón-Navarta, J.A., Konc, Z., Giaconia, F., Frets, E. and Marchesi, C. (2013) Backarc basin inversion and subcontinental mantle emplacement in the crust: kilometre-scale folding and shearing at the base of the proto-Alborán lithospheric mantle (Betic Cordillera, southern Spain). *Journal of the Geological Society, London*, **170**, 47–55.

Hidas, K., Tommasi, A., Garrido, C.J., Padrón-Navarta, J.A., Mainprice, D., Vauchez, A., Barou, F. and Marchesi, C. (2016) Fluid-assisted strain localization in the shallow subcontinental lithospheric mantle. *Lithos*, **262**, 636–650.

Hier-Majumder, S. and Kohlstedt, D.L. (2006) Role of dynamic grain boundary wetting in fluid circulation beneath volcanic arcs. *Geophysical Research Letters*, **33**.

Higgie, K. and Tommasi, A. (2012) Feedbacks between deformation and melt distribution in the crust-mantle transition zone of the Oman ophiolite. *Earth and Planetary Science Letters*, **359**, 61–72.

Higgie, K. and Tommasi, A. (2014) Deformation in a partially molten mantle: Constraints from plagioclase lherzolites from Lanzo, western Alps. *Tectonophysics*, **615**, 167–181.

Hirth, G. and Kohlstedt, D.L. (1995a) Experimental constraints on the dynamics of the partially molten upper-mantle – Deformation in the diffusion creep regime. *Journal of Geophysical Research – Solid Earth*, **100**, 1981–2001.

Hirth, G. and Kohlstedt, D.L. (1995b) Experimental constraints on the dynamics of the partially molten upper-mantle – Deformation in the dislocation creep regime. *Journal of Geophysical Research – Solid Earth*, **100**, 15441–15449.

Hirth, G. and Kohlstedt, D.L. (1996) Water in the oceanic upper mantle: Implications for rheology, melt extraction and the evolution of the lithosphere. *Earth and Planetary Science Letters*, **144**, 93–108.

Hirth, G. and Kohlstedt, D. (2003) Rheology of the upper mantle and the mantle wedge: a view from the experimentalists. Pp. 83–105 in: *Inside the Subduction Factory* (J.M. Eiler, editor). Geophysical Monograph, American Geophysical Union, San Francisco, USA.

Holland, T.J.B. and Powell, R. (2011) An improved and extended internally consistent thermodynamic dataset for phases of petrological interest, involving a new equation of state for solids. *Journal of Metamorphic Geology*, **29**, 333–383.

Holland, T.J.B., Green, E.C.R. and Powell, R. (2018) Melting of Peridotites through to Granites: A Simple Thermodynamic Model in the System KNCFMASHTOCr. *Journal of Petrology*, **59**, 881–900.

Holtzman, B.K., Groebner, N.J., Zimmerman, M.E., Ginsberg, S.B. and Kohlstedt, D.L. (2003a) Stress-driven melt segregation in partially molten rocks. *Geochemistry Geophysics Geosystems*, **4**, art. no.-8607.

Holtzman, B.K., Kohlstedt, D.L., Zimmerman, M.E., Heidelbach, F., Hiraga, T. and Hustoft, J. (2003b) Melt segregation and strain partitioning: Implications for seismic anisotropy and mantle flow. *Science*, **301**, 1227–1230.

Huang, Y., Nakatani, T., Nakamura, M. and McCammon, C. (2019) Saline aqueous fluid circulation in mantle wedge inferred from olivine wetting properties. *Nature Communications*, **10**, 5557.

Jaroslow, G.E., Hirth, G. and Dick, H.J.B. (1996) Abyssal peridotite mylonites: implications for grain-size sensitive flow and strain localization in the oceanic lithosphere. *Tectonophysics*, **256**, 17–37.

Jollands, M.C., Dohmen, R. and Padrón-Navarta, J.A. (2023) Hide and seek — Trace element incorporation and diffusion in olivine. *Elements*, **19**, 144–150.

Jung, H. and Karato, S.-I. (2001) Water-induced fabric transitions in olivine. *Science*, **293**, 1460–1463.

Jung, H., Katayama, I., Jiang, Z., Hiraga, I. and Karato, S. (2006) Effect of water and stress on the lattice-preferred orientation of olivine. *Tectonophysics*, **421**, 1-22.

Kakihata, Y., Michibayashi, K. and Dick, H.J.B. (2022) Heterogeneity in texture and crystal fabric of intensely hydrated ultramylonitic peridotites along a transform fault, Southwest Indian Ridge. *Tectonophysics*, **823**, 229206.

Karato, S.-I. (2004) Mapping water content in the upper mantle. Pp. 135–152 in: *Inside the Subduction Factory* (J. Eiler, editor). Geophysical Monograph Series, **138**, American Geophysical Union, Washington, D.C.

Karato, S.-I. (2008) *Deformation of Earth materials: An Introduction to the Rheology of Solid Earth.* Cambridge University Press, Cambridge, New York, 463 pp.

Karato, S.-I. (2010) Rheology of the Earth's mantle: A historical review. *Gondwana Research*, **18**, 17–45.

Karato, S.-I. (2013) *Rheological Properties of Minerals and Rocks. Physics and Chemistry of the Deep Earth.* John Wiley & Sons, Ltd, pp. 94–144.

Karato, S.-I. and Jung, H. (2003) Effects of pressure on high-temperature dislocation creep in olivine. *Philosophical Magazine*, **83**, 401–414.

Karato, S.-I. and Wu, P. (1993) Rheology of the Upper Mantle: A Synthesis. *Science*, **260**, 771–778.

Karato, S.-I., Jung, H., Katayama, I. and Skemer, P. (2008) Geodynamic significance of seismic anisotropy of the upper mantle: New insights from laboratory studies. *Annual Review of Earth and Planetary Sciences*, **36**, 59–95.

Katayama, I. and Karato, S.-I. (2006) Effect of temperature on the B- to C-type olivine fabric transition and implication for flow pattern in subduction zones. *Physics of The Earth and Planetary Interiors*, **157**, 33–45.

Katayama, I., Jung, H. and Karato, S.-I. (2004) New type of olivine fabric from deformation experiments at modest water content and low stress. *Geology*, **32**, 1045–1048.

King, D.S.H., Zimmerman, M.E. and Kohlstedt, D.L. (2010) Stress-driven Melt Segregation in Partially Molten Olivine-rich Rocks Deformed in Torsion. *Journal of Petrology*, **51**, 21–42.

Klemme, S. and Stalder, R. (2018) Halogens in the Earth's mantle: What we know and what we don't. Pp. 847–869 in: *The Role of Halogens in Terrestrial and Extraterrestrial Geochemical Processes: Surface, Crust, and Mantle* (D.E. Harlov and L. Aranovich, editors). Springer International Publishing.

Kneller, E.A., van Keken, P.E., Karato, S.-I. and Park, J. (2005) B-type olivine fabric in the mantle wedge: Insights from high-resolution non-Newtonian subduction zone models. *Earth and Planetary Science Letters*, **237**, 781–797.

Kohli, A.H. and Warren, J.M. (2020) Evidence for a deep hydrologic cycle on oceanic transform faults. *Journal of Geophysical Research: Solid Earth*, **125**, e2019JB017751.

Kohlstedt, D.L. and Hansen, L.N. (2015) Constitutive equations, rheological behavior, and viscosity of rocks. Pp. 441–472 in: *Treatise on Geophysics* (2nd edition). (G. Schubert, editor). Elsevier, Amsterdam.

Kohlstedt, D.L. and Zimmerman, M.E. (1996) Rheology of partially molten mantle rocks. *Annual Review of Earth and Planetary Sciences*, **24**, 41–62.

Kohlstedt, D.L., Evans, B. and Mackwell, S.J. (1995) Strength of the lithosphere - Constraints imposed by laboratory experiments. *Journal of Geophysical Research – Solid Earth*, **100**, 17587–17602.

Kohlstedt, D.L., Zimmerman, M.E. and Mackwell, S.J. (2010) Stress-driven melt segregation in partially molten feldspathic rocks. *Journal of Petrology*, **51**, 9–19.

Konzett, J. and Ulmer, P. (1999) The stability of hydrous potassic phases in lherzolitic mantle – An experimental study to 9.5 GPa in simplified and natural bulk compositions. *Journal of Petrology*, **40**, 629–652.

Kovács, I.J., Liptai, N., Koptev, A., Cloetingh, S.A.P.L., Lange, T.P., Matenco, L., Szakács, A., Radulian, M., Berkesi, M., Patkó, L., Molnár, G., Novák, A., Wesztergom, V., Szabó, Cs. and Fancsik, T. (2021) The 'pargasosphere' hypothesis: Looking at global plate tectonics from a new perspective. *Global and Planetary Change*, **204**, 103547.

Kumamoto, K.M., Warren, J.M. and Hansen, L.N. (2019) Evolution of the Josephine Peridotite shear zones: 2. Influences on olivine CPO evolution. *Journal of Geophysical Research: Solid Earth*, **124**, 12763–12781.

Lassak, T.M., Fouch, M.J., Hall, C.E. and Kaminski, É. (2006) Seismic characterization of mantle flow in subduction systems: Can we resolve a hydrated mantle wedge? *Earth and Planetary Science Letters*, **243**, 632–649.

Long, M.D. and Silver, P.G. (2008) The Subduction Zone Flow Field from Seismic Anisotropy: A Global View. *Science*, **319**, 315–318.

Mainprice, D., Bachmann, F., Hielscher, R. and Schaeben, H. (2014) Descriptive tools for the analysis of texture projects with large datasets using MTEX: strength, symmetry and components. Special Publications, **409**, Geological Society, London.

Mainprice, D., Barruol, G. and Ben Ismaïl, W. (2000) The seismic anisotropy of the Earth's mantle: from single crystal to polycrystal. Pp. 237–264 in: *Earth's Deep Interior: Mineral Physics and Tomography from the Atomic to the Global Scale* (S.I. Karato, editor). Geodynamics Series, AGU, Washington, D.C.

Mandler, B.E. and Grove, T.L. (2016) Controls on the stability and composition of amphibole in the Earth's mantle. *Contributions to Mineralogy and Petrology*, **171**, 68.

Manning, C.E. (2004a) The chemistry of subduction-zone fluids. *Earth and Planetary Science Letters*, **223**, 1–16.

Manning, C.E. (2004b) Polymeric silicate complexing in aqueous fluids at high pressure and temperature, and its implications for water-rock interaction. Pp. 45–49 in: *Water–Rock Interaction* (R.B. Wanty and R.R. Seal, II, editors). Taylor & Francis Group, London.

Mei, S. and Kohlstedt, D.L. (2000a) Influence of water on plastic deformation of olivine aggregates: 1. Diffusion creep regime. *Journal of Geophysical Research: Solid Earth*, **105**, 21457–21469.

Mei, S. and Kohlstedt, D.L. (2000b) Influence of water on plastic deformation of olivine aggregates: 2. Dislocation creep regime. *Journal of Geophysical Research: Solid Earth*, **105**, 21471–21481.

Mei, S., Bai, W., Hiraga, T. and Kohlstedt, D.L. (2002) Influence of melt on the creep behavior of olivine-basalt aggregates under hydrous conditions. *Earth and Planetary Science Letters*, **201**, 491–507.

Melekhova, E., Schmidt, M.W., Ulmer, P. and Pettke, T. (2007) The composition of liquids coexisting with dense hydrous magnesium silicates at 11–13.5 GPa and the endpoints of the solidi in the MgO–SiO$_2$–H$_2$O system. *Geochimica et Cosmochimica Acta*, **71**, 3348–3360.

Mengel, K. and Green, D.H. (1989) Stability of amphibole and phlogopite in metasomatized peridotite under water-saturated and water-undersaturated conditions. Pp. 571–581 in: *Fourth International Kimberlite Conference, Perth* (J. Ross, editor). Special Publication, Geological Society of America, Boulder, Colorado, USA.

Mibe, K., Fujii, T. and Yasuda, A. (1999) Control of the location of the volcanic front in island arcs by aqueous fluid connectivity in the mantle wedge. *Nature*, **401**, 259–262.

Mibe, K., Kanzaki, M., Kawamoto, T., Matsukage, K.N., Fei, Y. and Ono, S. (2007) Second critical endpoint in the peridotite-H$_2$O system. *Journal of Geophysical Research: Solid Earth*, **112**, B03201.

Mizukami, T., Wallis, S.R. and Yamamoto, J. (2004) Natural examples of olivine lattice preferred orientation patterns with a flow-normal a-axis maximum. *Nature*, **427**, 432–436.

Mysen, B. (2022) Fluids and physicochemical properties and processes in the Earth. *Progress in Earth and Planetary Science*, **9**, 54.

Newton, R.C. and Manning, C.E. (2002) Experimental determination of calcite solubility in H$_2$O-NaCl solutions at deep crust/upper mantle pressures and temperatures: Implications for metasomatic processes in shear zones. *American Mineralogist*, **87**, 1401–1409.

O'Reilly, S. and Griffin, W.L. (2013) *Mantle Metasomatism, Metasomatism and the Chemical Transformation of Rock.* Springer Berlin Heidelberg, pp. 471–533.

Otsuka, K. and Karato, S.-I. (2011) Control of the water fugacity at high pressures and temperatures: Applications to the incorporation mechanisms of water in olivine. *Physics of the Earth and Planetary Interiors*, **189**, 27–33.

Padrón-Navarta, J.A. (2019) Phase diagrams for Nominally Anhydrous Minerals, Goldschmidt2019. *Goldschmidt Abstracts*, Barcelona, Spain, p. 2548.

Padrón-Navarta, J.A. and Hermann, J. (2017) A subsolidus olivine water solubility equation for the Earth's upper mantle. *Journal of Geophysical Research: Solid Earth*, **122**, 9862–9880.

Padrón-Navarta, J.A., Tommasi, A., Garrido, C.J., López Sánchez-Vizcaíno, V. (2012). Plastic deformation and development of antigorite crystal preferred orientation in high-pressure serpentinites. *Earth and Planetary Science Letters*, **349–350**, 75–86.

Padrón-Navarta, J.A., Hermann, J., Garrido, C.J., López Sánchez-Vizcaíno, V. and Gómez-Pugnaire, M.T. (2010) An experimental investigation of antigorite dehydration in natural silica-enriched serpentinite. *Contributions to Mineralogy and Petrology*, **159**, 25–42.

Padrón-Navarta, J.A., Hermann, J. and O'Neill, H.S.C. (2014) Site-specific hydrogen diffusion rates in forsterite. *Earth and Planetary Science Letters*, **392**, 100–112.

Paterson, M.S. (1982) The determination of hydroxyl by infrared absorption in quartz, silicate glasses and similar materials. *Bulletin de Minéralogie*, **105**, 20–29.

Patterson, S.N., Lynn, K.J., Prigent, C. and Warren, J.M. (2021) High temperature hydrothermal alteration and amphibole formation in Gakkel Ridge abyssal peridotites. *Lithos*, **392–393**, 106107.

Pitzer, K.S. and Sterner, S.M. (1994) Equations of state valid continuously from zero to extreme pressures for H_2O and CO_2. *Journal of Chemical Physics*, **101**, 3111–3116.

Powell, R., Evans, K.A., Green, E.C.R. and White, R.W. (2019) The truth and beauty of chemical potentials. *Journal of Metamorphic Geology*, **37**, 1007–1019.

Précigout, J., Prigent, C., Palasse, L. and Pochon, A. (2017) Water pumping in mantle shear zones. *Nature Communications*, **8**, 15736.

Précigout, J., Stünitz, H. and Villeneuve, J. (2019) Excess water storage induced by viscous strain localization during high-pressure shear experiment. *Scientific Reports*, **9**, 3463.

Prigent, C., Guillot, S., Agard, P. and Ildefonse, B. (2018) Fluid-Assisted Deformation and Strain Localization in the Cooling Mantle Wedge of a Young Subduction Zone (Semail Ophiolite). *Journal of Geophysical Research: Solid Earth*, **123**, 7529–7549.

Prigent, C., Warren, J.M., Kohli, A.H. and Teyssier, C. (2020) Fracture-mediated deep seawater flow and mantle hydration on oceanic transform faults. *Earth and Planetary Science Letters*, **532**, 115988.

Raterron, P., Amiguet, E., Chen, J., Li, L. and Cordier, P. (2009) Experimental deformation of olivine single crystals at mantle pressures and temperatures. *Physics of The Earth and Planetary Interiors*, **172**, 74–83.

Raterron, P., Chen, J., Geenen, T. and Girard, J. (2011) Pressure effect on forsterite dislocation slip systems: Implications for upper-mantle LPO and low viscosity zone. *Physics of the Earth and Planetary Interiors*, **188**, 26–36.

Raterron, P., Girard, J. and Chen, J. (2012) Activities of olivine slip systems in the upper mantle. *Physics of the Earth and Planetary Interiors*, **200-201**, 105–112.

Rouméjon, S. and Cannat, M. (2014) Serpentinization of mantle-derived peridotites at mid-ocean ridges: Mesh texture development in the context of tectonic exhumation. *Geochemistry, Geophysics, Geosystems*, **15**, 2354 – 2379.

Rouméjon, S., Cannat, M., Agrinier, P., Godard, M. and Andreani, M. (2015) Serpentinization and fluid pathways in tectonically exhumed peridotites from the Southwest Indian Ridge (62–65°E). *Journal of Petrology*, **56**, 703–734.

Rutter, E.H. (1976) The kinetics of rock deformation by pressure solution. *Philosophical Transactions of the Royal Society of London*, **A283**.

Schwartz, S., Guillot, S., Reynard, B., Lafay, R., Debret, B., Nicollet, C., Lanari, P. and Auzende, A.L. (2013) Pressure–temperature estimates of the lizardite/antigorite transition in high pressure serpentinites. *Lithos*, **178**, 197–210.

Scott, T. and Kohlstedt, D.L. (2006). The effect of large melt fraction on the deformation behavior of peridotite. *Earth and Planetary Science Letters*, **246**, 177–187.

Shaw, C.S.J. (2009) Textural development of amphibole during breakdown reactions in a synthetic peridotite. *Lithos*, **110**, 215–228.

Soustelle, V., Tommasi, A., Demouchy, S. and Ionov, D.A. (2010) Deformation and Fluid–Rock Interaction in the Supra-subduction Mantle: Microstructures and Water Contents in Peridotite Xenoliths from the Avacha Volcano, Kamchatka. *Journal of Petrology*, **51**, 363–394.

Soustelle, V., Walte, N.P., Manthilake, M.A.G.M. and Frost, D.J. (2014) Melt migration and melt-rock reactions in the deforming Earth's upper mantle: Experiments at high pressure and temperature. *Geology*, **42**, 83–86.

Spear, F.S. (1995) *Metamorphic Phase Equilibria and Pressure-temperature-time Paths*. Mineralogical Society of America, Washington, D.C., 799 pp.

Spiers, C.J., De Meer, S., Niemeijer, A.R. and Zhang, X. (2004) Kinetics of rock deformation by pressure solution and the role of thin aqueous films. Pp. 129–158 in: *Physicochemistry of Water in Geological and Biological Systems: Structures and Properties of Thin Aqueous Films* (S. Nakashima, C.J. Spiers, L. Mercury, P.A. Fenter and M.F. Hocheller, editors). Universal Academy Press, Inc., Tokyo.

Takei, Y. and Holtzman, B.K. (2009a) Viscous constitutive relations of solid-liquid composites in terms of grain boundary contiguity: 1. Grain boundary diffusion control model. *Journal of Geophysical Research: Solid Earth*, **114**, B06205.

Takei, Y. and Holtzman, B.K. (2009b) Viscous constitutive relations of solid-liquid composites in terms of grain boundary contiguity: 2. Compositional model for small melt fractions. *Journal of Geophysical Research: Solid Earth*, **114**, B06206.

Takei, Y. and Holtzman, B.K. (2009c) Viscous constitutive relations of solid-liquid composites in terms of grain boundary contiguity: 3. Causes and consequences of viscous anisotropy. *Journal of Geophysical Research: Solid Earth*, **114**, B06207.

Tasaka, M., Michibayashi, K. and Mainprice, D. (2008) B-type olivine fabrics developed in the fore-arc side of the mantle wedge along a subducting slab. *Earth and Planetary Science Letters*, **272**, 747–757.

Tasaka, M., Zimmerman, M.E. and Kohlstedt, D.L. (2016) Evolution of the rheological and microstructural properties of olivine aggregates during dislocation creep under hydrous conditions. *Journal of Geophysical Research: Solid Earth*, 121, 92–113.

Tielke, J.A., Zimmerman, M.E. and Kohlstedt, D.L. (2017) Hydrolytic weakening in olivine single crystals. *Journal of Geophysical Research: Solid Earth*, **122**, 3465–3479.

Tollan, P.M.E., O'Neill, H.S.C. and Hermann, J. (2018) The role of trace elements in controlling H incorporation in San Carlos olivine. *Contributions to Mineralogy and Petrology*, **173**, 89.

Tommasi, A. and Vauchez, A. (2015) Heterogeneity and anisotropy in the lithospheric mantle. *Tectonophysics*, **661**, 11–37.

Tommasi, A., Tikoff, B. and Vauchez, A. (1999) Upper mantle tectonics: three-dimensional deformation, olivine crystallographic fabrics and seismic properties. *Earth and Planetary Science Letters*, **168**, 173–186.

Tommasi, A., Mainprice, D., Canova, G. and Chastel, Y. (2000) Viscoplastic self-consistent and equilibrium-based modeling of olivine lattice preferred orientations: Implications for the upper mantle seismic anisotropy. *Journal of Geophysical Research – Solid Earth*, **105**, 7893–7908.

Tommasi, A., Langone, A., Padrón-Navarta, J.A., Zanetti, A. and Vauchez, A. (2017) Hydrous melts weaken the mantle, crystallization of pargasite and phlogopite does not: Insights from a petrostructural study of the Finero peridotites, southern Alps. *Earth and Planetary Science Letters*, **477**, 59–72.

Urai, J.L., Means, W.D. and Lister, G.S. (1986) Dynamic recrystallization of minerals. Pp. 161–199 in: *Mineral and Rock Deformation: Laboratory Studies (the Paterson volume)* (B.E. Hobbs and H.C. Heard, editors). American Geophysical Union, Washington, D.C.

Vannucci, R., Piccardo, G. B., Rivalenti, G., Zanetti, A., Rampone, E., Ottolini, L., Oberti, R., Mazzucchelli, M. and Bottazzi, P. (1995) Origin of LREE-depleted amphiboles in the subcontinental mantle. *Geochimica et Cosmochimica Acta*, **59(9)**, 1763–1771.

Vieira Duarte, J.F., Kaczmarek, M.-A., Vonlanthen, P., Putlitz, B. and Müntener, O. (2020) Hydration of a Mantle Shear Zone Beyond Serpentine Stability: A Possible Link to Microseismicity Along Ultraslow Spreading Ridges? *Journal of Geophysical Research: Solid Earth*, **125**, e2020JB019509.

Vissers, R.L.M., Drury, M.R., Strating, E.H.H., Spiers, C.J. and Vanderwal, D. (1995) Mantle shear zones and their effect on lithosphere strength during continental breakup. *Tectonophysics*, **249**, 155–171.

Wallis, D., Hansen, L.N., Tasaka, M., Kumamoto, K.M., Parsons, A.J., Lloyd, G.E., Kohlstedt, D.L. and Wilkinson, A.J. (2019) The impact of water on slip system activity in olivine and the formation of bimodal crystallographic preferred orientations. *Earth and Planetary Science Letters*, **508**, 51–61.

Wang, C., Liang, Y. and Xu, W. (2021) Formation of amphibole-bearing peridotite and amphibole-bearing pyroxenite through hydrous melt-peridotite reaction and in situ crystallization: an experimental study. *Journal of Geophysical Research: Solid Earth*, **126**, e2020JB019382.

Wang, J., Takahashi, E., Xiong, X., Chen, L., Li, L., Suzuki, T. and Walter, M.J. (2020) The water-saturated solidus and second critical endpoint of peridotite: implications for magma genesis within the mantle wedge. *Journal of Geophysical Research: Solid Earth*, **125**, e2020JB019452.

Warren, J.M. and Hirth, G. (2006) Grain size sensitive deformation mechanisms in naturally deformed peridotites. *Earth and Planetary Science Letters*, **248**, 438–450.

Wassmann, S. and Stöckhert, B. (2013) Rheology of the plate interface — Dissolution precipitation creep in high pressure metamorphic rocks. *Tectonophysics*, **608**, 1–29.

Wassmann, S., Stöckhert, B. and Trepmann, C.A. (2011) Dissolution precipitation creep versus crystalline plasticity in high-pressure metamorphic serpentinites. *Geological Society, London, Special Publications*, **360**, 129–149.

Watson, E.B. and Brenan, J.M. (1987) Fluids in the lithosphere. 1: Experimentally-determined wetting characteristics of CO_2-H_2O fluids and their implications for fluid transport, host-rock physical properties, and fluid inclusion formation. *Earth and Planetary Science Letters*, **85**, 497–515.

Whitney, D.L. and Evans, B.W. (2010) Abbreviations for names of rock-forming minerals. *American Mineralogist*, **95**, 185–187.

Wintsch, R.P. and Yi, K. (2002) Dissolution and replacement creep: a significant deformation mechanism in mid-crustal rocks. *Journal of Structural Geology*, **24**, 1179–1193.

Wyllie, P.J. and Ryabchikov, I.D. (2000) Volatile components, magmas, and critical fluids in upwelling mantle. *Journal of Petrology*, **41**, 1195–1206.

Zanetti, A., Vannucci, R., Bottazzi, P., Oberti, R. and Ottolini, L. (1996) Infiltration metasomatism at Lherz as monitored by systematic ion-microprobe investigations close to a hornblendite vein. *Chemical Geology*, **134**, 113–133.

Zhang, M., Niu, Y. and Hu, P. (2009) Volatiles in the mantle lithosphere: Modes of occurrence and chemical compositions. Pp. 171–212 in: *The Lithosphere: Geochemistry, Geology and Geophysics* (J.E. Anderson, editor). Nova Science Publishers Inc.

Zhang, S., Karato, S.-I., Fitz Gerald, J., Faul, U.H. and Zhou, Y. (2000) Simple shear deformation of olivine aggregates. *Tectonophysics*, **316**, 133–152.

Zimmerman, M.E. and Kohlstedt, D.L. (2004) Rheological properties of partially molten lherzolite. *Journal of Petrology*, **45**, 275–298.

Zimmerman, M.E., Zhang, S.Q., Kohlstedt, D.L. and Karato, S.-I. (1999) Melt distribution in mantle rocks deformed in shear. *Geophysical Research Letters*, **26**, 1505–1508.

Kinetic controls on the thermometry of mantle rocks: A case study from the Xigaze Ophiolites, Tibet

L. Zhao and S. Chakraborty

Institut für Geologie, Mineralogie und Geophysik, Ruhr Universität Bochum, Bochum, Germany

Temperature-dependent equilibrium partitioning of elements between different mineral (or melt/glass) phases forms the basis of geothermometry. In natural rock systems it is necessary to determine whether equilibrium partitioning of a given element was obtained between two phases before calculating temperatures using the tool. With the improvement of spatial resolution of analytical tools and our understanding of solid-state kinetics it has become clear that compositional heterogeneities on different scales exist in mantle rocks because of incomplete equilibration, and a kinetic evaluation is necessary before application of geothermometers. This work summarizes the kinetic situations that may arise and provides some guidelines and criteria for testing whether partitioning equilibrium was obtained. A suite of dunites and harzburgites from an ophiolite suite in the Himalaya (Xigaze, Tibet) is used to illustrate the application of some of these concepts. It is shown that when compositions used for geothermometry are chosen bearing these kinetic considerations in mind, a systematic pattern of freezing temperatures is obtained from the geothermometers. These data provide insights into the cooling histories of these rocks with complex, multistage (e.g. melt percolation) histories. Some potential pitfalls for geospeedometry are also illustrated along the way.

1. Introduction

Petrological analysis has evolved from the study of rocks to understand the condition of formation to a recognition that rocks record a 'chain of processes' that took place in the course of their evolution through different pressure-temperature conditions. So, the question becomes: when does a given geothermometer/barometer 'freeze', and at what stage in the history of the rock does it record? Conventionally, thermobarometry has been studied largely in the domain of thermodynamics, but with this expanded view, kinetics plays as much of a role. As thermobarometry becomes a freezing issue rather than just a phase equilibrium problem, one has to ask, for example, the question: Which compositions were in equilibrium with each other (i.e. what should be measured and combined with each other) to get pressures and temperatures? Equilibrium calculations do not depend on the history of a system, and consideration of only the phases and components involved in the equilibrium process determine the system completely. Kinetics, on the other hand, does depend on pathways of processes (e.g. through initial- and boundary-conditions mechanisms of reaction). Therefore, the practice of thermobarometry becomes dependent on the nature and behaviour of the surrounding medium in addition to those involved directly in the equilibrium of interest, and the temperatures that are

DOI: 10.1180/EMU-notes.21.7

obtained from different element exchange geothermometers depend on the kinetics of element exchange between the mineral phases concerned.

Generally, the differences between freezing temperatures are amplified for slower cooling rates. Element exchange thermometry of mantle rocks is an important tool for reconstructing the thermal history of different processes in the mantle. Compared to rapidly ascended xenoliths, the role of kinetics is more pronounced in the slowly cooled and emplaced mantle segments in ophiolites. On the one hand, the spatial context provided by different lithological units within an ophiolite sequence provide additional information and an internal check of methods. On the other hand, mineral pairs within rocks in an ophiolite sequence are affected by multiple events (= thermal pulses) that result from percolating melts that affect the phase assemblages and compositional distributions; and these need to be considered in the determination of compositions for and interpretation of the results of thermometry. Thus mantle ophiolite sequences provide some challenges as well as unique opportunities to study the kinetic controls on thermometry. Here, after outlining the principles of element exchange thermometry and some of the kinetic controls that govern element exchange we present an illustration using samples from the Xigaze ophiolite in Tibet. The implications of the results also include aspects about the application of geospeedometry to such mantle rocks.

2. Thermodynamics of thermobarometry and the role of kinetics

If two elements (i and j) interchange between phases α and β by the reaction: i-α + j-β $\leftrightarrow$ j-α + i-β, then K_D is defined as the ratio of the two exchanged elements in one phase times the inverse ratio of the two exchanged elements in the other phase (equation 1).

$$K_D = \frac{X_i^\alpha * X_j^\beta}{X_j^\alpha * X_i^\beta}, \text{ where } X_i^\alpha = i/(i+j) \tag{1}$$

K_D at equilibrium is a function of only P and T in thermodynamically ideal systems, and is a function of composition as well in non-ideal systems. The temperature dependence of K_D in the form $\ln K_d = A/T + B + C(X)$, is used as a thermometer. Here A and B are constants the values of which depend on the identities of i, j, α and β; $C(X)$ appears only in non-ideal systems and may have complex functional forms. Conventionally, the compositions X_i^α are measured in a sample to calculate a K_D which is then compared to an experimental (or model) calibration of the same as a function of temperature to determine the temperature of the natural sample. The underlying assumption of this exercise is that the measured compositions in the natural sample represent a frozen equilibrium state at some point in the history of the sample. In the event where multiple compositions of the same mineral are found in a sample (either as compositional zoning within individual grains; or as different grains, often in different textural locations), a process of kinetic evaluation is necessary to determine which compositions to pair with which others, and whether any of the compositions were in equilibrium with each other.

Two relationships derived from the above, that are useful (Ganguly and Saxena, 1987) are:

$$X_i^\alpha = \frac{X_i^\beta}{K_D(1 - X_i^\beta) + X_i^\beta} \tag{2}$$

$$X_i^\beta = \frac{K_D * X_i^\alpha}{X_i^\alpha(K_D - 1) + 1} \tag{3}$$

These equations hold if both phases (α and β) are ideal solutions; again, additional terms appear in non-ideal systems but the general behaviour may still be studied with the help of these equations. Kinetic tests essentially involve checking if equations 2 and 3 are satisfied. Towards this end, Roozeboom diagrams that plot compositional pairs X_i^α vs. X_i^β are a valuable aid. Indeed, these have been used since the early days of applications of thermodynamics in petrology to demonstrate that phase equilibrium is obtained among natural mineral assemblages and thermodynamics may be applied at all (Ganguly and Saxena, 1987; see Ganguly, 2021, for a historical perspective). In such a plot, calibration curves at different temperatures that were used to develop a thermometer (e.g. from experiments or a model) are plotted. Compositional pairs X_i^α vs. X_i^β that plot on or adjacent to these curves indicate that those analyses could be either in or approaching equilibrium. On the other hand, a scatter (or plot outside the domain of calibrated curves) indicates that element exchange equilibrium was not achieved and these compositions should not be used for thermometry (examples are provided in the case study below). The Roozeboom diagram of X_i^α vs. X_i^β would be symmetrical with reference to the intersection diagonal in ideal solutions at a given constant P and T. Other criteria in addition to consistency in a Roozeboom diagram are necessary for a complete kinetic test; these are discussed in the next section.

Fe-Mg exchange between many mineral pairs have been calibrated as geothermometers and the following are used in the illustrative examples with dunites and harzburgites below: orthopyroxene-clinopyroxene (Ganguly *et al.*, 2013), olivine-clinopyroxene (Kawasaki and Ito, 1994), olivine-orthopyroxene (von Seckendorff and O'Neill, 1993), spinel-olivine (Ballhaus *et al.*, 1991) and spinel-orthopyroxene (Liermann and Ganguly, 2003). The thermometers are usually temperature dependent and are insensitive to pressure. For example, the temperature variation is <2°C/kbar and 4°C/kbar for spinel-olivine (Ballhaus *et al.*, 1991) and spinel-orthopyroxene (Liermann and Ganguly, 2003) Fe-Mg exchange thermometers respectively. In addition to these, the *REE* in the two-pyroxene thermometer of Liang *et al.* (2013) has been used in the present study.

3. Kinetics of Thermobarometry

The main change that occurs due to the entry of kinetics into the practice of thermobarometry is that spatial (= textural) context becomes important. This manifests itself in
various forms ranging from the need to identify reaction textures, through identification
of different generations of minerals, to the appropriate selection of compositions to use
for thermobarometry from a compositionally zoned mineral. As the spatial disposition
of mineral grains controls kinetics, different situations arise depending on the properties
and geometric disposition of the grains and their surrounding medium. Compositional
readjustment during cooling from some peak temperature in the absence of growth-dissolution have been considered in several models that considered primarily diffusion in
mineral grains (e.g. Dodson, 1973; Dodson, 1976; Dodson, 1986; Eiler et al., 1991;
Lasaga, 1983). Four broad types of situations may be recognized: (1) a mineral grain
is embedded in a matrix with which unrestrained (i.e. enough solubility and diffusivity
in the matrix) exchange is possible, and for thermobarometry/geospeedometry-only
analyses from this one mineral is necessary; (2) two mineral grains exchanging elements
are in contact with each other (and "leakage" via the contact grain boundary to other minerals is negligible/limited); (3) two mineral grains that are not in contact with each other
and exchange elements through a "grain boundary medium", but transport via this
medium is fast and unhindered (the 'fast grain boundary model'); and (4) the most
general case where the two mineral grains are not in contact with each other and exchange
elements through a grain boundary, but the properties of the grain boundary (ability to
contain the element of interest, transport rates, length, etc.) are also considered explicitly
(Fig. 1).

(i) **Kinetics of compositional preservation during cooling in a mineral grain
embedded in a matrix with unrestrained transport:** The formulations for this situation
are due to Dodson (1973, 1976, 1986). There is a general misconception that these relate
to diffusion-controlled processes only. The general form of the Dodson equation
(Dodson, 1976) is

$$T_c = \frac{E}{R \ln\left(-\tau \frac{k_0}{a}\right)} \tag{4}$$

where T_c is the temperature at which a mineral composition freezes, R is the gas constant,
k_0 is a pre-exponential factor and E the activation energy for 'any kinetic process that
follows an Arrhenius type of law' of temperature dependence, a is a geometric parameter
and τ is a characteristic (cooling) time constant. The more commonly seen formulation is
where the kinetic parameters have been replaced explicitly by parameters related to
diffusion. As can be seen from the expression, there are no provisions for accounting
for the behaviour of the matrix (no terms that describe the behaviour in the matrix) or
for different initial conditions (i.e. conditions far removed from the initial condition so
that the initial condition is not preserved anywhere in the crystal). Modifications that

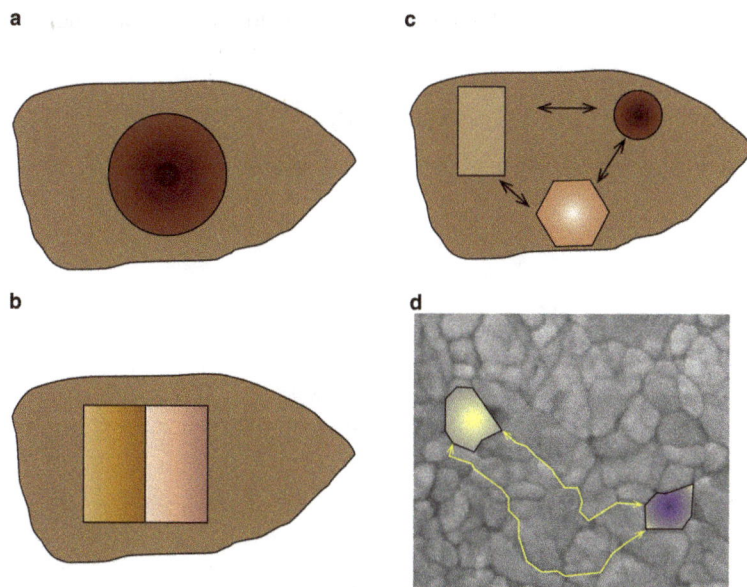

Fig. 1. Schematic representation of various models of kinetics of element exchange. (*a*) The Dodson model, where a crystal (dark brown sphere) exchanges elements with a matrix (lighter brown) where transport is unrestricted. (*b*) Geospeedmetry element exchange model of Lasaga (1983) where two crystals (rectangular prisms of different colours in the middle), with colour gradation to represent compositional zoning that develops during cooling. The surrounding matrix, isolated in this case from the exchanging crystal-pair, is depicted in the same light brown shade as in (*a*). (*c*) Fast grain boundary model of Eiler *et al.* (1991, 1992) and Jenkins *et al.* (1994) where multiple crystals sit in a matrix (light brown) physically separated from each other and communicate (= exchange elements) with each other via grain boundaries through which transport is fast and unrestricted. Colour gradation represents compositional zoning. (*d*) The Dohmen and Chakraborty model where two crystals that are physically separated from each other (yellow and blue) exchange elements with each other via a grain boundary network which has its own properties (solubility, diffusion rates for different elements as well as geometrical features such as length and surface area of contact). Colour gradation represents compositional zoning again.

account for some of these aspects have been developed (e.g. Ganguly and Tirone, 1999) that allow the formulation to be used in some cases. But the fact that the exchange partner is not considered explicitly makes the formulation of somewhat limited use and some of the formulations discussed below are more easily adaptable for describing the kinetics of geothermometry. However, the generalized form of the expression above is very useful in particular for cases where diffusion is not the only kinetic process at play (e.g. diffusion + dissolution/precipitation).

(ii) Kinetics of compositional preservation during cooling from a peak temperature for two mineral grains exchanging elements that are in contact with each other: The formulation for this situation is due to Lasaga (1983) and related works. For a system cooling linearly from a peak temperature obeying $T \approx T_P - s\,t$ where T is temperature, T_P the peak temperature, s the initial cooling rate and t is time, the partition coefficient (K_D) and diffusion coefficient are considered to decay exponentially with time obeying

$$K_D(t) = K_D^o e^{-\beta' t}$$

and

$$D(t) = D_o e^{-\gamma t}$$

The parameters controlling the variations, β' and γ, are given by

$$\beta' \equiv \frac{\Delta H^o s}{RT_P^2}$$

and

$$\gamma \equiv \frac{E s}{RT_P^2}$$

The role of the exchange partner of a crystal was described by a parameter, β/γ, where

$$\beta = \frac{\Delta H^o s \sqrt{D_B^o/D_A^o}}{RT_P^2 \left\{ \sqrt{D_B^o/D_A^o} [1 + (C_{1A}^I/C_{2A}^I)] + [(C_{1A}^I/C_{1B}^I) + (C_{1A}^I/C_{2B}^I)] \right\}}$$

which depends on the concentrations of various exchanging elements, the rate of diffusion in the two minerals, and most notably, the enthalpy change of the element exchange reaction, ΔH^o. C_{1A}^I refers to the 'initial' composition of component 1 in mineral A; other C_{XX}^I terms may be interpreted analogously.

Using these quantities, ultimately the kinetics of compositional resetting in a given crystal, for a particular value of β/γ, is described by a parameter, γ', given by equation 5:

$$r' = \frac{Esa^2}{R \cdot (T_P)^2 \cdot D_P} \tag{5}$$

where E is the activation energy, D_p is the diffusion coefficient at the peak temperature, T_p, s is the cooling rate, a is a relevant length scale and R is the universal gas constant. Two values of γ' are particularly relevant (for a β/γ of 0.15 or so which is typical of many Fe-Mg exchange geothermometers) for $\gamma' > 10$, the composition at the peak temperature is preserved at the core of a crystal, and for $\gamma' > 100$ most of the crystal retaians its peak composition and only the very rims are reset. These provide important and useful guidance for the choice of systems and compositions for geothermometry, geospeedometry and geochronology. It is seen easily from equation 5 that different minerals (with

different values of relevant diffusion parameters e.g. E, D_p) in the same rock would preserve compositions to different extents, as would different grain sizes of the same
mineral. It is interesting to note that aside from the role of the exchange medium, the
expressions for γ' in the Lasaga formulations and T_c in the Dodson formulations
depend on similar variables and in a similar manner. Even though the work was
focused on garnets, Chakraborty and Ganguly (1991, their fig. 19) provided a discussion
and an easy-to-use graphical representation of these relationships, that can be used to
evaluate the extent of compositional resetting for any element exchange thermometer
as long as β/γ lies close to 0.15.

(iii, iv) Kinetics of element exchange mediated by a grain boundary: If two grains
are physically separated from each other, properties of the intervening matrix come into
play as well in determining the extent of compositional readjustment. Early treatments of
this effect include the fast grain boundary diffusion models of Eiler *et al.* (1992) and a
similar treatment by Jenkin *et al.* (1994). As the name indicates, these models operate
under the assumption that transport in the grain boundary for all elements is infinitely
fast and effective. However, it has been shown, specifically for mantle minerals, that
not all elements are incorporated in a grain boundary to the same extent (e.g. Hiraga
et al., 2003; Hiraga *et al.*, 2004). This phenomenon is well known in the materials
science literature and is described by the grain-boundary segregation factor. Moreover,
rates of grain boundary transport of different elements also differ by several orders of
magnitude (e.g. see a review by Dohmen and Milke, 2010).

The role of these factors (i.e. concentration of an element in a grain boundary, its
diffusion rate in the grain boundary, as well as the rate at which the element is
exchanged between a crystal and the grain boundary) have been considered in a quantitative model by Dohmen and Chakraborty (2003). Implications for thermobarometry
have been considered in that work as well as by Chakraborty and Dohmen (2001). The
main finding was that the behaviour of such systems (two minerals exchanging
elements/isotopes via an intervening medium) depend on three non-dimensional parameters (β, γ and δ) that consist of: (1) diffusivity of the relevant elements/isotopes
in the mineral grains; (2) diffusivity of the same elements/isotopes in the grain boundary of the intervening medium; (3) solubility of the element/isotope in the grain
boundary region ('segregation factor' in the Materials Science literature); (4) grain
sizes of the grains; (5) surface area of the minerals (related to grain size); (6)
surface exchange reaction rate; and (7) distance between the mineral grains, and
surface area of the grain boundary network. Depending on the values of these
various parameters, six different kinds of situations (rather than the classical
'diffusion-control' or 'interface-control') may obtain. These have been depicted in a
reaction mechanism map by Dohmen and Chakraborty (2003) (Fig. 2, where the
non-dimensional parameters γ, δ and β are defined as well). For many situations
related to geothermobarometry in relatively coarse-grained high-temperature rocks
(igneous rocks, metamorphic rocks, mantle samples), three of these domains, the
'solid-diffusion control' domain, the 'fluid diffusion-control' domain, and a mixture
of the two, are relevant. A property of particular importance is the solubility of the
element of interest in the intervening medium; this changes substantially in melt-

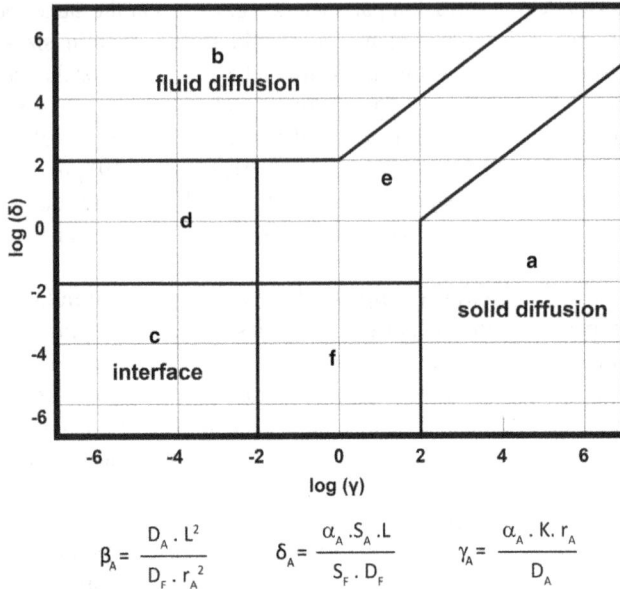

$$\beta_A = \frac{D_A \cdot L^2}{D_F \cdot r_A^2} \qquad \delta_A = \frac{\alpha_A \cdot S_A \cdot L}{S_F \cdot D_F} \qquad \gamma_A = \frac{\alpha_A \cdot K \cdot r_A}{D_A}$$

Fig. 2. Reaction mechanism map of Dohmen and Chakraborty (2003). Instead of a binary classification of kinetics into 'diffusion control' and 'interface control', for diffusive element exchange kinetics, we have several possible situations that are labeled a–f, depending on two non-dimensional parameters, γ_A and δ_A while the rate depends on a third non-dimensional parameter, β_A.

The parameters are defined in the figure. Symbols used in the definition are: D_A: Diffusivity of the element in mineral grain A, D_F: Diffusivity of the element in the grain boundary 'fluid', L: distance along the grain boundary between the two grains, r_A: radius/half width of grain A, S_A: surface area of grain A, S_F: surface area along grain boundary between the two grains, A and B, that are exchanging the element, α_A: kinetic constant for surface reaction for exchange of the element between mineral A and grain boundary 'fluid', and K: Partition coefficient of element in the grain boundary 'fluid', defined as K = (Concentration in 'fluid')/ (Concentration in solid A). For details, see Dohmen and Chakraborty (2003).

The reaction mechanism map shows the different mechanisms that arise for different values of γ_A and δ_A. (*a*) Classical diffusion control, with exchange controlled by diffusion rates in the solid. (*b*) Fluid diffusion control, where diffusion is controlled by the rate of transport (diffusion, solubility) in the grain boundary fluid medium. The 'fluid' can be physical (e.g. aqueous fluid, melt) or virtual (a dry grain boundary with its own transport properties. (*c*) Classical interface control where element exchange is controlled by the kinetic rate of exchange at the mineral–grain boundary interface. (*d*) Mixed interface–fluid diffusion control, which has characteristics of both (b) and (c), (*e*) a zone of mechanism that is a mixed control of solid and fluid diffusion control, or even all three end-member mechanisms: solid-, fluid- and interface-diffusion control, (*f*) a mixed zone of solid diffusion and interface control. The behaviour of crystals are different in each of these cases, and must be evaluated on a case-by-case basis to decide which compositions may be suitable for geothermometry. In several cases (e.g. fields (b), (d) or (c) even the rims of two crystals are not always in equilibrium with each other; in other cases the rims may be in equilibrium, but not the cores (depending on the value of the third parameter, the rate β_A).

present vs. melt-absent situations. The presence of melt facilitates attainment of equilibrium between at least the rims of two grains that are physically separated from each other.

Based on all of these considerations, there are a few criteria that may be used to determine whether compositions of two such grains are suitable for geothermobarometry. If the rim compositions of all grains of a mineral, irrespective of distance from the exchange partners, are the same then there is a high likelihood that the composition is in equilibrium. If this applies to both elements involved in an exchange process, then it is possible that the core compositions of the two minerals may be paired to obtain the temperature/pressure at an earlier stage of evolution (see above for the parameter γ' that provides guidance in this matter). In addition, of course, chemical criteria that have been mentioned above (e.g. Roozeboom diagram), need to be tested. This highlights the importance of compositional measurements in a textural context, with attention to initial and boundary conditions under which the element/isotope exchange process occurred, rather than in isolated grains. For example, as seen from the above discussion, distance between the grains exchanging the element may be relevant, depending on the mechanism of reaction that operated. This requirement relates to single-mineral thermobarometry as well, where the 'single mineral' aspect arises from certain assumptions about the nature of the exchange partner/medium (e.g. fixed activity of certain components). Dohmen and Chakraborty (2003) demonstrated how it is possible to get artefacts and even temperatures higher than any that were ever attained if these criteria are not properly applied.

The overall outcome of the above analysis is that a simple analysis of diffusion using a $x^2 \sim Dt$ kind of framework, or a simple calculation of closure temperature using the conventional textbook expression of the Dodson equation, is inadequate for evaluating which thermometers are reset to what extents, or even what the temperature given by a given thermometer means. We note, however, that the general expressions of Dodson (1973) or its later extensions (e.g. Ganguly and Tirone, 1999), used judiciously do permit such evaluations.

Before considering the specific natural system of interest, we summarize below a few lessons/corollaries that follow from the mathematical formulations of the kinetics of element exchange that have been described above:

Lesson 1: Multiple processes in series or parallel; the special status of diffusion + growth/dissolution

As rocks are subjected to changing conditions (change in pressure or temperature, change in oxygen or water fugacity due to hydration/dehydration, or due to melt–rock interaction, as discussed in the example below) minerals adjust their compositions by a combination of diffusion processes and growth/dissolution of existing crystals (Fig. 3). As illustrated in Fig. 3, dissolution/precipitation can reduce/increase the size of a crystal, but compositional changes must involve diffusion (defined as the relative motion of a particle relative to another particle) in the fluid as well as the solid. As these processes with rather different kinetic rates occur together, expressions that consider the kinetics of processes occurring in series (i.e. sequentially after one after the other) or in parallel need to be considered. For processes that occur in parallel, the fastest sub-process is rate controlling; for processes that occur in series, it is the

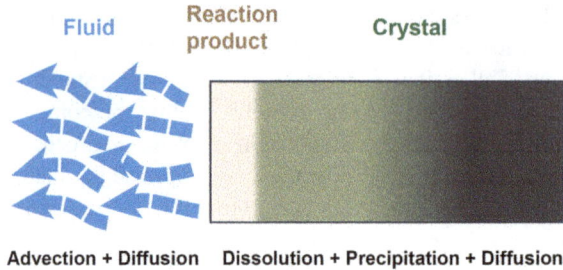

Fig. 3. Schematic depiction of a dissolution–precipitation reaction involving a mineral solid-solution. The process of dissolution and precipitation is, by itself, only a removal/addition of crystalline material. For a solid solution, the reactivity (solubility/saturation limit in the fluid) is invariably different for the different components of the solid solution. This leads to the production of concentration gradients in the crystal, with resulting diffusion (shown by colour gradation), and in the fluid where advection of the fluid and diffusion in the fluid act together.

slowest one. Therefore, in spite of the occurrence of dissolution/precipitation, diffusion rates continue to play a central role in the overall evolution. See fig. 2 and the discussion of the Dohmen and Chakraborty (2003) model above for examples of how diffusion in the solid as well as the fluid phases may be coupled and play a role in the overall exchange process. It is worthwhile noting, in the context of application of the generalized Dodson equation (equation 1), that in many cases the process of dissolution/precipitation + diffusion may be described by an effective diffusion equation and the process even depends on the same parameter, δ, that has been described above in the Dohmen and Chakraborty formulation (see Lasaga, 1986, 2014 for derivations and discussion).

There are important implications of this for thermobarometry. During diffusive exchange, the core composition of a crystal may retain the 'memory' of a previous stage, the rim composition relates to a later stage in the evolution of the rock (e.g. dependent on the values of the parameters β/γ and γ' discussed above). But if adjustment of composition occurs by growth-dissolution where grains may recrystallize completely, this is not the case and the composition at the core may reflect the point in time when the dissolution/precipitation process occurred. The challenge lies in the fact that in real systems it is rarely an either/or situation; more commonly the two processes occur in conjunction. The first outcome is that one may have compositionally zoned minerals, bearing signatures of diffusion, but for thermobarometry (as well as geospeedometry/diffusion chronometry) one needs to decide, which, if any, compositions of other minerals may be paired with compositions in the zoned mineral for thermobarometry. Second, depending on the relative rates of growth/dissolution vs. diffusion, one may obtain homogeneous grains which are not in equilibrium with each other (see Dohmen and Chakraborty, 2003, for examples). Therefore, it is important to carry out chemical tests for (local) equilibrium (including the Roozeboom diagrams mentioned above) before calculating temperatures and pressures rather than rely on compositional zoning or homogeneity (see example below).

Lesson 2: Lifetime of a crystal vs. lifetime of a phase; the role of textural maturation

The important role of the textural context of compositional evolution leads to additional effects that arise from textural maturation. Even when a system is not chemically out of equilibrium, the need to minimize surface/interfacial free energy, and hence surface area of minerals, can drive grain growth and dissolution. A well-known example of such a process is Ostwald ripening; an outcome of this aspect is that existing grains (e.g. finer-grained ones) may dissolve to permit growth of other (e.g. coarser) grains. There may be other situations as well. Two common processes of particular relevance to the study of mantle samples are the following: (1) permeation of externally derived melt through a matrix may facilitate textural maturation and compaction processes, leading to dissolution of existing grains/parts of grains and precipitation of new material (with modified compositions); and (2) plastic deformation involves the motion of dislocations and this may lead to the annihilation of existing old grains the birth of new grains with new compositions (e.g. Hackl and Renner, 2013).

These have important consequences for geothermobarometry (and geochronology, geospeedometry) in that: (1) the evolution of compositional zoning in minerals, (2) the stage of evolution of a rock that is preserved in the compositional record of a mineral (= the last stage after such textural maturation process) are affected. Examples of how this may influence geological inferences from such data were provided by Beyer and Chakraborty (2021).

Lesson 3: Multiple diffusion mechanisms, particularly of trace elements, and impact on closure temperatures

It is being found increasingly that the diffusion of trace elements in minerals occurs by more than one mechanism, and hence rate, depending on their concentrations and other factors (e.g. oxygen fugacity, concentration in the surrounding medium). This effect was first shown for diffusion of Li in olivine (Dohmen *et al.*, 2010) and since recorded for many systems such as REE in olivine (Dohmen *et al.*, 2016), Al in olivine (Zhukova *et al.*, 2017), Li in zircon (Cisneros de León and Schmitt, 2019; Tang *et al.*, 2017), *REE* in garnet (Bloch *et al.*, 2020) and Nb, Ta, Hf, Zr in rutile (Dohmen *et al.*, 2019). In addition, major elements may occur in multiple sites in different minerals (such as Ca in garnets as grossularite or andradite components, Al in IV and VI coordination in several ferromagnesian minerals) and may therefore diffuse by different mechanisms (with different activation energies and other parameters of diffusion, see above for their role in kinetics). These lead to two consequences: (1) closure temperatures for different mechanisms may be different, and (2) either multicomponent coupling (e.g. Borinski *et al.*, 2012) or a reaction-diffusion equation involving homogeneous reaction between different "species" (= an element in different sites/part of different components, e.g. Dohmen and Chakraborty, 2010) is necessary to model the freezing behaviour accurately. Simplifications are possible, but the assumptions made in such simplifications should be borne in mind. The implication for geothermobarometry is that depending on which mechanism and rate was operative in a given case (= different pre-exponential factors and activation energies), quite different stages of evolution of a rock may be recorded by the same trace element composition of a mineral; the results need to be interpreted accordingly.

In the following section we report data from an ophiolite sequence in Tibet (Xigaze/Luqu Ophiolite) to illustrate that simply pairing mineral compositions from such rocks without attention to kinetic factors leads to a large, indiscriminate scatter of calculated temperatures that cannot be interpreted in any meaningful manner. However, consideration of the kinetic factors not only produces a very systematic and kinetically meaningful pattern; it provides important insights into the evolution of the sequence as a whole.

4. A case study from mantle samples from Xigaze Ophiolite, Tibet

Mantle rocks are a multiphase system typically consisting of the phases olivine, orthopyroxene, clinopyroxene and spinel/garnet in different proportions, connected to each other via a grain boundary network. In addition, a melt phase may be present at different stages of evolution of a rock. Commonly, mantle peridotites experience several events of melt percolation, and depending on the exact location (e.g. which grain boundaries/contacts between minerals), magnitude (e.g. extent of melt pockets) and duration of such events, different thermometers may be reset to different extents. Minerals in direct contact under mantle conditions would maintain elemental exchange equilibrium. However, this equilibrium could be disturbed by many factors in ophiolitic mantle rocks before or during the emplacement process. Therefore, identification of element exchange equilibria for elements of interest between the phases that are present, and determination of whether they are at equilibrium or disequilibrium are crucial for acquiring meaningful temperature data and to further understanding the mechanisms affecting the processes endured by the rocks.

As discussed above, geothermometry in such systems requires the simultaneous application of thermodynamics and kinetics to determine temperatures and evaluate their significance. Additionally, it is necessary to evaluate whether concentration profiles are suitable for the determination of cooling rates before applying tools of geospeedometry.

We begin by introducing the rock suites used in the present study, followed by justifying the choice of analytical points for the determination of temperatures and then presenting the results. As will be seen, different thermometers record different sets of temperatures, more or less consistent with expectations based on diffusion rates of elements in the minerals that are involved in a geothermometer. However, a detailed analysis of concentration profiles reveal that the resetting was not entirely by diffusive processes (i.e. dissolution/precipitation played a role also; see above), and caution needs to be exercised in extracting cooling rates from these data.

4.1. Geological setting and samples used in this study

The Yarlung-Zangbo Suture zone (YZS), the eastern segment of the Indus-Yalung Zangbo Suture (IYS) from the Nanga Parbat syntaxis (NPS) to the Namcha Barwa syntaxis (NBS), is the youngest suture among four major well-defined sutures in the Tibetan Plateau (Fig. 4a) (Dewey and Bird, 1970; Gansser, 1977; Molnar and Tapponnier, 1975; Yin *et al.*, 1988). Geographically, it is divided into three segments: the western segment (from Kiogar to Saga), the central segment (from Sangsang to Dazhuqu) and the eastern segment (from Zedong to the very east) (Hébert *et al.*, 2012). It marks the collision zone between the Indian (Greater Indian block) and Eurasian (Lhasa block) plates, in the wake of the Neotethyan ocean closure lasting from late Cretaceous to early Tertiary (Fig. 4A') (Allégre *et al.*, 1984; Dai *et al.*, 2013; Hébert *et al.*, 2012; Yin and Harrison, 2000).

The Luqu ophiolite, also called Beimarang or Xigaze ophiolitic massif (Girardeau *et al.*, 1985; Huot *et al.*, 2002), covers an area of ~150 km^2 at an elevation ranging from 3800 to 5000 m. It is located in the central segment (from Sangsang in the west to Dazhuqu in the east) and is sandwiched by the Xigaze Group flysch to the north and the Upper Jurassic-Lower Cretaceous red radiolarite and Trias-Lias flysch in fault contact to the south (Fig. 4a, b).

At Luqu, a continuous ophiolitic section is preserved exhibiting the complete lithospheric sequence (Huot *et al.*, 2002; Nicolas *et al.*, 1981). In the upper mantle section of the ophiolite, based on the mineralogical, petrographic and field relation characteristics, the ultramafic massifs can be divided into three groups (Fig. 4b). Group A: ultramafic massifs mainly consisting of fresh harzburgites, dunites and pyroxenites. Group B: ultramafic massifs mainly consisting of serpentinized harzburgites, dunites, pyroxenites and abundant mafic dykes. Group C: ultramafic massifs mainly consisting of serpentinites, highly serpentinized Cpx-harzburgites and lherzolites at the base of the mantle section.

In this paper, the studied samples are mainly selected from fresh harzburgite and dunite in group A (and chosen specific fresh samples in group B) as they maximally retain the magmatic features eliminating the effects of secondary alteration such as serpentinization.

4.2. Petrography

The fresh ultramafic rocks (harzburgite and dunite) generally have coarse to porphyroclastic textures. The latter texture is marked by the inequigranular mineral grains and the development of smaller neoblasts of olivine, orthopyroxene, clinopyroxene and spinel between pyroxene or olivine porphyrocalsts. Based on the petrographic study, minerals in the rocks can be classified into different types and generations.

4.2.1. Fresh Dunite

The fresh dunite consists of olivine (~99 vol.%), spinel (~0.5 vol.%) and Cpx (~0.5 vol.%) and a minor amount of sulfide, e.g. pentlandite. No orthopyroxene occurs in the dunite.

Olivine can be subdivided into three types according to the crystal size and mineral assemblage: (1) fine-grained olivine (~20–100 μm) occurs forming interstitial mineral

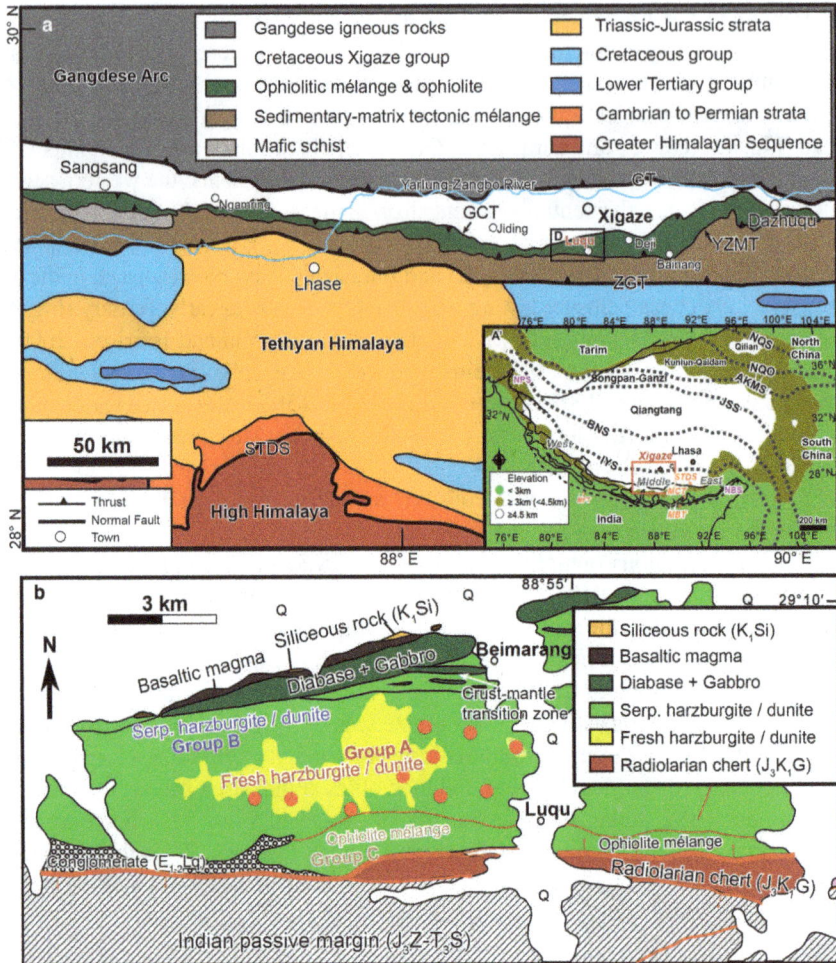

Fig. 4. (*a*) Geological map of the central segment of Yarlung-Zangbo Suture Zone (modified after Bédard *et al.*, 2009; Dai *et al.*, 2013; Ding *et al.*, 2005). Abbreviations: STDS, south Tibetan detachment system; ZGT, Zhongba-Gyangze thrust; YZMT, Yarlung-Zangbo Mantle thrust; GCT, Greater Counter thrust; GT, Gangdese thrust. (A') Tectonic framework of Tibetan Plateau (modified after DeCelles *et al.*, 2002; Yin and Harrison, 2000). Abbreviations: MFT, Main Frontal Thrust; MBT, Main Boundary Thrust; MCT, Main Central Thrust; STDS, South Tibetan Detachment System; IYS, Indus-Yalung Zangbo Suture; BNS, Banggong-Nujiang Suture; JSS, Jinsha Suture; AKMS, Anyimaqen-Kunlun-Muztagh Suture; NQO, North Qaidam Orogen; NQS, North Qilian Suture; NPS, Nanga Parbat syntaxis; NBS, Namcha Barwa syntaxis, (*b*) Geological sketch map of Luqu ophiolite in Xigaze ophiolites (modified after Girardeau *et al.*, 1985; Nicolas *et al.*, 1981; Zhang *et al.*, 2017).

aggregates (Cpx+spinel+olivine) between the former two types of olivine (Fig. 5a). All the symplectic minerals of the aggregates are anhedral in shape and show no undulose extinction. (2) Coarse-grained olivine as porphyroclastic crystals (∼1–3 mm) show

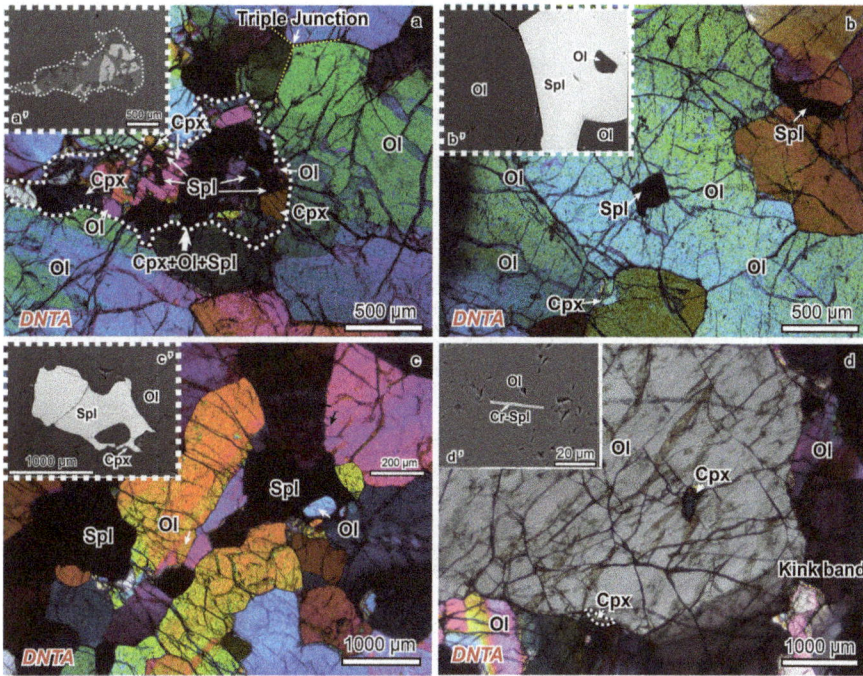

Fig. 5. Optical photomicrographs and (inset) backscattered electron images (BSE) of fresh dunite (Sample DNTA). The minerals are labelled in the individual frames, showing the different textural modes of occurrence of a mineral.

undulose extinction and kink bands. Some crystals host spare inclusions of spinel and Cpx (Fig. 5b). (3) Medium- to coarse-grained olivine crystals (–100–1000 μm) exist between the porphyroclastic ones. Triple junction texture is well-developed indicating recrystallization effects (Fig. 5c). Fewer amounts of mineral inclusions are observed in this type of olivine, but kink bands are common.

Spinel can be classified into the following types: (1) fine-grained crystals (~30–100 μm) in anhedral shape occur in the interstitial mineral aggregate (Fig. 5a). (2) Fine-grained inclusions (~30–100 μm) in olivine which are mostly euhedral (Fig. 5b). Occasionally, there are olivine inclusions in the spinel inclusion, and altogether enclosed in massive olivine (Fig. 5b). (3) Fine- to medium-grained anhedral crystals form an interstitial phase (Fig. 5b). (4) Medium-grained crystals (~500–2000 μm) in subhedral to euhedral shape occur in the matrix of olivine (Fig. 5c). (5) Needle-shaped chromian spinel (~30 μm long) exists as inclusions in olivine (Fig. 5d).

Cpx generally occurs in the following four types. (1) Fine-grained crystals (~30–100 μm) in anhedral shape occur in the interstitial mineral aggregates (Cpx + olivine + spinel) (Fig. 5a). (2) Fine-grained interstitial single crystals (~100 μm) exist between olivine in subhedral to euhedral olivine (Fig. 5b). (3) Fine-grained interstitial crystal slices occur at the rim of medium-grained spinel (Fig. 5c). (4) Fine-grained crystals (~100 μm) in anhedral shape are included in the porphyroclastic olivine (Fig. 5d).

4.2.2. *Fresh Harzburgite*

Olivine in the Cpx-bearing harzburgites is divided into five types. (1) Porphyroclastic olivine (~500–5000 μm) contains euhedral spinel inclusions and are surrounded by fine- to medium-grained olivine, Cpx and spinel. Textures of kink and undulose extinction are well developed in the coarse-grained olivine type (Fig. 6a, b). (2) Fine- to medium-grained inclusions in spinel or in assemblage with anhedral spinel (Fig. 6b). (3) Medium-grained euhedral crystals are included Cpx (Fig. 6c). (4) Fine- to medium-grained crystals generally in subhedral shape recrystallize between the porphyroclastic olivine and opx. Triple junctions are prevalent (Fig. 6d). (5) Fine-grained olivine neoblasts occur at the embayments of porphyroclastic Opx as a part of the interstitial polyphase aggregates (Ol+Cpx+Spl±Opx±Pn) (Fig. 6b, d).

Opx can be classified into three types according to petrographic features. (1) Porphyroclastic crystals (500–3000 μm) bearing Cpx exsolutions (anhedral) and spinel inclusions (subhedral to euhedral) may be surrounded by the interstitial polyphase aggregates (Fig. 6a, d). (2) Fine- to medium-grained crystals exist in the interstitial polyphase

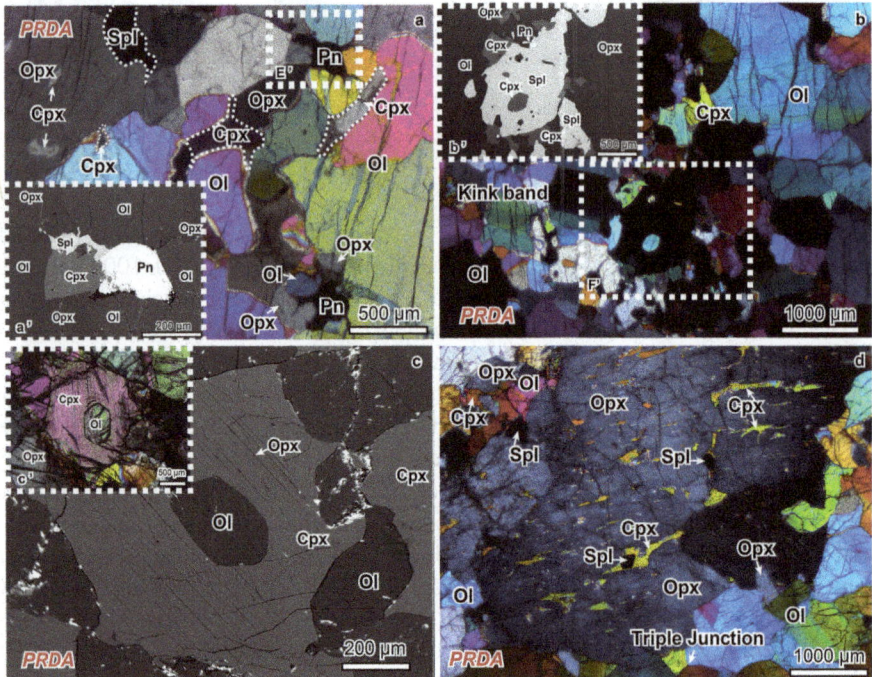

Fig. 6. Optical photomicrographs and (inset) backscattered electron images (BSE) of fresh harzburgite (Sample PRDA). Minerals are labelled in the individual frames, showing the different textural modes of occurrence of a mineral.

aggregate at the embayment of porphyroclastic opx (Fig. 6a, d). (3) Exsolution lamellae exsolved internally in medium-grained Cpx (Fig. 6c).

Cpx mainly occurs forming three types: (1) medium-sized crystals in subhedral shape, some of these crystals bear exsolution lamellae of Opx (thickness: ~1–2 µm) in their core (Fig. 6a, d). The width of Opx lamellae in Cpx is much thinner than the Cpx exsolution in Opx (thickness: ~3–10 µm) in the Opx porphyroclasts. (2) Fine-grained interstitial crystals in anhedral shape in the interstitial polyphase aggregate (Fig. 6b, d). (3) Fine-grained anhedral crystals in layered shape on the boundaries of spinel (Fig. 6b). (4) Cpx exsolution lamellae (thickness: ~3–10 µm) in the core of porphyroclastic Opx (Fig. 6a, b).

Spinel is identified as forming three types. (1) Fine-grained anhedral spinel occurs in interstitial polyphase aggregates $(Ol+Cpx+Spl \pm Opx \pm Pn)$ (Fig. 6a, b, c). (2) Medium- to coarse-grained subhedral to euhedral crystals occur between porphyroclastic olivine and Opx. Some spinel host inclusions of Cpx and Opx (Fig. 6b). (3) Fine-grained euhedral crystals as inclusions in porphyroclastic Opx (Fig. 6d).

The fine-grained interstitial phases are interpreted, based on their textural mode of occurrence as well as major- and trace-element (e.g. REE) compositional characteristics, to be products of crystallization from interstitial melt that must have percolated through the rock matrices.

4.3. Estimation of equilibrium temperatures

Compositional profiles across various mineral pairs in different textures were measured in an electron microprobe in the wavelength dispersive mode at Ruhr-Universität Bochum (Cameca SX Five-FE). Well established silicate and oxide standards were used with a 1 µm beam spot size and optimized step size. Data from such profiles, rather than isolated individual spot analyses, were used for evaluation of suitability for thermometry using the criteria discussed above.

As dunites and harzburgites in the Xigaze ophiolite that are studied here are both in the spinel-stable field, the pressure was set at 1.5 Gpa (15 Kbar) in all thermometric calculations for comparison. Mineral pairs were selected according to (a) the kinetic considerations outlined above, (b) bearing in mind the different textural modes of occurrence of a given mineral that has been described above, and (c) the nature of compositional zoning in the minerals. Next, the chemical analyses from such selected pairs are marked on the compositional profiles and projected on the Roozeboom diagram to show the compositional variation trends and test for approach to element partitioning equilibrium. Temperature ranges are reported only when mineral pairs pass these tests. It is seen that the same mineral pairs from different textural locations record different stages in the history of evolution of the rocks.

4.3.1. Thermometry with spinel–olivine pairs as an example

The approximate linear ranges traversing the curves on the Roozeboom diagram indicate that compositions of each mineral pair are in equilibrium in such conditions (Fig. 7). Different temperatures (ranging from ~560 to ~690°C) were estimated in different

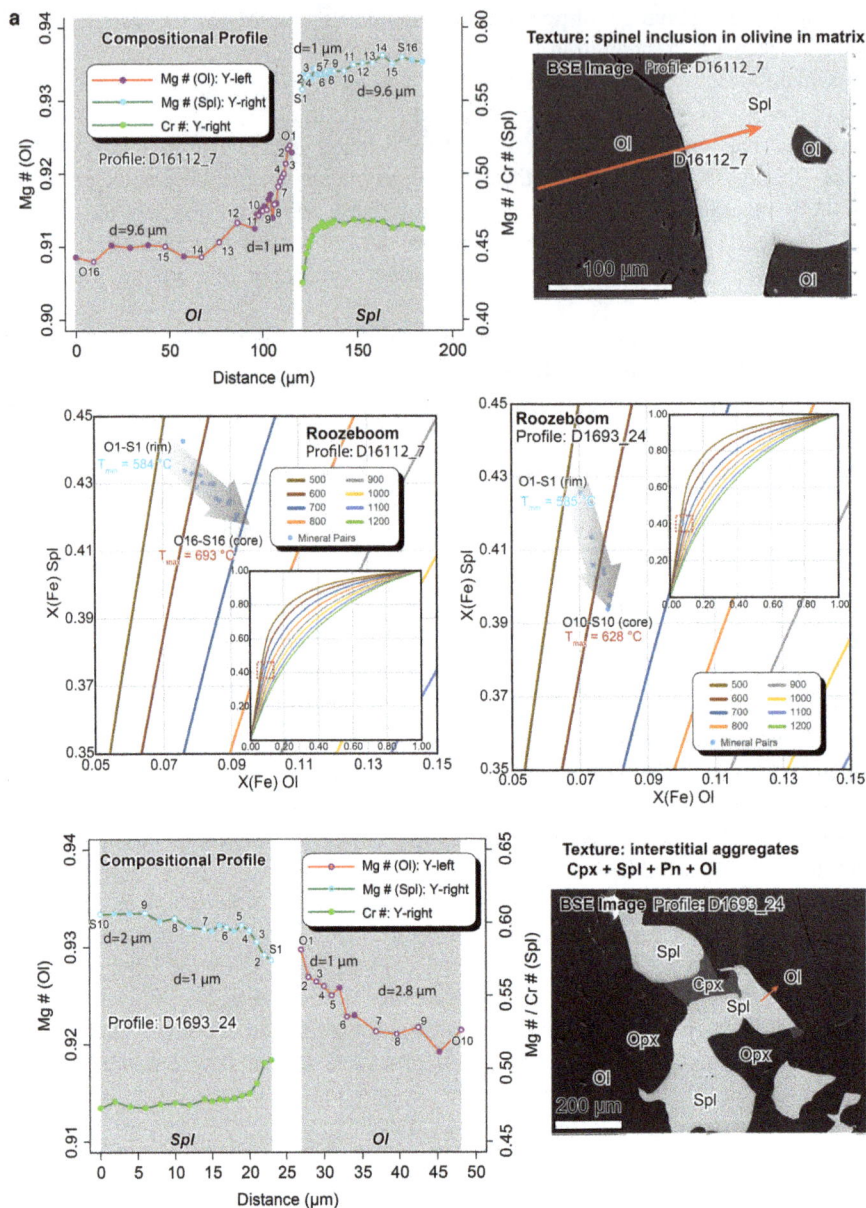

Fig. 7. Four selected compositional profiles across spinel–olivine for equilibrium temperature estimation. The BSE images show the location of the line measurements. Roozeboom diagrams indicate the equilibrium conditions among mineral pairs and the approximate temperature ranges. The T_{max} values at the core and T_{min} values at the interface are calculated using the thermometer of Ballhaus *et al.* (1991). Gaps indicate mixed analyses/cracks with alteration (i.e. not olivine/spinel stoichiometry) at interfaces. Images and profiles from fresh dunite (DNTA, sample 16112) and fresh harzburgite in group A (PRDA, Sample 1693).

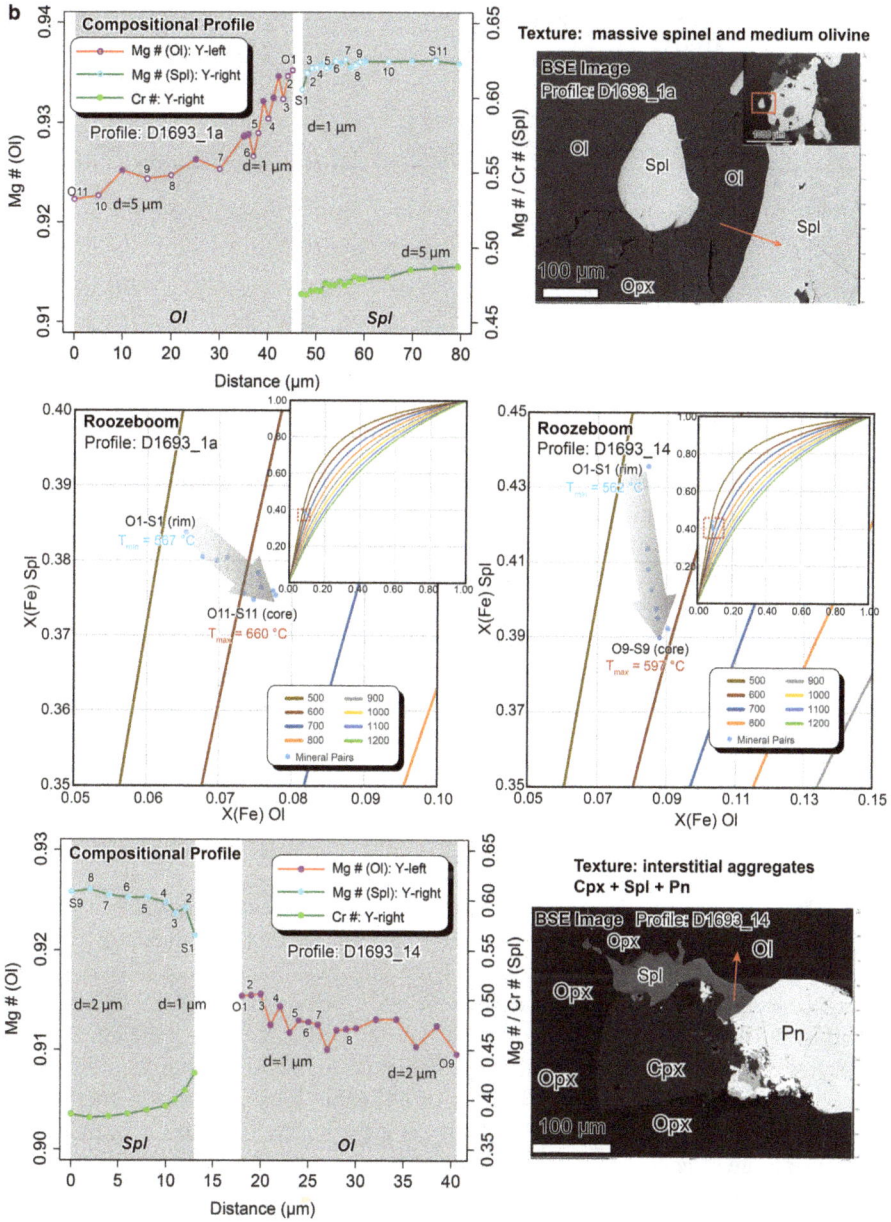

Fig. 7b. *Continued.*

spinel-olivine pairs among various textures (Table 1, Fig. 7). Analytical data points either from porphyroclastic olivine or massive spinel (larger grain size) record the higher T_{max} at the core part, and points from mineral pairs in the aggregates with smaller grain sizes

Table 1. Summary of the estimated temperatures (spinel–olivine) at the interface (T_{min}) and the core (T_{max}).

Profiles	D16112_7	D1693_24	D1693_1a	D1693_14
T_{max} (°C)	693	628	660	598
T_{min} (°C)	584	586	567	563
Texture	Ol: massive	Ol: interstitial	Ol: medium	Ol: interstitial
	SPl: inclusion	Spl: interstitial	Spl: massive	Spl: interstitial

Profile D16112_7: Fresh dunite in Group A (DNTA).
Profile D1693_24, 1a & 14: Fresh harzburgite in Group A (PRDA1).
See Fig. 7 for more details.

record the lower T_{max} (Fig. 7). On the other hand, temperatures (T_{min}) estimated at the interface of the mineral pairs are quite similar among occurrences in different textural modes. For example, spinel (massive)–olivine (medium) in profile D1693_1a and spinel (inclusion)–olivine (massive) in profile 16112_7 present T_{max} values of 660°C and 693°C respectively. The T_{max} values of the large grains are generally ~50°C greater than that calculated from the mineral pairs in the polyphase aggregates. However, the temperatures at the interface (T_{min}) have a more minor variation (20°C) despite the various textures and mineral grain sizes. This overall distribution demonstrates clearly a history of progressive compositional resetting with resetting at rims at all textural locations > cores of smaller grains > cores of larger grains.

4.3.2. Thermometry of spinel–orthopyroxene pairs

Two selected spinel and Opx pairs for equilibrium temperature estimates are listed below as examples (Fig. 8 & Table 2). The temperature estimated from the massive pairs with large grain size yield a higher T_{max} (835°C) compared with the T_{max} (783°C) from interstitial mineral pairs with very small grain sizes. Similar to the spinel-olivine pairs, the T_{min} estimated from the rims of the two different scale gain sizes present only minor temperature differences (760°C in large grain sizes, 774°C in small grain sizes).

4.3.3. Thermometry of clinopyroxene (Cpx)–orthopyroxene (Opx) pairs

The estimated temperatures among three different textures (massive, medium and exsolution) yield similar T_{max} (890–904°C) using the analytical points obtained from near the mineral cores (Table 3, Fig. 9). The wide range of T_{min} difference (768–893°C) might be caused by the loss of the measurement points bordering the interface of the mineral pairs (gradients at pyroxene–pyroxene interfaces are expected to be sharper than the others. This is partly a consequence of Fick's law of diffusion: Flux = Diffusion coefficient x concentration gradient. So, for smaller [slower] diffusion coefficients, as in pyroxenes, the gradients are steeper, making it more difficult to measure the exact rim compositions which may be in equilibrium with each other).

4.3.4. Disequilibrium Mineral pairs

Equilibrium among different mineral pairs were not always attained in these ophiolite samples. Cpx–olivine pairs are a typical example of mineral pairs showing disequilibrium relationships. For example, the massive Cpx in the matrix is surrounded by medium-sized other phases, including olivine (Fig. 10). The equilibrium condition was tested by the

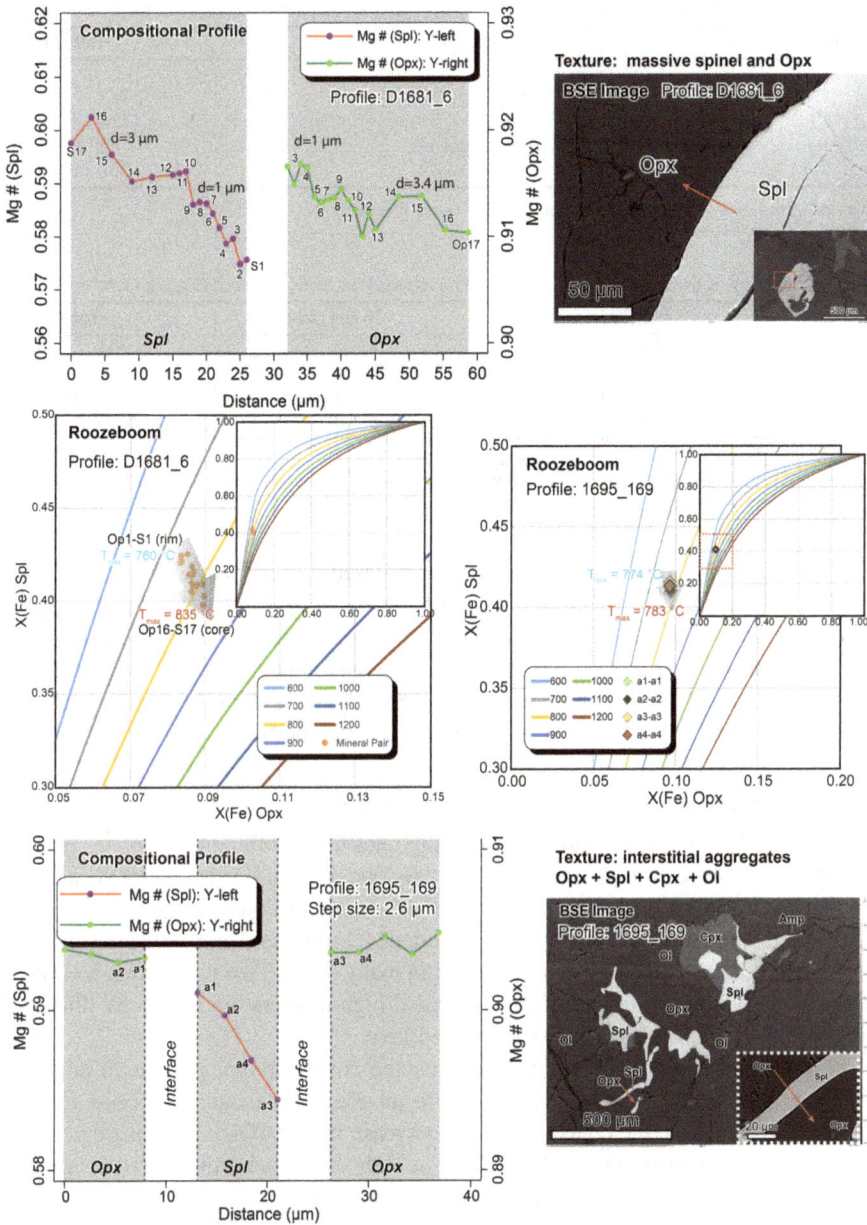

Fig. 8. Two selected compositional profiles across spinel–Opx pairs for equilibrium temperature estimation. The BSE images show the location of the line measurements. Roozeboom diagrams indicate the equilibrium conditions among mineral pairs and the approximate temperature ranges. The T_{max} values at the core and T_{min} values at the interface are calculated using the thermometer of Liermann and Ganguly (2003). Images and profiles from the harzburgite in group B (PRDB, Samples 1681 & 1695).

Table 2. Summary of the estimated temperatures (spinel–Opx) at the interface (T_{min}) and the core (T_{max}).

Profiles	1681_6	1695_169
T_{max} (°C)	835	783
T_{min} (°C)	760	774
Textures	Opx: massive	Opx: interstitial
	SPl: massive	Spl: interstitial

Profile 1681_16 & 1695_169: Serpentinized harzburgite in Group B (PRDB1).
See Fig. 8 for more details.

Table 3. Summary of the estimated temperatures (Cpx–Opx) close to the interface (T_{min}) and the core (T_{max}).

Profiles	16104c_273	16104_294	1681_14
T_{max} (°C)	901	890	904
T_{min} (°C)	893	782	768
Textures	Cpx: massive	Cpx: massive	Cpx: exsolution
	Opx: medium	Opx: massive	Opx: massive

Profile 16104c_273 & 294: Fresh harzburgite in Group A (PRDA2).
Profile 1681_14: Serpentinized harzburgite in Group B (PRDB1).
See Fig. 9 for more details.

mineral compositions in the Roozeboom diagram based on Kawasaki and Ito (1994). The compositional points are all projected above the curve of 1200°C in the Roozeboom diagram, indicating that equilibrium was not obtained in this case. Caution needs to be exercised when one estimates temperatures using such analytical data points from the Cpx–olivine pairs. If the test for equilibrium step is ignored, the estimated temperature, would be ~1300°C, which would be a meaningless number in this case.

4.4. Concentration profiles and Diffusion modelling

Compositional zoning is used to obtain cooling rates using different techniques of geospeedometry. Often, compositional zoning in only one mineral is used, under the assumption that the composition at the core represents the composition at the peak temperature and then the retrograde zoning at the rim is modelled. From the discussion above it should be apparent that this condition is not fulfilled in many cases. Two pitfalls that arise commonly in many mantle (and other slowly cooled high-temperature rocks) are illustrated here through examples from the rocks studied here.

The first case in Fig. 11 demonstrates that the compositional profiles in adjoining mineral grains do not represent those due to an element exchange process. Element exchange would cause the concentration to decrease in one mineral and increase in the other. However, as seen in Fig. 11, the Mg# in both spinel and orthopyroxene increases towards the rim, indicating that there was an external source of the element (e.g. leakage/supply via the grain boundary between the two). Such supply may have occurred, for example, through a percolating melt phase. Clearly, the profiles are not suitable for modelling to obtain cooling/heating rates unless the boundary conditions for diffusion can be properly recognized and refined.

The second case in Fig. 12 shows a situation where the disposition of the profiles are in the right direction (i.e. gain in one, loss in the other), but a closer inspection reveals that

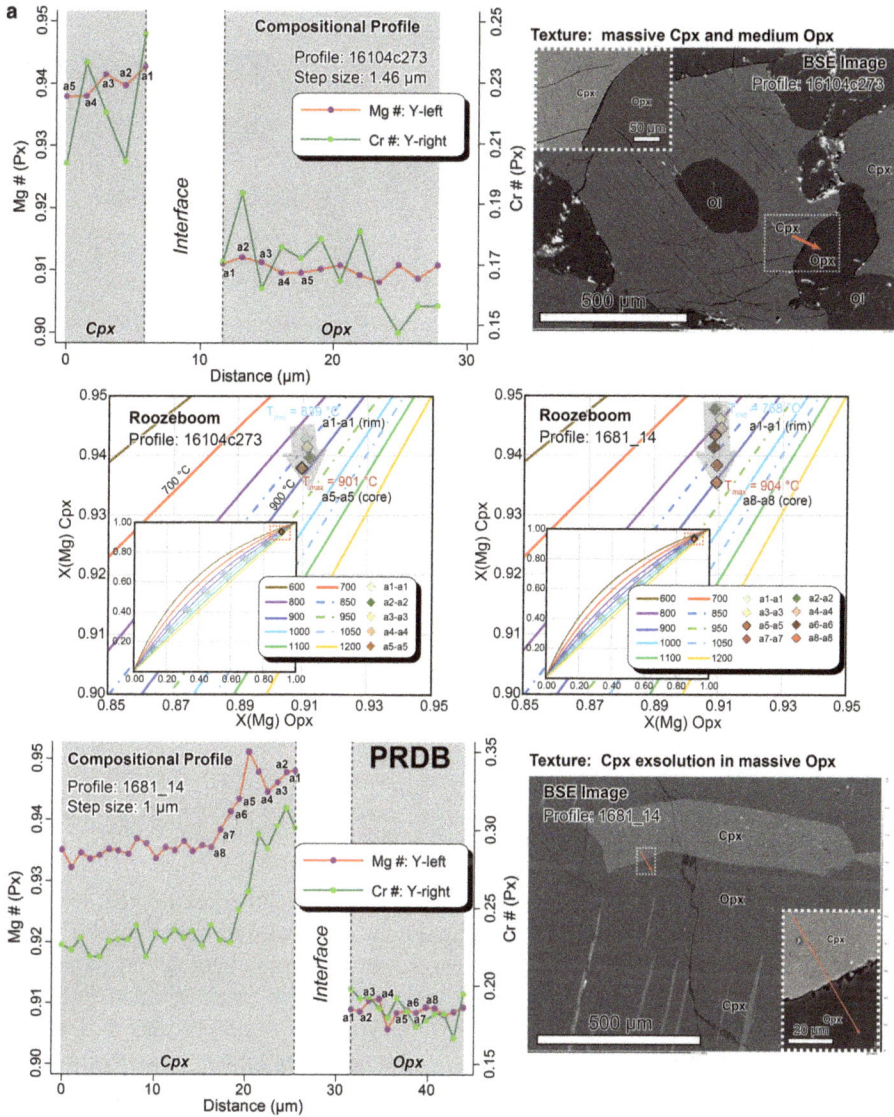

Fig. 9. Three selected compositional profiles across Cpx–Opx for equilibrium temperature estimation. The BSE images shows the location of the line measurements. Roozeboom diagrams indicate the equilibrium conditions among mineral pairs and the approximate temperature ranges. The T_{max} at the core and T_{min} at the interface are calculated using the thermometer of Ganguly *et al.* (2013). Images and profiles from fresh harzburgite (DNTA, sample 16104) in group A and harzburgite (PRDB, Sample 1681) in group B.

the area under the two profile curves are not the same. This is a necessity for mass balance and diffusion modelling – the amount of element lost from one mineral (given by the area under the profile curve) must be compensated exactly by the gain in the other mineral.

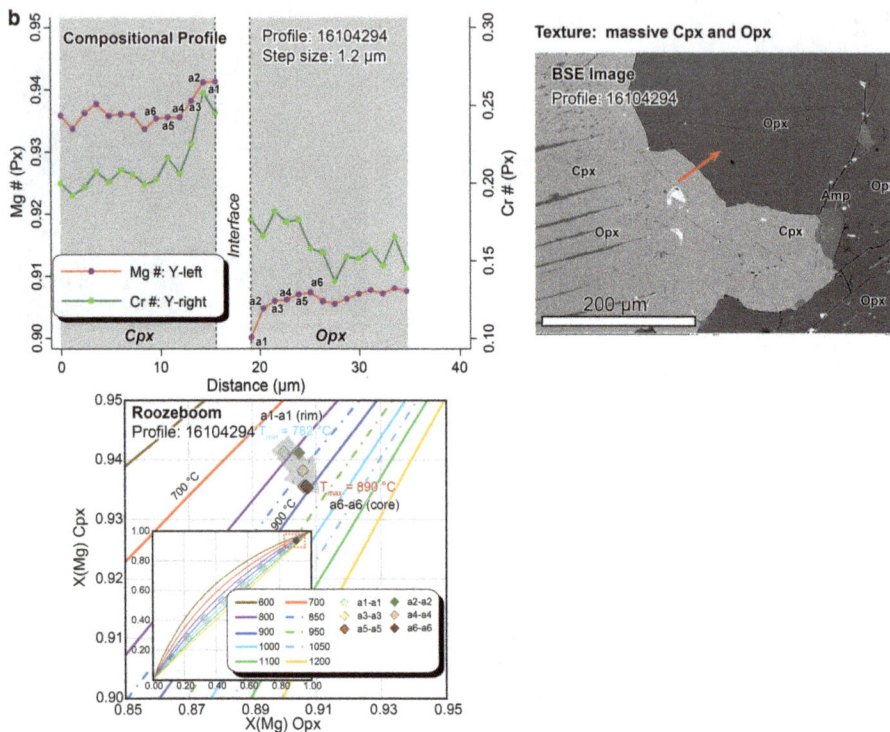

Fig. 9b. *Continued.*

Once again, even if the profiles show the right trends these are not suitable for modelling to obtain cooling/heating rates.

Both situations point to the operation of processes other than just element exchange by diffusion in these rocks. The petrographic and textural data reported above, that point clearly to the effects of melt percolation/infiltration and formation of multiple generations of minerals, provide an indication of possible processes that may have occurred: dissolution, precipitation and textural maturation (see lessons 1–3 above). On the whole, these underscore the need for evaluating the kinetic processes involved and the boundary conditions before applying diffusion models to determine cooling/heating rates; not all concentration profiles that have the appearance of diffusion profiles are amenable to diffusion modelling. Note that the shapes of the profiles themselves may be fit by diffusion models if single minerals are used; it is just that the cooling/heating rates that would be obtained would lack physical significance in these cases.

4.5. Putting it together: implications for cooling history

Figure 13a summarizes the results of geothermometry carried out using: (a) all the different systems where equilibrium conditions were fulfilled (*REE* between Cpx-Opx

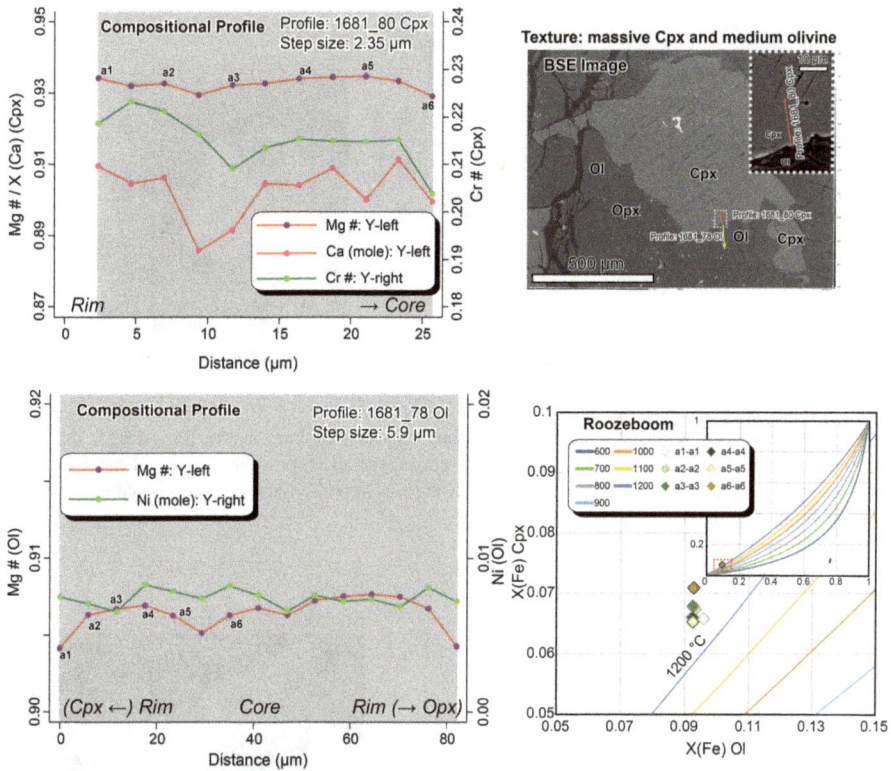

Fig. 10. One selected compositional profile across Cpx–olivine for equilibrium temperature estimation from harzburgite in group B (PRDB, sample 1681). The BSE image shows the location of the line measurement. The Roozeboom diagram indicates the equilibrium conditions among mineral pairs and the approximate temperature ranges (Kawasaki and Ito, 1994). The analytical points plot clearly outside the limits permitted by equilibrium at reasonable temperatures, indicating disequilibrium.

and Fe-Mg exchange between Opx-Cpx, Opx-spinel and Olivine-spinel) for (b) minerals occurring in the different textural modes in (c) all the different rock types. One observes several regularities in the pattern of the data.

First, temperatures from the same mineral pair from different rocks lie in the same range, emphasizing the kinetic control. Here, the usefulness of an ophiolite sequence compared to isolated mantle xenoliths becomes obvious; the spatial context provided by the field relations require that the different lithologies have gone through similar thermal histories (so that the same mineral pair should not yield widely different temperatures in these different rocks). At the same time, the different lithologies help to restrict/eliminate variabilities that may arise from bulk compositional control on the thermometers. The similarity of temperature obtained from different lithologies demonstrate that these, and the cooling histories, relate to the ophiolite sequence as a whole and not just to a particular rock type or outcrop.

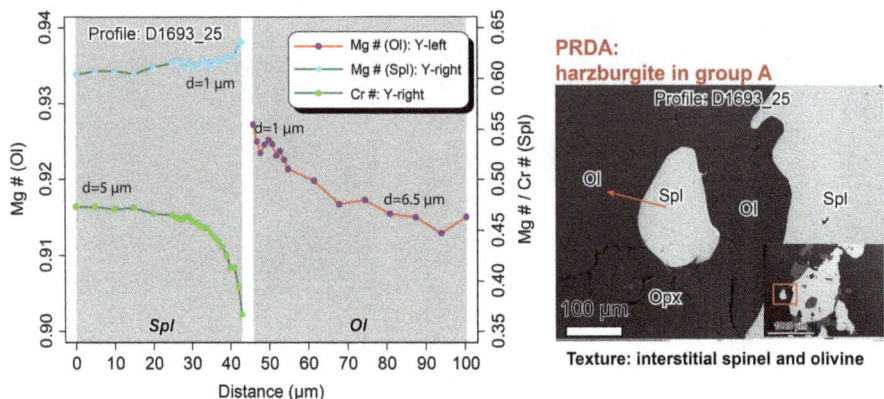

Fig. 11. Compositional profiles from an olivine–spinel contact (red line on the BSE image) that looks apparently clean but the profiles (left panel) do not correspond to shapes expected by element exchange between the two grains alone (Mg# increases towards the rim in both minerals). The sample is harzburgite PRDA in group A, sample number 1693.

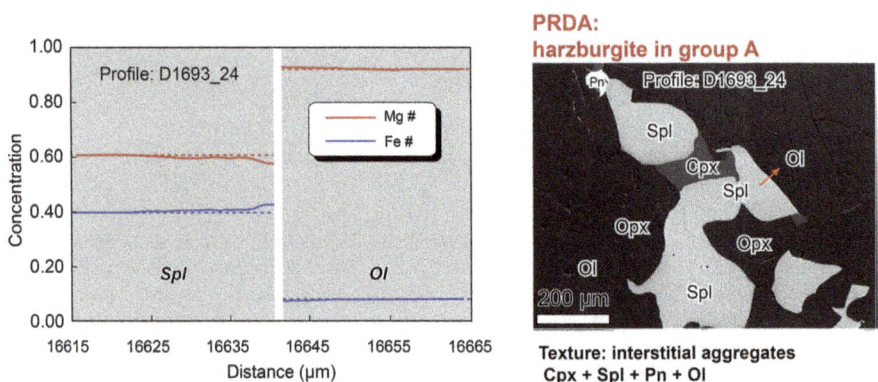

Fig. 12. Compositional profiles at an olivine–spinel contact (left panel) from a location shown by the red line in the BSE image (right panel). The areas under the profile curves show that mass balance is not obtained, and element exchange must have occurred between more partners than the adjacent olivine and spinel grains. This situation is unsuitable for geospeedometry modelling. The sample is harzburgite PRDA in group A, sample number 1693.

Second, the temperatures are sorted in the sequence expected from the diffusion parameters of the slower diffusing mineral in a pair (Table 4). Thus, REE from the two pyroxenes freeze at the highest temperature (1050–1250°C), followed by Fe-Mg in pyroxenes (750–1050°C), the spinel- orthopyroxene pairs (750–850°C), and the olivine–spinel pair was reset until freezing at the lowest temperatures below 750°C (600–750°C). Moreover, for a given mineral pair (e.g. olivine–spinel) and textural setting (e.g. matrix minerals) it is found that coarser spinel sizes yield greater temperatures compared to finer spinel grains

Table 4. Summary of the estimated temperatures using different thermometers among different coexisting mineral pairs.

ID	Sample ID	Profile ID	Label	Pairs	Grain Size	T_{min}	T_{max}	Texture
1	16112	16112_23	DNTA	Spl-Ol	50	620.10	683.37	3
2	16112	16112_73	DNTA	Spl-Ol	200	632.10	648.67	3
3	16112	16112_74	DNTA	Spl-Ol	200	614.60	646.60	3
4	16112	16112_22	DNTA	Spl-Ol	50	663.32	689.86	1
5	16112	16112_76	DNTA	Spl-Ol	80	680.63	695.22	1
6	16112	16112_72	DNTA	Spl-Ol	1000	677.10	748.97	0
7	16112	16112_9	DNTA	Spl-Ol	1500	714.03	742.93	0
8	16562	16562_1	PRDA2	Spl-Ol	200	659.90	679.29	1
9	1693	1693_2	PRDA1	Spl-Ol	800	685.44	688.59	1
10	1693	1693_2	PRDA1	Spl-Opx		828.00	846.00	1
11	16562c	16562_1	PRDA2	Spl-Opx		777.79	782.20	1
12	1695	1695_169	PRDB1	Spl-Opx		772.79	783.59	1
13	1681	1681_85	PRDB1	Spl-Opx		802.04	820.77	0
14	1681	1681_18	PRDB1	Spl-Opx		791.89	805.35	0
15	1693	1693_2	PRDA1	Cpx-Opx		903.73	924.25	2
16	1693	1693_266	PRDA1	Cpx-Opx		888.26	905.58	2
17	16562c	16562_154	PRDA2	Cpx-Opx		806.38	860.70	2
18	1693	1693_2	PRDA1	Cpx-Opx		946.36	991.05	1
19	16562c	16562_1	PRDA2	Cpx-Opx		901.46	959.35	1
20	16104c	16104_273	PRDA2	Cpx-Opx		839.24	900.69	0
21	1681	1681_18	PRDB1	Cpx-Opx		902.66	993.27	0
22	1681	1681_4	PRDB1	Cpx-Opx		909.25	1053.56	3
23	1681	1681_14	PRDB1	Cpx-Opx		768.97	903.93	3
24	16104v	16104v294	PXA1	Cpx-Opx		782.73	891.51	0
25	1317	1317_36	PXA2	Cpx-Opx		792.79	842.84	0
26	1317	1317_38	PXA2	Cpx-Opx		878.72	916.72	3
27	16562v	16562v80	PXA3	Cpx-Opx		859.49	912.60	0
28	16562v	16562v81	PXA3	Cpx-Opx		892.88	962.50	0
29	1693	1693_20	PRDA1	REE_2Px		1198.00	1240.00	2
30	1682v	1682v37	PXA3	REE_2Px		1022.00	1068.00	0
31	1695	1695_05	PRDB1	REE_2Px		1087.00	1151.00	0
32	1695	1695_03	PRDB1	REE_2Px		1106.00	1256.00	0
33	1681	1681_23	PRDB1	REE_2Px		1051.00	1073.00	2
34	1681	1681_24	PRDB1	REE_2Px			1075.00	2
35	16111	16111_11	PRDC	REE_2Px		1088.00	1118.00	0

0	Matrix
1	Aggregate
2	Exsolution
3	Inclusion

DNTA: Fresh dunite in Group A.
PRDA1: Fresh harzburgite in Group A (Mode (Cpx) >3 vol. wt.%).
PRDA2: Fresh harzburgite in Group A (Mode (Cpx) <3 vol. wt.%).
DNTB: Serpentinized dunite in Group B.
PRDB1: Serpentinized harzburgite in Group B (Mode (Cpx) >3 vol. wt.%).
PRDB2: Serpentinized harzburgite in Group B (Mode (Cpx) <3 vol. wt.%).
PRDC: Serpentinized Cpx-bearing harzburgite in Group C.
PXA1: Clinopyroxenite in Group A.
PXA2: Websterite in Group A.
PXA3: Orthopyroxenite in Group A.
PXB1: Orthopyroxenite in Group B.

(temperatures obtained from interstitial spinel grains fall totally outside of the trend, emphasizing the role of textural location and hence, reaction history; it is not just the grain size that controls the variability). This pattern highlights the important role of diffusion in the freezing process, even if it has been clearly demonstrated that other processes such as dissolution, precipitation and textural maturation also played a role (such as through interaction with percolating melt; see the previous section).

This aspect may be seen more quantitatively by considering the non-dimensional parameters in the various models that were introduced earlier. In the context of the reaction mechanism map of Dohmen and Chakraborty (2003), the parameter $\log \delta$ is ~ 9.0 and $\log \gamma > 13.0$ for all combinations of diffusion coefficients, grain sizes and other geometric parameters (e.g. distance between grains, surface area calculated based on average grain size; the surface reaction rate, α, has been considered to be 1.0 in the absence of any available experimental data). This implies that the kinetics of element exchange in the samples were controlled essentially by a 'solid-diffusion control' mechanism; in some cases, a 'mixed solid-fluid diffusion control' may have operated, consistent with the inference in the previous paragraph. Given such 'solid diffusion control', the γ' parameter of Lasaga (1983) may be used to estimate approximately the freezing temperature range of different minerals. The calculation is approximate because: (a) the temperature depends on the exchange partner as well, and any given mineral in the sample (e.g. olivine) exchanged Fe and Mg with several other minerals (e.g. orthopyroxene, clinopyroxene, spinel), and (b) the mathematical analysis is based on a linear cooling history. In a thermally evolving rock system, for values of $\gamma' < 10$, even the composition at the core is reset while for $\gamma' > 100$, even the rim compositions remain frozen (Section 2.ii above). Thus, the range of temperatures obtained from the core and the rim compositions correspond roughly to the range $10 > \gamma' > 100$. This is illustrated in Fig. 13a, b. Figure 13b illustrates schematically how the different temperatures might relate to the overall thermal evolution of the rock package. The purple line in the figure, including pulses at the higher temperature partly resulting from melt percolation events, shows a possible thermal history followed by the ophiolite sequence. There is an overall continuous (but not necessarily constant) cooling trend. Superposed on this trend, the ranges of temperatures obtained from different geothermometers, which are expected to freeze at different temperatures based on the above considerations, are shown as different coloured bands (the ranges are the same in Figs 13a and 13b, and are based on measurements). It is seen that there is some overlap between freezing ranges of different thermometers (depending, for example, on the grain sizes of minerals used). Diffusion modelling can yield cooling rates from zoning profiles between cores and rims of crystals – this is illustrated through red arrows for spinel-orthopyroxene and spinel-olivine pairs. The depiction underscores that concentration profiles in different mineral pairs would potentially record cooling rates at different segments of the overall thermal history, and that these need not be the same – a mineral grain has no memory of temperatures where $\gamma' < 10$, and is not affected by the thermal history once $\gamma' > 100$.

Third, as discussed above, processes such as growth and dissolution of crystals also operated (e.g. related to melt percolation) and element transport occurred along grain boundaries (external supply/leakage in mineral exchange pairs) so that it was not possible to carry out quantitative geospeedometry using the profiles reported here. Nevertheless, the systematic pattern of freezing of the thermometers point to a relatively slow, and more or less continuous cooling of the ophiolite sequence during emplacement (the

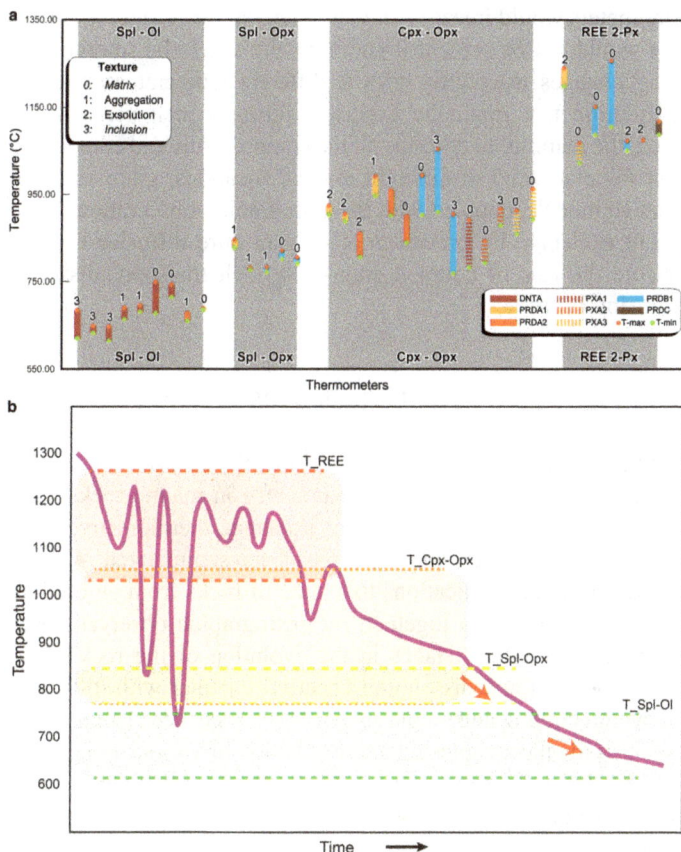

Fig. 13. (*a*) Summary of the estimated temperatures using different thermometers among different coexisting mineral pairs from different rocks (different colours) and different textural locations (numbers in legend). Temperatures obtained from different thermometers are different, but the behaviour is systematic across different rocks and mineral pairs, and are therefore physically meaningful. Please see text for more details and discussion of implications.

Mineral pairs in different colours are taken from DNTA: Dunite in Group A, PRDA1: Harzburgite in group A (Mode (Cpx) >3 vol. wt.%), PRDA2: Harzburgite in group A (Mode (Cpx) <3 vol. wt.%), PRDB1: Harzburgite in group B (Mode (Cpx) >3 vol. wt.%), PRDC: Cpx-bearing harzburgite in Group C, PXA1: Clinopyroxenite in group A, PXA2: Websterite in group A, PXA3: Orthopyroxenite in group A respectively.

(*b*) A schematic depiction of a possible thermal history for the ophiolite package (purple line) with the various temperature ranges recorded by different thermometers (various colours) in different textural settings (grain size, location of chemical analysis such as core or rim of a crystal as well as inclusion or matrix phase). The ranges recorded by a given thermometer are indicated by dashed lines of the same colour. Oscillations at high temperatures indicate the effect of melt percolation events. Temperatures may drop to quite low values between events, but the thermometers would be reset in successive events. And they freeze when the freezing temperature range is crossed finally (see text for details of kinetic parameters that control this). Red arrows indicate the segment of cooling history that may be accessed by diffusion modelling of compositional zoning in a given mineral pair, if suitable boundary conditions are fulfilled (see text for details). Note that these ranges are different for different mineral pairs, and that the cooling rates need not be the same over different segments of the thermal history.

different thermometers would freeze at similar temperatures for fast cooling, and the systematic pattern would not be present if cooling were not to be smooth). Note that this excludes neither changes in cooling rates at different temperatures, nor short thermal pulses (such as those that might be expected during a melt percolation event); see Fig. 13b. Indeed, the inability to describe all the compositions and compositional profiles in the different minerals (and in different textural locations, grain sizes) using simple diffusive exchange models, coupled with the petrographic observations of melt percolation and element exchange by processes other than pure diffusive-exchange between grains, underscore the role of coupled grain-scale melt transport, dissolution/precipitation, and diffusion processes.

5. Conclusions

We have outlined the importance of kinetics, and thereby chemical analysis with spatial–textural context, for performing geothermobarometry in mantle rocks. Some common formulations for kinetic treatment of element exchange thermometry, along with their implications for the practice of thermometry, have been summarized. We have illustrated the role of kinetics through applications to a suite of rocks from an ophiolite sequence (Xigaze ophiolite, Tibet). Taken together, the petrographic observations and mineral-chemical data point to multiple stages in the evolution of the rocks in the ophiolite sequence. Mineral growth and dissolution occurred together with diffusion during this evolution and overall cooling and emplacement of the sequence. Thermometry and geospeedometry of such mantle samples require evaluation of various kinetic aspects related to these processes. However, once such factors are considered, a very systematic pattern emerges pointing to a history of relatively slow and continuous cooling during emplacement of the ophiolite body.

Acknowledgements

This research was supported financially by grants from the National Natural Science Foundation of China (41572044 and U1906207). Additional funds from the Ruhr Universität Bochum to SC helped fund the research. We acknowledge the help of the following through many discussions: Prof. Xuping Li, Dr Arne Willner, Dr Hans-Peter Schertl and Dr Ralf Dohmen.

References

Allégre, C.J., Courtillot, V., Tapponnier, P., Hirn, A., Mattauer, M., Coulon, C., Jaeger, J.J., Achache, J., Schärer, U., Marcoux, J., Burg, J.P., Girardeau, J., Armijo, R., Gariépy, C., Göpel, C., Li, Tindong, Xiao, Xuchang, Chang, Chenfa, Li, Guangqin, Lin, Baoyu, Teng, Jiwen, Wang, Naiwen, Chen, Guoming, Han, Tonglin, Wang, Xibin, Den, Wanming, Sheng, Huaibin, Cao, Yougong, Zhou, Ji, Qiu, Hongrong, Bao, Peisheng, Wang, Songchan, Wang, Bixiang, Zhou and Yaoxiu & Ronghua, Xu (1984) Structure and evolution of the Himalaya–Tibet orogenic belt. *Nature*, **307**(5946), 17–22.

Ballhaus, C., Berry, R. and Green, D.H. (1991) High pressure experimental calibration of the olivine–orthopyr-oxene-spinel oxygen geobarometer: implications for the oxidation state of the upper mantle. *Contributions to Mineralogy and Petrology*, **107**(1), 27–40.

Bédard, É., Hébert, R., Guilmette, C., Lesage, G., Wang, C.S. and Dostal, J. (2009) Petrology and geochemistry of the Saga and Sangsang ophiolitic massifs, Yarlung Zangbo Suture Zone, Southern Tibet: Evidence for an arc–back-arc origin. *Lithos*, **113**(1), 48–67.

Beyer, C. and Chakraborty, S. (2021) Internal stress-induced recrystallization and diffusive transport in $CaTiO_3–PbTiO_3$ solid solutions: A new transport mechanism in geomaterials and its implications for ther-mobarometry, geochronology, and geospeedometry. *American Mineralogist*, **106**(12), 1940–1949.

Bloch, E.M., Jollands, M.C., Devoir, A., Bouvier, A.-S., Ibañez-Mejia, I. and Baumgartner, L.P. (2020) Multi-species Diffusion of Yttrium, Rare Earth Elements and Hafnium in Garnet. *Journal of Petrology*, **61**(7), egaa055.

Borinski, S.A., Hoppe, U., Chakraborty, S., Ganguly, J. and Bhowmik, S.K. (2012) Multicomponent diffusion in garnets I: general theoretical considerations and experimental data for Fe–Mg systems. *Contributions to Mineralogy and Petrology*, **164**(4), 571–586.

Chakraborty, S. and Dohmen, R. (2001) Some aspects of the role of intergranular fluids in the compositional evolution of metamorphic rocks. *Journal of Earth System Science*, **110**(4), 293–303.

Chakraborty, S. and Ganguly, J. (1991) Compositional zoning and cation diffusion in garnets. Pp. 120–175 in: *Diffusion, Atomic Ordering, and Mass Transport*. (J. Ganguly, editor). Selected Topics in Geochemistry. Springer US, New York.

Cisneros de León, A. and Schmitt, A.K. (2019) Reconciling Li and O diffusion in zircon with protracted mag-matic crystal residence. *Contributions to Mineralogy and Petrology*, **174**(4), 28.

Dai, J., Wang, C., Polat, A., Santosh, M., Li, Y. and Ge, Y. (2013) Rapid forearc spreading between 130 and 120 Ma: evidence from geochronology and geochemistry of the Xigaze ophiolite, southern Tibet. *Lithos*, **17**, 1–16.

DeCelles, P.G., Robinson, D.M. and Zandt, G. (2002) Implications of shortening in the Himalayan fold-thrust belt for uplift of the Tibetan Plateau. *Tectonics*, **21**(6), 121–12-25.

Dewey, J.F. and Bird, J.M. (1970) Mountain belts and the new global tectonics. *Journal of Geophysical Research*, (1896–1977), **75**(14), 2625–2647.

Ding, L., Kapp, P. and Wan, X. (2005) Paleocene–Eocene record of ophiolite obduction and initial India-Asia collision, south central Tibet. *Tectonics*, **24**(3), 1–18.

Dodson, M.H. (1973) Closure temperature in cooling geochronological and petrological systems. *Contributions to Mineralogy and Petrology*, **40**(3), 259–274.

Dodson, M.H. (1976) Kinetic processes and thermal history of slowly cooling solids. *Nature*, **259**(5544), 551–553.

Dodson, M.H. (1986) Closure Profiles in Cooling Systems. *Materials Science Forum*, **7**, 145–154.

Dohmen, R. and Chakraborty, S. (2003) Mechanism and kinetics of element and isotopic exchange mediated by a fluid phase. *American Mineralogist*, **88**(8-9), 1251–1270.

Dohmen, R. and Milke, R. (2010) Diffusion in Polycrystalline Materials: Grain Boundaries, Mathematical Models, and Experimental Data. Reviews in Mineralogy and Geochemistry, **72**(1), 921–970.

Dohmen, R., Kasemann, S.A., Coogan, L. and Chakraborty, S. (2010) Diffusion of Li in olivine. Part I: Exper-imental observations and a multi species diffusion model. *Geochimica et Cosmochimica Acta*, **74**(1), 274–292.

Dohmen, R., Marschall, H., Wiedenbeck, M., Polednia, J. and Chakraborty, S. (2016) Trace element diffusion in minerals: the role of multiple diffusion mechanisms operating simultaneously. AGU Fall Meeting Abstracts, pp. MR51A–2682.

Dohmen, R., Marschall, H.R., Ludwig, T. and Polednia, J. (2019) Diffusion of Zr, Hf, Nb and Ta in rutile: effects of temperature, oxygen fugacity, and doping level, and relation to rutile point defect chemistry. *Physics and Chemistry of Minerals*, **46**(3), 311–332.

Eiler, J.M., Baumgartner, L.P. and Valley, J.W. (1991) Numerical modeling of diffusive exchange of stable iso-topes in the lithologic setting, Abstracts with Programs-Geological Society of America. Geological Society of America, pp. A447–A447.

Eiler, J.M., Baumgartner, L.P. and Valley, J.W. (1992) Intercrystalline stable isotope diffusion: a fast grain boundary model. *Contributions to Mineralogy and Petrology*, **112**, 543–557.

Ganguly, J. (2021) Academic Reminiscences and Thermodynamics–Kinetics of Thermo–Barometry–Chronology. *Geochemical Perspectives*, **10**(1), 1–3.

Ganguly, J. and Saxena, S.K. (1987) Exchange equilibrium and inter-crystalline fractionation, mixtures and mineral reactions. Pp. 131–165 in: *Mixtures and Mineral Reactions* (J. Ganguly and S.K. Saxena, editors). Minerals and Rocks Series, Springer.

Ganguly, J. and Tirone, M. (1999) Diffusion closure temperature and age of a mineral with arbitrary extent of diffusion: theoretical formulation and applications. *Earth and Planetary Science Letters*, **170**(1), 131–140.

Ganguly, J., Tirone, M., Chakraborty, S. and Domanik, K. (2013) H-chondrite parent asteroid: A multistage cooling, fragmentation and re-accretion history constrained by thermometric studies, diffusion kinetic modeling and geochronological data. *Geochimica et Cosmochimica Acta*, **105**, 206–220.

Gansser, A. (1977) The great suture zone between Himalaya and Tibet: A Preliminary account. Himalaya-sciences de la terra Colloqes International, 7–10 December 1976, Editions du Centre National de la Researche Scientifique, Paris, 268, pp. 181–192.

Girardeau, J., Mercier, J. and Yougong, Z. (1985) Structure of the Xigaze ophiolite, Yarlung Zangbo suture zone, southern Tibet, China: Genetic implications. *Tectonics*, **4**(3), 267–288.

Hackl, K. and Renner, J. (2013) High-temperature deformation and recrystallization: A variational analysis and its application to olivine aggregates. *Journal of Geophysical Research: Solid Earth*, **118**(3), 943–967.

Hébert, R., Bezard, R., Guilmette, C., Dostal, J., Wang, C.S. and Liu, Z.F. (2012) The Indus–Yarlung Zangbo ophiolites from Nanga Parbat to Namche Barwa syntaxes, southern Tibet: First synthesis of petrology, geo-chemistry, and geochronology with incidences on geodynamic reconstructions of Neo-Tethys. *Gondwana Research*, **22**(2), 377–397.

Hiraga, T., Anderson, I.M. and Kohlstedt, D.L. (2003) Chemistry of grain boundaries in mantle rocks. *American Mineralogist*, **88**(7), 1015–1019.

Hiraga, T., Anderson, I.M. and Kohlstedt, D.L. (2004) Grain boundaries as reservoirs of incompatible elements in the Earth's mantle. *Nature*, **427**(6976), 699–703.

Huot, F., Hébert, R., Varfaly, V., Beaudoin, G., Wang, C., Liu, Z., Cotten, J. and Dostal, J. (2002) The Beimar-ang mélange (southern Tibet) brings additional constraints in assessing the origin, metamorphic evolution and obduction processes of the Yarlung Zangbo ophiolite. *Journal of Asian Earth Sciences*, **21**(3), 307–322.

Jenkin, G., Farrow, C., Fallick, A. and Higgins, D. (1994) Oxygen isotope exchange and closure temperatures in cooling rocks. *Journal of Metamorphic Geology*, **12**(3), 221–235.

Kawasaki, T. and Ito, E. (1994) An experimental determination of the exchange reaction of Fe^{2+} and Mg^{2+} between olivine and Ca-rich clinopyroxene. *American Mineralogist*, **79**(5–6), 461–477.

Lasaga, A.C. (1983) Geospeedometry: An Extension of Geothermometry. Pp. 81–114 in: *Kinetics and Equilibrium in Mineral Reactions* (S.K. Saxena, editor). Springer, New York.

Lasaga, A.C. (1986) Metamorphic reaction rate laws and development of isograds. *Mineralogical Magazine*, **50** (357), 359–373.

Lasaga, A.C. (2014) *Kinetic Theory in the Earth Sciences*. Princeton University Press, Princeton, New Jersey, USA.

Liang, Y., Sun. C., and Yao, L. (2013) A REE-in-two-pyroxene thermometer for mafic and ultramafic rocks. *Geochimica et Cosmochimica Acta*, **102**, 246–260.

Liermann, H.-P. and Ganguly, J. (2003) Fe^{2+}–Mg fractionation between orthopyroxene and spinel: experimental calibration in the system $FeO–MgO–Al_2O_3–Cr_2O_3–SiO_2$, and applications. *Contributions to Mineralogy and Petrology*, **145**(2), 217–227.

Molnar, P. and Tapponnier, P. (1975) Cenozoic Tectonics of Asia: Effects of a Continental Collision. Features of recent continental tectonics in Asia can be interpreted as results of the India-Eurasia collision. *Science*, **189**(4201), 419–426.

Nicolas, A. (1981) The Xigaze ophiolite (Tibet): a peculiar oceanic lithosphere. *Nature*, 294(5840): 414.

Tang, M., Rudnick, R.L., McDonough, W.F., Bose, M., Goreva, Y. (2017) Multi-mode Li diffusion in natural zircons: Evidence for diffusion in the presence of step-function concentration boundaries. *Earth and Planetary Science Letters*, **474**, 110–119.

von Seckendorff, V. and O'Neill, H.S.C. (1993) An experimental study of Fe-Mg partitioning between olivine and orthopyroxene at 1173, 1273 and 1423 K and 1.6 GPa. *Contributions to Mineralogy and Petrology*, **113**(2), 196–207.

Yin, A. and Harrison, T.M. (2000) Geologic evolution of the Himalayan-Tibetan orogen. *Annual Review of Earth and Planetary Sciences*, **28**(1), 211–280.

Yin, J., Xu, J., Liu, C. and Li, H. (1988) The Tibetan plateau: regional stratigraphic context and previous work. *Philosophical Transactions of the Royal Society of London. Series A, Mathematical and Physical Sciences*, **327**(1594), 5–52.

Zhang, C.. Liu, C.Z., Wu, F.-Y., Ji, W.-B., Liu, T. and Xu, Y. (2017) Ultra-refractory mantle domains in the Luqu ophiolite (Tibet): Petrology and tectonic setting. *Lithos*, **286–287**, 252–263.

Zhukova, I., O'Neill, H. and Campbell, I.H. (2017) A subsidiary fast-diffusing substitution mechanism of Al in forsterite investigated using diffusion experiments under controlled thermodynamic conditions. *Contributions to Mineralogy and Petrology*, **172**(7), 53.

www.ingramcontent.com/pod-product-compliance
Lightning Source LLC
Chambersburg PA
CBHW061411210326
41598CB00035B/6180